Springer Tracts in Modern Physics

Volume 294

Springer Tracts in Modern Physics provides comprehensive and critical reviews of topics of current interest in physics. The following fields are emphasized:

– Particle and Nuclear Physics
– Condensed Matter Physics
– Light Matter Interaction
– Atomic and Molecular Physics

Suitable reviews of other fields can also be accepted. The Editors encourage prospective authors to correspond with them in advance of submitting a manuscript. For reviews of topics belonging to the above mentioned fields, they should address the responsible Editor as listed in "Contact the Editors".

Eric Stachura

Mathematical Methods in Liquid Crystal Optics and Lens Design

Springer

Eric Stachura
Dept of Mathematics, Building 850
Kennesaw State University
Marietta, GA, USA

ISSN 0081-3869 ISSN 1615-0430 (electronic)
Springer Tracts in Modern Physics
ISBN 978-3-031-46613-7 ISBN 978-3-031-46614-4 (eBook)
https://doi.org/10.1007/978-3-031-46614-4

This Springer imprint is published by the registered company Springer Nature Switzerland AG
The registered company address is: Gewerbestrasse 11, 6330 Cham, Switzerland

To Lucas

Preface

This book arose during my time with the Basque Center for Applied Mathematics (BCAM) in Bilbao, Spain, during the Spring of 2021. I quickly realized that there is a vast physics and engineering literature devoted to various aspects related to optical phenomena in liquid crystals, and more generally the optics of anisotropic media. However, there is much less available in the mathematics literature. I found myself collecting notes and various calculations I had done, and this collection continued to grow. Some parts of this book are well known; for instance, Chap. 3, which provides a brief background on liquid crystals, would be quite well understood by the liquid crystal experts. However, this chapter could be particularly useful to someone with a mathematical optics background, but not necessarily having experience in liquid crystals. Other parts of this book are new, such as the lens design problem considered in Sect. 6.2. Many of the new results, though, are limited to simple cases. Largely this is an artifact of the challenges of understanding light propagation in general inhomogeneous, anisotropic media. In such general media, light rays are not necessarily straight lines; it is this case that has been mainly studied in the freeform lens design/mathematical optics community.

Liquid crystals offer unique capabilities for imaging due to their tunability and flexibility: simply applying a voltage across a liquid crystal can change the alignment of the liquid crystal molecules. As such, my hope here is to attract attention to liquid crystals for use in mathematical freeform lens design and imaging problems. These lens design problems typically go as follows: design a surface or lens (sandwiched between two surfaces) to create a desired illumination effect. This can involve either reflection or refraction of light rays, and the aim could be to focus in the far field (send rays into some prescribed set of directions) or the near field (focus rays to a single point). If the lens design problem involves refraction, one of the keys in the development of a refracting lens or surface is Snell's Law of Refraction. With this in mind, I have spent a substantial amount of text devoted to this principle (in particular, throughout Chap. 4). Since mathematically, a variational perspective is often taken to study liquid crystals (e.g., in trying to obtain minimizers of some free energy), we formulate Snell's Law in a "variational framework," which becomes equivalent to the classical Weierstrass-Erdmann corner conditions for parametric variational integrals.

This book may be interesting for researchers in mathematical optics, especially those who are interested in exploring optics in general anisotropic and inhomogeneous media. Moreover, advanced graduate students or postdoctoral students in mathematics who want to understand the basis of light propagation in liquid crystals will find this text useful. Those physicists who may be interested in some of the mathematical intricacies of light propagation in liquid crystals, especially using variational methods or the language of Riemannian geometry, would also find the text helpful. Each chapter is fairly independent, so readers with different backgrounds I hope will find this to be a useful reference.

A solid liquid crystal background is not required for this text; we have provided a review in Chap. 3 and provided plenty of references for those interested in diving deeper into the subject. A firm knowledge of light propagation in anisotropic media is also not required—Chap. 2 sets the framework starting from the equations of macroscopic electromagnetism. A basic knowledge of calculus of variations is helpful, especially for Chaps. 4 and 5. Additionally, a basic background in analysis and differential geometry is helpful for Chap. 5. That being said, we provide a basic review of the essentials from Riemannian geometry (see also Appendix A.1), at least what is needed to formulate the geodesic equations. Some understanding of integrable and superintegrable systems is useful for the discussion of absolute instruments, but again, in Chap. 7, we provide a basic review of the essentials. In Sect. 8.3, a boundary value problem is solved using functional analytic techniques; this particular section is aimed more toward those interested in functional analytic methods for partial differential equations. Overall, I have tried to keep each chapter relatively self-contained, so a thorough background in any given topic is not strictly necessary.

The outline of this book is the following. The first three chapters provide an introduction to the text and a background on liquid crystals and ray propagation in general anisotropic media. Of course, one small chapter on liquid crystals does not do the topic justice, and I have not pretended to provide an expansive discussion, but rather just a taste of some key ideas that are particularly useful here. Chapter 4 is devoted to the classical Fermat principle and Snell's Law in various settings, with the last section focusing on liquid crystals. Chapter 5 is perhaps more mathematically rigorous than the previous sections; variational ideas and tools from Riemannian geometry are used in this chapter. In Chap. 6, I do not assume the reader is familiar with optical aberrations, so this is reviewed before moving to a lens design problem. Chapter 7 addresses absolute instruments; it is in this chapter where connections to dynamical systems (in particular, superintegrable systems) are made. Chapter 8 considers electromagnetics, and in particular optics, of helical media. Snell's Law appears here again, and mathematical methods such as semigroup theory, Hilbert space methods, and Floquet theory are used throughout this chapter. Finally, Chap. 9 concludes with a number of possible research directions. More details of each chapter are provided in Chap. 1.

This is not a typical book in the sense that I provide both new and well-known results, the goal of which is to provide the reader with a balanced overview of the state of the art. Moreover, instead of solving a fundamental problem, many times I allude to challenges and obstructions in advancing the state of the art (e.g., the challenge of designing absolute instruments). Ultimately, it is my hope to bring together seemingly disjoint areas of mathematics and liquid crystal physics to facilitate the use of liquid crystals for study and use in imaging and freeform lens design problems.

Marietta, Georgia
December 2023

Eric Stachura

Acknowledgments

I would first like to acknowledge the Visiting Fellow program at BCAM where this work started during the Spring of 2021. I am also grateful to my institution, Kennesaw State University, for the time given to me during Spring 2021 to work with BCAM and, in particular, Dr. Arghir Zarnescu. I am particularly indebted to Dr. Zarnescu for motivating me to write this book and sharing his endless knowledge on the mathematical theory of liquid crystals. I would also like to thank those who took time to discuss their wealth of liquid crystal knowledge with me, especially Prof. Dr. Miha Ravnik. I would also like to thank Dr. Jeremy Gulley for a careful reading of the manuscript and his insightful comments. I am also grateful to Dr. Tim Faver for an initial perusal of a draft manuscript, as well as Dr. Maxim Gilula for interesting discussions related to oscillatory integrals, especially related to the content in Sect. 9.2.6. I am also pleased to thank one undergraduate student, in particular Elizabeth Ehme, for her hard work on topics related to this book during the Summer of 2023. Finally, many thanks to my wife Tamara for her patience as I worked on this book.

Contents

Chapter 1
Introduction

Liquid crystals are a state of matter which have some fluid-like properties and some solid-like properties [116, 329]. The most studied liquid crystals from the mathematical perspective are the so-called nematic liquid crystals, where molecules are shaped like rods and the center of mass is distributed isotropically, yet there is long-range orientational order (i.e., locally, the molecules have a preferred direction).

A classical starting point for a liquid crystal variational problem is the following (see the review [26]). Suppose a liquid crystal fills a container $\Omega \subset \mathbb{R}^3$. The configuration of the liquid crystal at $\mathbf{x} \in \mathbb{R}^3$ is given by the *director* $\mathbf{d}(\mathbf{x})$, a unit vector describing the average orientation of the liquid crystal molecules at the point $\mathbf{x}$. Then, the basic energy minimization problem is to find a director $\mathbf{d}$ which minimizes the energy

$$I[\mathbf{d}] = \int_{\Omega} W(\mathbf{d}, \nabla\mathbf{d}) d\mathbf{x}$$

subject to the constraint that $\mathbf{d}$ be a unit vector field, plus some appropriate boundary conditions. Here $W(\mathbf{d}, \nabla\mathbf{d})$ denotes the so-called Oseen–Frank energy

$$W(\mathbf{d}, \nabla\mathbf{d}) = K_1(\nabla\cdot\mathbf{d})^2 + K_2(\mathbf{d}\cdot\nabla\times\mathbf{d})^2 + K_3|\mathbf{d}\times\nabla\mathbf{d}|^2 + (K_2+K_4)(\mathrm{tr}(\nabla\mathbf{d})^2 - (\nabla\cdot\mathbf{d})^2)$$

where K_1, K_2, K_3, K_4 are the so-called Franck constants [131, 422]. This model assumes $\mathbf{d} : \Omega \to S^2$, the unit sphere in $\mathbb{R}^3$. Other competing models exist as well, including the Ericksen theory [123] which also uses a scalar function s, the degree of orientation, so that $(\mathbf{d}, s) : \Omega \to S^2 \times [-1/2, 1]$. There is also the Landau–de Gennes theory [98], which uses Q-tensors $Q : \Omega \to S_0 = \{Q \in \mathbb{R}^{3\times3} : Q = Q^T; \mathrm{tr}(Q) = 0\}$. We discuss this in more detail in Chap. 3.

Most of the works studying liquid crystals using variational methods are aimed at so-called thermotropic nematic liquid crystals, which change phase under a

E. Stachura, *Mathematical Methods in Liquid Crystal Optics and Lens Design*,
Springer Tracts in Modern Physics 294,
https://doi.org/10.1007/978-3-031-46614-4_1

temperature change. We will not attempt to write a comprehensive reference list but rather refer the interested reader to a brief review in Chap. 3, the aforementioned reviews, as well as just mention the substantial body of work that exists in the mathematical community.[1]

At the same time, another class of problems is being studied in the realm of mathematical optics. The equations that describe the solutions to these problems are typically fully nonlinear partial differential equations of the Monge–Ampère type:

$$\det\left(D^2\phi(x) + A(x, \nabla_x\phi(x))\right) = B(x, \nabla_x\phi(x))$$

for some potential function $\phi(x)$, which is usually given by $\phi(x) = \log\rho(x)$, where ρ is a defining function for an optical surface. This is due to the fact that solutions to the optics problems are typically converted into optimal mass transport problems, where there is a rich theory on existence and uniqueness of solutions and their regularity, see, e.g., [101, 128, 156, 264, 411, 416, 421], just to name a few references on the vast subject. The function B above is usually given by

$$B(x, \nabla_x\phi(x)) = |\det\left(\nabla_{x_k}\nabla_{y_\ell}c(x, Y(x, \nabla_x\phi(x)))\right)|\frac{f(x)}{f^*(Y(x, \nabla_x\phi(x)))}$$

where $c(x, y)$ is the cost function associated with the optimal mass transport problem, $Y(x, \nabla_x\phi(x)) \in S^{n-1}$, and f is a function defined in a neighborhood of a point $x \in S^{n-1}$, where f^* denotes the Legendre transform of f. Finally, the matrix A is given by

$$A(x, \nabla_x\phi(x)) = \nabla^2_{x_k x_\ell}c(x, Y(x, \nabla_x\phi(x)))$$

and in the optics problem, one typically needs to determine the A and B. This is done by explicit determination of the cost function.

Some of the first optics problems that were studied in this context were related to reflection [61–63, 151, 204, 225, 316, 317, 320, 354, 425, 426]. A typical reflection type problem is the following: Given two domains in the unit sphere $D, D^* \in S^{n-1} \subset \mathbb{R}^n$, suppose monochromatic light emanates from a point O in an isotropic medium with some intensity $f(x)$, with $x \in D$. Then, find a perfectly reflecting surface $\mathcal{R}$ such that all light rays reflected by $\mathcal{R}$ have directions in D^*, with some prescribed illumination intensity $f^*(m)$ with $m \in D^*$.

Since one of our main focuses in this book is on variational methods, it is worth pointing out in [170] that a reflection problem was also turned into an energy minimization problem. Indeed, a solution u of the so-called reflector problem which minimizes the energy

[1] According to MathSciNet, the number of papers that mention liquid crystals just from 2017 to 2022 is more than 1,000.

$$I[u] = \int_{\Omega} |\nabla u(x) - F(x)|^2 dx$$

among all functions $u \in C^2(\Omega) \cap C^1(\overline{\Omega})$ with appropriate boundary data is the solution of the Poisson equation

$$\Delta u = \nabla \cdot F \qquad \text{in } \Omega$$

In this setting, F is a C^1 field that depends on where the rays are reflected.

Similar problems were then considered for refraction, see, e.g., [157, 158, 161, 171, 205, 298, 315, 319], just to name a few. Since then, extensions have been made in various directions [113, 166, 206, 318] , such as to exotic media including negative refractive index materials [168] and metasurfaces [160, 164], as well as to including an inverse square law [165] or chromatic aberration [167] and some numerical methods [57, 58, 99, 162, 214]. Of course, this list is not exhaustive, but rather a sampling of more recent problems studied.

Typically, such problems start by developing an appropriate version of Snell's Law of refraction in vector form [155]. This vector formulation is particularly useful for lens design problems, see, e.g., [145, 146]. In the most basic setting of an isotropic material, one supposes that Γ is a surface in $\mathbb{R}^3$ which separates two homogeneous, isotropic materials. The refractive index in material I is given by $n_1 = c/v_1$, and the refractive index in material II is given by $n_2 = c/v_2$, where c is the speed of light in vacuum and v_1, v_2 are the velocities of light in the respective media. If a light ray with direction $\mathbf{a} \in S^2$ travels through medium I and strikes Γ at a point P, then the ray is refracted into a direction $\mathbf{m} \in S^2$ according to the Snell Law in vector form:

$$n_1(\mathbf{a} \times \nu) = n_2(\mathbf{m} \times \nu)$$

where ν is the unit normal at P pointing from medium I to medium II; see Fig. 1.1 below.

The vector form implies the usual scalar version of Snell's Law [48] by taking norm on both sides of the vector equation:

$$n_1 \sin(\theta_1) = n_2 \sin(\theta_2)$$

where θ_1 is the angle between $\mathbf{a}$ and the normal, and θ_2 is the angle between $\mathbf{m}$ and the normal (again see Fig. 1.1). From here one then asks: Given these two media, what is the surface having the uniform refraction property? For instance, the near field problem is as follows: Given a point O inside medium I and a point Q inside medium II, what is the surface interface between the two media which refracts all monochromatic light rays emanating from O into the point Q? The far field version of the refraction problem is the following, which can typically be treated with optimal mass transport methods. Given a direction $\mathbf{m} \in S^2$, what is the surface interface between two materials which refracts all rays into rays parallel to $\mathbf{m}$? For

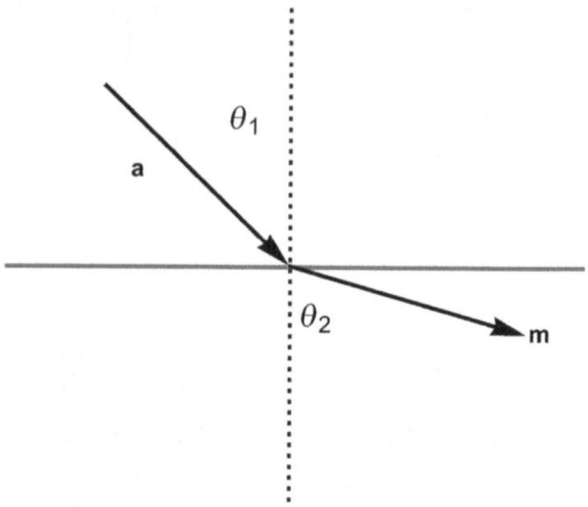

Fig. 1.1 Illustration of the classical version of Snell's Law. The horizontal line separates the top medium with refractive index n_1 from the bottom medium with refractive index n_2. The normal vector $\boldsymbol{\nu}$ points in the direction orthogonal to the separating line

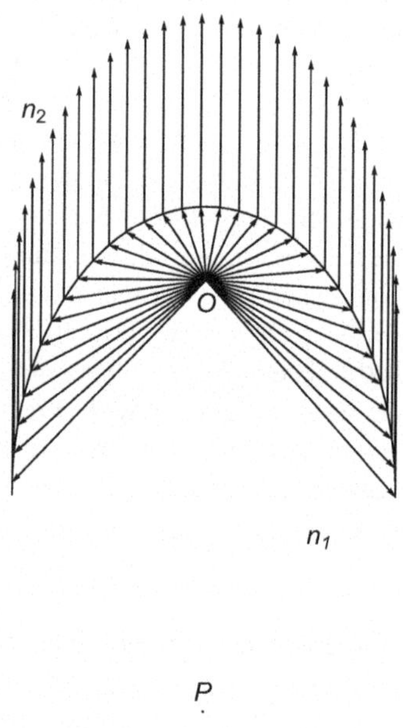

Fig. 1.2 Far field surface with uniform refraction property

instance, in homogeneous, isotropic media with refractive indices $n_1 > 0$ and $n_2 < 0$, this surface was found in [168] to be either a semiellipsoid or a hyperboloid of revolution, depending on the ratio n_2/n_1; see Fig. 1.2.

Determination of such uniformly refracting surfaces in the far field leads to a variational framework of optimal transport with a relevant cost function, which then leads to the explicit PDE of Monge–Ampère type as discussed above. Let us illustrate this framework by means of the results from [157], see also [421].

The idea is to appeal to the so-called Kantorovich dual problem [201–203], in which the key player is *price* rather than *cost* in the original Monge–Kantorovich optimal mass transport problem [303]. Thus one is interested in a maximization problem rather than minimization: while one usually wants to minimize cost, the dual problem concerns maximizing profit.

So, consider two domains $\Omega, \Omega^* \subset S^{n-1}$ in the sphere such that $|\partial\Omega| = 0$ (measure in the sphere). Let $\mathcal{N}$ be a multivalued map, $\mathcal{N} : \overline{\Omega} \to \overline{\Omega^*}$ such that $\mathcal{N}(x)$ is single valued a.e. Let $F \subset \overline{\Omega^*}$, and set

$$\mathcal{T}(F) = \mathcal{N}^{-1}(F) = \left\{x \in \overline{\Omega} : \mathcal{N}(x) \cap F \neq \emptyset\right\}$$

Given some nonnegative, measurable function $g \in L^1(\Omega)$ and some finite Radon measure μ on $\overline{\Omega^*}$ such that

$$\int_{\overline{\Omega}} g(x)dx = \mu(\overline{\Omega^*}) > 0$$

$\mathcal{N}$ is said to be measure preserving from $\eta = g(x)dx$ to μ if for any Borel set $F \subset \overline{\Omega^*}$, there holds

$$\int_{\mathcal{T}(F)} g(x)dx = \mu(F).$$

Now, consider a general Lipschitz cost function $c(x, m)$ and a class of continuous admissible functions

$$\mathcal{K} = \left\{(u, v) : u \in C(\overline{\Omega}),\ v \in C(\overline{\Omega^*}),\ u(x) + v(m) \le c(x, m)\ \forall\ x \in \Omega,\ \forall\ m \in \Omega^*\right\}$$

We can then define the Kantorovich dual functional $I(u, v)$ by

$$I(u, v) = \int_{\Omega} u(x)g(x)dx + \int_{\Omega^*} v(m)d\mu(m)$$

as well as the c and c^* transforms of u and v:

$$u^c(m) = \inf_{x \in \overline{\Omega}} (c(x, m) - u(x)),\ m \in \overline{\Omega^*},\quad v_c(x) = \inf_{m \in \overline{\Omega^*}} (c(x, m) - v(m)),\ x \in \overline{\Omega}$$

Now, a continuous function ϕ is called c-concave if for $x_0 \in \overline{\Omega}$ there exist some $m_0 \in \overline{\Omega^*}$ and some real number b such that

$$\phi(x) \le c(x, m_0) - b, \qquad \phi(x_0) = c(x_0, m_0) - b$$

Under appropriate assumptions on the cost function, one can show (see, e.g., Lemma 3.12 in [157]) that there exists some c-concave function ϕ which maximizes the dual functional $I(u, v)$ in $\mathcal{K}$.

For example, if two media separated by a refracting surface have refractive indices $0 < n_1 < n_2$, then the cost function that transforms the so-called refractor problem into this optimal mass transport framework is

$$c(x, m) = \log\left(\frac{1}{1 - (n_2/n_1)x \cdot m}\right) \tag{1.1}$$

where x is the incident ray direction and m is the refracted ray direction.

To illustrate the framework more clearly, we provide an example from classical convex analysis. Given a convex function u, take the map $\mathcal{N} = \partial u$, the subdifferential of u. Recall that the subdifferential of a convex function is given by

$$\partial u(x) = \left\{y \in \mathbb{R}^n : u(z) \geq u(x) + y \cdot (z - x)\ \forall z\right\}$$

and can be thought of as the set of supporting hyperplanes to u at x (i.e., the collection of all subgradients of u at x). This $\mathcal{N}$ is measurable since one can show that $m \in \partial u(x_0)$ if and only if $x_0 \in \partial u^*(m)$, where $u^*(m)$ is the Legendre transform of u. Thus for a Borel set F, $(\partial u)^{-1}(F) = \partial u^*(F)$ and so $\mathcal{N} = \partial u$ is measurable. One typically has to be careful about measurability since it is not *a priori* guaranteed in this optimal mass transport framework (e.g., if c is not continuous). Notice that there were no measurability assumptions in the transform definitions.

Now, what about c-convexity? This is defined similar to c-concavity as mentioned above. If $c(x, m) = -x \cdot m$, then the c-transform coincides with the usual Legendre transform and c-convexity is the usual notion of convexity in $\mathbb{R}^n$.

Now, what is the challenge with combining the two seemingly distinct research areas of liquid crystals and mathematical optics (naively stated)? As it turns out, the previous geometric optics results generally used the fact that light rays traced out *linear* trajectories. That is, the media are typically assumed homogeneous and isotropic (although recently some of these optics ideas were extended to anisotropic media [159], but still with linear light trajectories). However, liquid crystals are generally nonhomogeneous [229] (the director $\mathbf{d}$ depends on position) and anisotropic. In such materials, the trajectories of the light rays are not necessarily straight lines. For example, there is a large class of materials known as gradient index or GRIN materials, see, e.g., [143, 293, 305, 389]. In such materials, the refractive index is variable, and optical phenomena are produced by the gradient of the refractive index. In a classical Luneburg lens, a refracting structure is made with a variable refractive index which is spherically symmetric (more precisely, the refractive index of such a lens is $n(r) = \sqrt{2 - r^2/R^2}$, where R is the radius of the lens). Luneburg himself [262] found a particular solution of how such a lens focuses, and later Morgan [307] obtained a general expression for a radially symmetric refractive index which performs similar focusing. In such a lens, the light rays have been shown to trace out curved paths which follow an elliptical trajectory [172].

The question then becomes: If light rays are not necessarily following a linear trajectory, what is the appropriate starting point for analyzing the geometric optics problems similar to those mentioned before? Does it make sense to develop some kind of Snell's Law at the interface of a homogeneous material and a nematic liquid crystal? We will address this question in part in Chap. 4. Largely, we take a variational perspective using Fermat's Principle. It is an elementary exercise in basic geometry to show that for simple materials, Fermat's Principle actually implies Snell's Law. Thus, instead of directly considering Snell's Law itself, our goal here is to form a variational framework to study *optics* problems in liquid crystals.

There is of course a vast physics and engineering literature on optical phenomena in liquid crystals, see, for instance, [35, 210–212, 220, 323, 344, 362, 453]. Again we have not attempted to exhaust the enormous list of references.

There is great interest lately in optical phenomena in liquid crystals. For instance, a version of Snell's Law has recently been obtained for a frustrated[2] cholesteric liquid crystal confined between two plates with a topological soliton inside the liquid crystal sample [186]. Generally speaking, a soliton is a solution of a nonlinear partial differential equation describing some physical system and has some interesting topological properties [387]. This is particularly intriguing in light of the optics problems considered here as this can open a brand new avenue where so-called one-dimensional twist walls [31], two dimensional skyrmions [388], and even three dimensional Hopfions [2] can be studied for use in optical imaging. Indeed, recent mathematical research related to these topological solitons has seen work on knotted solutions of various equations (such as the Maxwell and Dirac equations) with intrinsic topological properties [44] and the development of solvable models of magnetic skyrmions [366], among many other directions. There is distinct room, however, in the realm of mathematical optics to consider topological solitons as well; more details on these topological objects are provided in Sect. 9.2.7. As we will see, the topological objects mentioned above can be described using algebraic topology (namely, the first and second homotopy groups). Interestingly enough, there are connections to knot theory here as well [208] since knotted and linked colloidal particles can induce knots and links of singular defects [387].

Moreover, the idea of controlling light via liquid crystals has seen recent interest for numerous applications [81, 111, 180, 339, 378, 462], such as broad-wavelength light control with little chromatic dispersion [309]. Indeed, such control could even lead to all-optical computing [186]. What is more, optical devices based on waveguides using liquid crystals can be used for optical switching or filtering [96].

We thus find it to be a unique time to set a firm foundation for studying optical phenomena in liquid crystals by creating a precise variational framework and to illustrate a number of applications in the process. To this end, in Chap. 2, we recall the basic physical background necessary to understand optical phenomena in liquid crystals. Starting from macroscopic electromagnetics, we recall the different types of anisotropy possible in a material (uniaxial vs. biaxial). This leads to one of the key players in the analysis to come: the optical indicatrix, see equation (2.17).

[2] Here we mean *geometric frustration*; that is, the material has an ideal local structure which cannot fill up space [218] since it violates some global constraint.

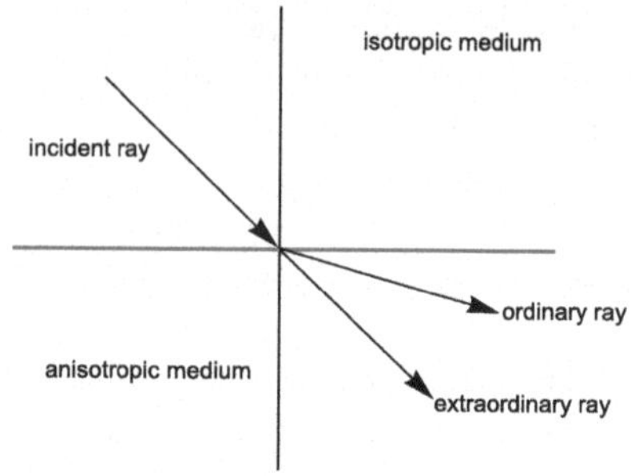

Fig. 1.3 At the interface of isotropic and anisotropic media, an incoming light ray splits into an ordinary and an extraordinary ray which travel in two different directions

One of the main subtleties which we will frequently highlight is how one should *define* a light ray. In isotropic media, there is no issue in defining a light ray as an orthogonal trajectory to the wavefront [48]. However, in general anisotropic media, there are two ways one can define a light ray: as an orthogonal trajectory to the wave normal **p** or as an integral curve of the time averaged Poynting vector. In isotropic media, these vectors are parallel, so there is no need to distinguish them. However, in general anisotropic media, these directions are typically different, which is due to the fact that the electric field **E** and the electric displacement **D** are not necessarily parallel. We then move to another key player for liquid crystals, the effective refractive index. At the interface of an isotropic and anisotropic medium, a light ray is split into the so-called ordinary and extraordinary rays. See Fig. 1.3.

The extraordinary ray feels an *effective* refractive index which depends on both the "ordinary" refractive index and the "extraordinary" refractive index. We discuss the effective indices in both cases, depending on the definition of light ray.

In Chap. 3, we give a brief background on liquid crystals themselves. For the liquid crystal experts, this chapter can be skipped. It is meant to serve as an introduction to the materials and some of the basic mathematical theories which exist for liquid crystals. We start by discussing the different liquid crystal phases, where orientational order can be characterized by certain local symmetries. There is a nice connection to representation theory here (which we do not explore) as one can obtain an irreducible representation of the gradient of a particular vector field. From here we explore the different order parameters that are used in the various mathematical theories for nematic liquid crystals: the Q-tensor and the director **d**. We also briefly review some very basic results on liquid crystal defects and see in particular how they relate to various imaging/optical phenomena. Finally, we review three of the main mathematical theories for nematics: Oseen–Frank, Ericksen, and Landau–de Gennes.

In Chap. 4 we discuss the classical Fermat Principle and Snell's Law as used in geometric optics. Starting from a precise statement of Fermat's Principle, we move to parametric functionals of the form

$$\mathcal{F}(\mathbf{x}) = \int F(\mathbf{x}(t), \dot{\mathbf{x}}(t))dt$$

which turns out to be convenient to analyze the various light ray trajectories in the context of liquid crystals. Indeed, since the effective refractive index for the extraordinary ray will depend on both $\mathbf{x}$ and $\dot{\mathbf{x}}$, it is necessary to consider functionals of the above type.

To address the seemingly ambiguous relationship between the variational form of Fermat's principle and Snell's Law of geometric optics, we appeal next to the so-called Weierstrass–Erdmann corner condition [139], which turns out to be the correct path from Fermat to Snell. We then discuss what Snell's Law becomes in various types of media, as the law follows from the corner condition. As mentioned previously, this is useful if the light rays trace out linear trajectories (as will be made clear). As an application of the Snell Law, we compute various critical angles of refraction, in both the ray and wave directions. Here we both recover known results [263] and provide new results on multiple critical angles at a liquid crystal interface, assuming an appropriate boundary condition for the director $\mathbf{d}$. Finally, we consider Snell's Law with absorption. The idea of absorption is briefly discussed at the end of Chap. 2; namely, in lossy materials there is a complex permittivity ε, which leads to a complex refractive index. That is, in lossy media, the conductivity $\sigma \neq 0$, and a wave will lose energy by heating the material. Starting from Maxwell's equations, looking for certain plane wave solutions leads to a fourth order equation for the real part of an effective refractive index in lossy media, see equation (4.50). Snell's Law holds with such a refractive index; this is obtained by matching phases of incident and transmitted waves at the interface of air with an absorbing medium. We also discuss the uniaxial and biaxial anisotropic cases and the corresponding effective refractive indices.

In Chap. 5 we take a differential geometric perspective and consider light rays as geodesics in an appropriate space. We start at the beginning and consider extremals of an appropriate functional and their relation to the classical Euler–Lagrange equation. We take our main (parametric) functional for liquid crystals and explicitly compute the Euler–Lagrange equation in terms of the director $\mathbf{d}$, assuming $\mathbf{d}$ is sufficiently smooth. However, it is often the case that $\mathbf{d}$ is not smooth (for instance, in the presence of defects). We then show how to go between the Lagrangian and Hamiltonian formulation of the optics problem, which turns out to be possible via the Legendre transform. In particular, we obtain a more general set of Hamilton's equations than those computed in [343] as we do not make any smallness assumptions on the wave vector $\mathbf{p}$. The resulting system can be solved with basic Runge–Kutta methods.

Then we move from the variational approach (where light rays, i.e., extremals of a functional, satisfy the Euler–Lagrange equations) to an equivalent geometric approach, where light rays satisfy a geodesic equation in a Finsler space. In principle, such spaces are the natural place to study such anisotropic geometric optics problems, since the relevant underlying structure is a Finsler function $F(x, dx)$, as opposed to the natural metric $g_{ij}(x)$ in classical Riemannian geometry. However, as we note, for the effective refractive index for the extraordinary ray, the metric is indeed independent of direction, so one can essentially use Riemannian geometry rather than going through Finsler spaces. We also provide a number of examples including the 2D hedgehog director. We will also show how implementing simple deformations of a given director, one can obtain modified lensing effects of light rays, observed for undeformed directors in [359, 360].

In Sect. 5.6 we consider broken extremals, that is, solutions of the Euler–Lagrange equation which are in some sense only piecewise C^1. Such extremals are to be expected in geometric optics when there is a surface discontinuity where rays pass through different media. In particular, solutions of certain Euler–Lagrange equations can be only piecewise continuously differentiable, i.e., can have corners. We review some of the basic theory on broken extremals, such as the excess function and indicatrix. Then, we apply the theory to liquid crystals in 2D. In particular, we show when it is possible to have a corner in the $x - y$ plane and the possibility of discontinuous extremals in 2D. We then provide an example of a director field in which corners could exist.

In Chap. 6, we discuss perfect imaging through the notions of absolute instruments and optical aberrations. Absolute instruments are, at least informally, instruments which can eliminate all optical aberrations [48, 415]. Thus we begin by reviewing the different types of aberrations that can occur in an optical system. We then proceed to solve a simple lens design problem: Given the first surface of a lens in 2D, what is the second surface, such that the lens containing a liquid crystal layer between the surfaces refracts the extraordinary ray into a given fixed point? This second surface is constructed to correct for aberrations generated by the first surface. We also introduce an "inverse" type director design problem: Namely, given a light ray trajectory, what director can produce this trajectory? We show in particular that in the simplest case of a linear trajectory in 2D, a constant director does the job.

In Chap. 7, we discuss the challenge of constructing absolute instruments. We begin by reviewing a classical analogy between optics and mechanics, wherein an optics problem with refractive index n can be related to a classical mechanics problem under the influence of a potential V. One challenge in the context of liquid crystals is that the effective refractive index is *anisotropic*, i.e., velocity dependent. This leads to a velocity dependent potential V, which is difficult to handle in the context of superintegrable systems [423], which we discuss next. Indeed, it turns out that the notion of perfect (stigmatic) imaging is related to separable Hamiltonians and maximally superintegrable Hamiltonian systems. Starting from a separable Hamiltonian, one can construct an absolute instrument [413]. To this end we review some basics of separability and superintegrability and explore the possibility of obtaining quadratically maximal superintegrable systems in the context of liquid crystals. We determine Hamilton's equations of motion in 2D and show what it takes to obtain even a linear integral of motion (let alone a quadratic integral of motion). Again, the challenge we face in even constructing linear equations of motion is that the corresponding mechanical potential is velocity dependent and depends on the director $\mathbf{d}$. Actually, the main difficulty is that the mechanical potential is *quadratic* in the momenta variables. We then consider an alternative approach to stigmatic systems: by using a particular homotopy. Finally, we mention an interesting connection to Riemannian geometry: the cut locus and the so-called Wiedersehenmannigfaltigkeit theorem. The solution of Blaschke's conjecture regarding Wiedersehen manifolds, in the language of optics, implies that spheres image perfectly.

In Chap. 8, we discuss the electromagnetics, and of course optics, of helical media. We begin with a very brief background on cholesteric liquid crystals, including a brief discussion on defects and how they differ from nematic liquid crystals. We then begin with some of the mathematics of chiral media, starting from Maxwell's equations. Chiral media, from the viewpoint of Maxwell's equations, can be characterized by two material parameters in the frequency domain (see Definition 8.1), depending on whether or not the material is reciprocal. Roughly, reciprocal media should be "the same going backward as forward" [66]. More precisely, reciprocity is based on time reversal symmetry: That is, one should ask if the appropriate (here, electromagnetic) quantity is symmetry under time reversal symmetry or is time reversal antisymmetric. It is a straightforward exercise to show the time reversal symmetry of Maxwell's equations, but it is indeed possible to break time reversal symmetry [22]. We review a classical example of reciprocal chiral media, where the constitutive relations are the Drude–Born–Fedorov relations [134].

Next we show a new result on H^s Sobolev regularity for solutions of a chiral Maxwell system on a Lipschitz domain.[3] Here, we give an abstract, functional theoretical argument to show that supposing the material parameters satisfy a certain multiplier property, one can find a solution to the Maxwell system (understood in a generalized sense) with certain Sobolev regularity; see Theorem 8.3. This particular section illustrates how certain Hilbert space/functional analysis methods can be used to study chiral media from the perspective of analysis and partial differential equations.

Then we move to the geometric optics regime. The periodic nature of cholesteric liquid crystals leads to interesting optical behavior. We discuss two phenomena in particular: selective reflection and Bragg reflection. Cholesteric liquid crystals can reflect light of a certain wavelength (selective reflection). Indeed, due to the constructive interference of light being reflected, light with the same handedness[4] as the helical structure of the cholesteric liquid crystal is reflected. At the same time, light with the opposite handedness to the cholesteric helix is not reflected due to destructive interference. In particular, we review here some of the ideas from [114, 115], which provides a rigorous mathematical discussion on selective reflection using ideas from Floquet theory. Due to the prevalence of Snell's Law in this book, we next find it appropriate to derive a vector form of Snell's Law for stable waves at the interface of a cholesteric liquid crystal. Here, stable waves are related to the relevant Floquet exponent: If this exponent is real, then the wave can propagate, and if the exponent is complex, the wave is damped.

---

[3] Roughly, this is an open connected set in $\mathbb{R}^n$ whose boundary can be viewed as locally the graph of a Lipschitz continuous function. This will be made precise in Definition 8.2.

[4] Chiral media have a "handedness" associated with them. Mathematically, this is distinguished as the sign of $q = 2\pi/p$ in the director (8.1), where p denotes the pitch of the cholesteric helix.

Regarding Bragg reflection, we review the Bragg–Snell law for photonic crystals; this is a classical result [216]. Indeed, for a certain frequency range, photonic crystals prevent light from passing through and are reflected through diffraction. We then discuss the corresponding law for cholesteric liquid crystals and show through a rigorous Fourier transform calculation using tempered distributions that Bragg reflection (at normal incidence) occurs when the wavelength λ satisfies $p = \lambda$. We then consider an incident light at arbitrary angles and ultimately derive a Bragg condition which is written only in terms of the incident ray and the refracted ray; see equation (8.62). This is in line with the theme of vector formulations of Snell's Law as seen throughout this book.

Finally, in Chap. 9 we provide a brief summary and provide a few open directions of research. The topic of each section has various open directions to pursue, ranging from more theoretical analysis related to the Monge–Ampère equation to more concrete lens design problems and controlling light via topological solitons.

Chapter 2
Physical Background

We begin with a discussion of macroscopic electromagnetics and then obtain the geometric optics regime as a limiting instance. From here we are able to move on to optics in various media. Since we are largely interested in liquid crystals, it is necessary to develop the relevant ideas in general anisotropic, inhomogeneous media. We will discuss both uniaxial anisotropy and biaxial anisotropy, although the main focus throughout these next few chapters is just on uniaxially anisotropic media (in terms of the liquid crystal director **d**, this means there is only one position dependent preferred direction, namely **d**).

We then hone in on a key definition, as simple as it may seem. Namely, we will be careful in how we *define* a light ray. Indeed, light rays could be defined in different ways depending on the anisotropy of the material. This will be discussed in detail in Sect. 2.3. We then discuss a key function which will be used frequently: the effective refractive index. In general, for a liquid crystal, the refractive index for a given direction of propagation can take two values (physically corresponding to the fast and slow polarizations of the light beam). As the orientation of a liquid crystal is changed, the first (ordinary) refractive index is constant, and the second (extraordinary) refractive index changes. This leads us to the notion of *effective* refractive index, which is determined from both the ordinary and extraordinary refractive indices. Light traveling along the "ordinary" vibration direction just feels the "ordinary" refractive index, whereas light traveling along the "extraordinary" vibration direction feels the effective refractive index, see, e.g., [362, Chapter 2].

2.1 Electromagnetism

The starting point is the macroscopic Maxwell equations, as these describe the electromagnetic fields associated with light propagation:

E. Stachura, *Mathematical Methods in Liquid Crystal Optics and Lens Design*,
Springer Tracts in Modern Physics 294,
https://doi.org/10.1007/978-3-031-46614-4_2

$$\nabla \times \mathbf{H}(\mathbf{x}, t) - \frac{\partial \mathbf{D}(\mathbf{x}, t)}{\partial t} = \mathbf{J}(\mathbf{x}, t) \tag{2.1}$$

$$\nabla \times \mathbf{E}(\mathbf{x}, t) + \frac{\partial \mathbf{B}(\mathbf{x}, t)}{\partial t} = 0 \tag{2.2}$$

$$\nabla \cdot \mathbf{D}(\mathbf{x}, t) = \rho(\mathbf{x}, t) \tag{2.3}$$

$$\nabla \cdot \mathbf{B}(\mathbf{x}, t) = 0 \tag{2.4}$$

where:

- $\mathbf{E}(\mathbf{x}, t), \mathbf{H}(\mathbf{x}, t)$ denote the electric and magnetic fields
- $\mathbf{D}(\mathbf{x}, t), \mathbf{B}(\mathbf{x}, t)$ denote the electric flux density and magnetic induction vector
- $\rho(\mathbf{x}, t), \mathbf{J}(\mathbf{x}, t)$ denote the electric charge density and current density

for $\mathbf{x} = (x, y, z) \in \mathbb{R}^3$ and $t > 0$. The divergence and curl operators at the moment are defined classically, i.e., with $\mathbf{E} = (E_1, E_2, E_3)$ and similar for the other fields,

$$\nabla \times \mathbf{E}(\mathbf{x}, t) = \left(\frac{\partial E_3}{\partial y} - \frac{\partial E_2}{\partial z}, \frac{\partial E_1}{\partial z} - \frac{\partial E_3}{\partial x}, \frac{\partial E_2}{\partial x} - \frac{\partial E_1}{\partial y} \right)$$

$$\nabla \cdot \mathbf{D}(\mathbf{x}, t) = \frac{\partial D_1}{\partial x} + \frac{\partial D_2}{\partial y} + \frac{\partial D_3}{\partial z}$$

That is, we have implicitly assumed that all fields are continuously differentiable in each variable. Coupled with the Maxwell equations are the constitutive relations, which relate the fields **E**, **H** to **D**, **B**:

$$\mathbf{D} = \mathbf{D}(\mathbf{E}, \mathbf{H}), \qquad \mathbf{B} = \mathbf{B}(\mathbf{E}, \mathbf{H}) \tag{2.5}$$

Some typical physical assumptions governing these relations include (see [350], Section 2.3):

- *Determinism*: For every cause there is precisely one effect.
- *Linearity*: The effect is linearly related to its cause.
- *Causality*: The effect cannot precede its cause.[1]
- *Spatial locality*: A cause at any point in space produces an effect at this point only.
- *Time invariance*: If the cause is translated by some time interval, the same time shift occurs for the effect.

Let us interpret these assumptions mathematically. To this end, consider a general operator $T : X \to Y$ where X, Y are vector spaces over $\mathbb{R}$ (or some other field). The determinism assumption means that T is a single valued operator which is not

[1] We have implicitly assumed this above by saying that $t > 0$, since causality means that we may neglect the behavior of the fields for $t < 0$.

everywhere zero. The operator T is linear if for any two elements $u, v \in X$ and any scalar $\lambda \in \mathbb{R}$, we have $T(u+v) = T(u) + T(v)$ and $T(\lambda u) = \lambda T(u)$. Regarding causality, if $u = u(x,t)$ and X is an appropriate function space, then causality means that if $u(x,t) = 0$ for $t \leq t_0$ then $T(u(x,t)) = 0$ for $t \leq t_0$. Spatial locality means that T is a local operator in the spatial variables, i.e., $T(u(x,\cdot))(x,\cdot) = f(x,\cdot)$ for some local functional f (for instance, a differential operator in x; i.e., $T(u(x,\cdot))(x,\cdot) = \frac{\partial u}{\partial x}(x,\cdot)$). Finally, time invariance means that

$$T \circ \tau_s = \tau_s \circ T$$

for all $s \geq 0$, where τ_s is the right shift operator: $\tau_s u(\cdot,t) = u(\cdot, t+s)$. More clearly, the above function composition means that

$$T(\tau_s(u(x,t))) = \tau_s(T(u(x,t))) \qquad \forall\, s \geq 0$$

The above assumptions regarding the constitutive relations lead to a mathematical relationship of the form

$$\begin{pmatrix} \mathbf{D} \\ \mathbf{B} \end{pmatrix} = \begin{pmatrix} \varepsilon & \xi \\ \eta & \mu \end{pmatrix} \begin{pmatrix} \mathbf{E} \\ \mathbf{H} \end{pmatrix} + \int_0^t \begin{pmatrix} \varepsilon_d & \xi_d \\ \eta_d & \mu_d \end{pmatrix} \begin{pmatrix} \mathbf{E} \\ \mathbf{H} \end{pmatrix} ds \tag{2.6}$$

for the four fields $\mathbf{D}, \mathbf{B}, \mathbf{E}, \mathbf{H}$. Here we will assume the second term in (2.6) is zero; physically this corresponds to ignoring dispersion effects. Using the previous relation we can classify media in the following way:

Definition 2.1 A material is called:

1. **Isotropic** if ε, μ are scalar multiplies of $I_{3\times 3}$, the 3×3 identity matrix, and $\eta = \xi = 0$
2. **Anisotropic** if at least one of ε, μ is not a scalar multiple of $I_{3\times 3}$ and $\eta = \xi = 0$
3. **Bi-isotropic** if $\varepsilon, \mu, \eta, \xi$ are scalar multiples of $I_{3\times 3}$ with $\eta \neq 0, \xi \neq 0$
4. **Bi-anisotropic** else

Spatially inhomogeneous media are characterized by ε or μ being functions of the position vector. Above, ε is called the (electric) permittivity of a material, μ the (magnetic) permeability, and η and ξ are coupling parameters which couple the electric and magnetic fields.

For general inhomogeneous, bi-anisotropic media, the second term in (2.6) is really a (temporal) convolution term. The nonlocal in time part of this term models dispersion, and the kernel is called the susceptibility kernel.

Remark 2.1 Typically one supplements the Maxwell system on a domain[2] $\Omega \subset \mathbb{R}^3$ with initial and boundary conditions. The boundary conditions, which describe the behavior of the electric and magnetic fields on the boundary of $\partial\Omega$ of the domain

[2] An open, connected set.

under consideration, depend on the physical situation. For instance, the "perfect conductor" boundary condition is given by prescribing the tangential component of the electric field on the boundary is zero:

$$\nu(\mathbf{x}) \times \mathbf{E}(\mathbf{x}, t) = 0, \qquad \mathbf{x} \in \partial\Omega, t > 0$$

where $\nu(\mathbf{x})$ denotes the outer unit normal at $\mathbf{x} \in \partial\Omega$. But for non-perfect conductors, one can impose a nonhomogeneous boundary condition. Moreover, if the domain is not smooth (i.e., Lipschitz[3] or worse), then one considers the previous cross product in the sense of traces—this is because the unit normal may not be defined everywhere. Here we mean trace in the mathematical sense of "restriction to the boundary." Indeed, on a Lipschitz domain, the tangential trace map $\gamma_t : \mathbf{E} \mapsto \nu \times \mathbf{E}|_{\partial\Omega}$ can be extended to a continuous linear map from $\mathbf{H}(\text{curl}, \Omega)$ into $\left(H^{-1/2}(\partial\Omega)\right)^3$, where

$$\mathbf{H}(\text{curl}, \Omega) = \left\{U \in \left(L^2(\Omega)\right)^3 : \nabla \times U \in \left(L^2(\Omega)\right)^3\right\}$$

and $H^{-1/2}(\partial\Omega)$ denotes the dual of the fractional Sobolev space $H^{1/2}(\partial\Omega)$, which can be defined on a Lipschitz domain [283, 397] as follows: The space $H^s(\partial\Omega)$ for $0 < s < 1$ is the space of distributions u on $\partial\Omega$ such that

$$\int_{\partial\Omega} |u(x)|^2 d\sigma(x) + \int_{\partial\Omega}\int_{\partial\Omega} \frac{|u(x) - u(y)|^2}{|x - y|^{n-1+2s}} d\sigma(x) d\sigma(y) < \infty$$

where σ denotes surface measure and n is the spatial dimension. For more in this direction, see, e.g., [59, 304, 396] among many other references.

We will neglect magnetic effects, so we shall assume that μ is just the identity matrix; thus we will only allow for electric anisotropy. Now we seek a solution of the Maxwell system in the form of plane waves:

$$\begin{aligned} \mathbf{E}(\mathbf{x}, t) &= \widetilde{\mathbf{E}}(\mathbf{x}) e^{i(k_0\psi(\mathbf{x}) - \omega t)}, \\ \mathbf{H}(\mathbf{x}, t) &= \widetilde{\mathbf{H}}(\mathbf{x}) e^{i(k_0\psi(\mathbf{x}) - \omega t)} \end{aligned} \tag{2.7}$$

where ψ denotes the Eikonal function, k_0 is the wave number in vacuum, and ω is the frequency of oscillation. Then, by plugging this ansatz into the Maxwell system, we obtain the system (see, e.g., [48], pg. 111)

[3] Roughly, the boundary is the graph of a Lipschitz function. For more precise definitions, see, e.g., Definition 8.2 and [304].

$$\nabla\psi \times \widetilde{\mathbf{H}} + c\varepsilon_0\varepsilon\widetilde{\mathbf{E}} = -\frac{1}{ik_0}\nabla \times \widetilde{\mathbf{H}}$$

$$\nabla\psi \times \widetilde{\mathbf{E}} - c\mu_0\widetilde{\mathbf{H}} = -\frac{1}{ik_0}\nabla \times \widetilde{\mathbf{E}}$$

$$\nabla\psi \cdot \varepsilon\widetilde{\mathbf{E}} = \frac{-1}{ik_0}\nabla \cdot \varepsilon\widetilde{\mathbf{E}}$$

$$\nabla\psi \cdot \widetilde{\mathbf{H}} = -\frac{1}{ik_0}\nabla \cdot \widetilde{\mathbf{H}} \tag{2.8}$$

where c denotes the speed of light in vacuum, ε_0 is the permittivity of free space, and μ_0 is the permeability of free space. The geometric optics regime corresponds to large values of k_0 (i.e., small wavelengths), which translates to the requirement that

$$\frac{|\nabla \cdot \varepsilon\widetilde{\mathbf{E}}|}{k_0} \ll 1 \tag{2.9}$$

2.2 The Optical Indicatrix

Below, we follow the ideas and notation mainly from [384].

For an inhomogeneous, anisotropic medium, the permittivity

$$\varepsilon(\mathbf{x}) = \begin{pmatrix} \varepsilon_{xx} & \varepsilon_{xy} & \varepsilon_{xz} \\ \varepsilon_{yx} & \varepsilon_{yy} & \varepsilon_{yz} \\ \varepsilon_{zx} & \varepsilon_{zy} & \varepsilon_{zz} \end{pmatrix}$$

depends on position with respect to the standard basis $\{\mathbf{e}_x, \mathbf{e}_y, \mathbf{e}_z\}$ in $\mathbb{R}^3$. We assume that ε is Hermitian;[4] thus, from the Spectral Theorem, we can find an orthonormal basis $\{\mathbf{e}_u, \mathbf{e}_v, \mathbf{e}_w\}$ where ε is diagonal:

$$\varepsilon(\mathbf{x}) = \begin{pmatrix} \varepsilon_u(\mathbf{x}) & 0 & 0 \\ 0 & \varepsilon_v(\mathbf{x}) & 0 \\ 0 & 0 & \varepsilon_w(\mathbf{x}) \end{pmatrix} \tag{2.10}$$

Note that, since ε is assumed to be Hermitian, the elements of the diagonal are real; thus for the moment we are considering only non-absorbing material (a general absorbing material has permittivity with complex eigenvalues). The basis $\{\mathbf{e}_u, \mathbf{e}_v, \mathbf{e}_w\}$ is called the *principal basis* and in fact depends on position. When

[4] Thus $\varepsilon_{ij} = \overline{\varepsilon_{ji}}$ where bar denotes complex conjugate.

the diagonal elements in (2.10) are different, the material is anisotropic. When two elements of (2.10) are equal, the material is called *uniaxially anisotropic*, and if all three elements are distinct, the material is *biaxially anisotropic*.

The $u-v-w$ axes are the local principal axes and define the principal coordinate system. Without loss of generality, we will assume

$$\varepsilon_u < \varepsilon_v < \varepsilon_w \tag{2.11}$$

The principle refractive indices n_u, n_v, n_w are defined by

$$n_u = \sqrt{\varepsilon_u}, \qquad n_v = \sqrt{\varepsilon_v}, \qquad n_w = \sqrt{\varepsilon_w} \tag{2.12}$$

Thus, a biaxially anisotropic medium is one with all distinct principle refractive indices.

We introduce the vector $\mathbf{p} = \nabla\psi$, where ψ is the Eikonal function; $\mathbf{p}$ is called the wave normal. Eliminating $\mathbf{H}$ from (2.8) yields the (anisotropic) Eikonal equation

$$\mathbf{p} \times (\mathbf{p} \times \widetilde{\mathbf{E}}) + \varepsilon\widetilde{\mathbf{E}} = 0 \tag{2.13}$$

From (2.13) we can recover the more familiar form of the Eikonal equation as follows. Using the identity

$$\mathbf{A} \times (\mathbf{B} \times \mathbf{C}) = \mathbf{B}(\mathbf{A} \cdot \mathbf{C}) - \mathbf{C}(\mathbf{A} \cdot \mathbf{B})$$

with $\mathbf{A} = \mathbf{p}$, $\mathbf{B} = \mathbf{p}$, $\mathbf{C} = \mathbf{E}$, we see that (2.13) implies that

$$(\mathbf{p} \cdot \widetilde{\mathbf{E}})\mathbf{p} - \widetilde{\mathbf{E}}|\mathbf{p}|^2 + \varepsilon\widetilde{\mathbf{E}} = 0$$

But from the geometric optics assumption (2.9) we see that $\mathbf{p} \cdot \widetilde{\mathbf{E}} \approx 0$, so that $\widetilde{\mathbf{E}}\left(\varepsilon - |\mathbf{p}|^2\right) = 0$. Assuming $\widetilde{\mathbf{E}}$ is never zero, we obtain the more familiar form of the Eikonal equation:

$$|\mathbf{p}| = \sqrt{\varepsilon} \tag{2.14}$$

By definition $\mathbf{p}$ is the wave vector $\mathbf{k}$ scaled by k_0; thus $|\mathbf{p}| = n$, the refractive index of the medium. This can also be seen from (2.14) since for isotropic media, $|\mathbf{p}| = \sqrt{\varepsilon} = \sqrt{\varepsilon\mu} = n$.[5]

Now, for the Eikonal equation (2.13) to hold, the quadratic equation

$$\mathcal{H}(|p_u|^2, |p_v|^2, |p_w|^2) = 0$$

[5] Again, for simplicity, in this chapter we consider non-magnetic media, so that the magnetic permeability μ is the identity matrix.

must hold; its solutions are two 3D surfaces in **p** space called the *optical indicatrix*. To see this, we write (2.13) as a matrix equation

$$\mathscr{P}(\mathbf{p})\widetilde{\mathbf{E}} = 0$$

by writing (2.13) as $\mathbf{p}(\widetilde{\mathbf{E}} \cdot \mathbf{p}) - |\mathbf{p}|^2\widetilde{\mathbf{E}} + \varepsilon\widetilde{\mathbf{E}} = 0$. In component form, we obtain $(p_i p_k + \varepsilon_{ik} - |\mathbf{p}|^2\delta_{ik})\widetilde{E}_k = 0$. This matrix equation has nontrivial solutions if the determinant of $\mathscr{P}$ vanishes, leading to the quadratic equation $\mathcal{H}(|p_u|^2, |p_v|^2, |p_w|^2) = 0$. Here, p_u, p_v, p_w denote the components of **p** in the basis $\{\mathbf{e}_u, \mathbf{e}_v, \mathbf{e}_w\}$.

Now, the form of the optical indicatrix depends on the type of anisotropy, which we distinguish in the sections below.

2.2.1 Uniaxial Anisotropy

For unaxially anisotropic media, the direction of the optical axis (the axis of symmetry of the material) depends on position; we call this position dependent optical axis the director **d**. The permittivity ε can be written in terms of **d** (see [386, Equation 23] or [228, Equation 14]):

$$\varepsilon_{ij}(\mathbf{x}) = n_o^2\delta_{ij} + (n_e^2 - n_o^2)d_i(\mathbf{x})d_j(\mathbf{x}), \qquad i, j = u, v, w \tag{2.15}$$

where the ordinary and extraordinary refractive indices are given by

$$n_o = \sqrt{\varepsilon_\perp}, \qquad n_e = \sqrt{\varepsilon_\parallel} \tag{2.16}$$

where $\varepsilon_\parallel$ is the permittivity in the direction of the director and $\varepsilon_\perp$ is the permittivity in the direction perpendicular to **d**. This is because the electric flux density **D** can be written as

$$\mathbf{D} = n_e^2(\mathbf{E} \cdot \mathbf{d})\mathbf{d} + n_o^2(\mathbf{E} - (\mathbf{E} \cdot \mathbf{d})\mathbf{d})$$

where $(\mathbf{E} \cdot \mathbf{d})\mathbf{d}$ denotes the electric field in the direction of the director; again, the permittivity in this direction is given by n_e^2. The last equation is the macroscopic material equation in terms of **d**. Then the constitutive relation $D_i = \varepsilon_{ij}E_j$ holds with ε given by (2.15).

The optical indicatrix for uniaxial symmetry can be written as

$$\mathcal{H} = \left[\left(|p_u|^2 + |p_v|^2\right)n_o^2 + |p_w|^2 n_e^2 - n_o^2 n_e^2\right]\left(|\mathbf{p}|^2 - n_o^2\right) = 0 \tag{2.17}$$

Recall that this arises from the determinant of $\mathscr{P}$ vanishing:

$$\det \begin{pmatrix} n_o^2 - p_v^2 - p_w^2 & p_u p_v & p_u p_w \\ p_u p_v & n_o^2 - p_u^2 - p_w^2 & p_v p_w \\ p_u p_w & p_v p_w & n_e^2 - p_u^2 - p_v^2 \end{pmatrix} = 0$$

The surface (2.17) gives rise to ordinary waves, corresponding to

$$\mathcal{H}_o = |\mathbf{p}|^2 - n_o^2 = 0 \tag{2.18}$$

and extraordinary waves, corresponding to

$$\mathcal{H}_e = \left(|p_u|^2 + |p_v|^2\right) n_o^2 + |p_w|^2 n_e^2 - n_o^2 n_e^2 = 0 \tag{2.19}$$

The Eikonal equation (2.13) with ε given by (2.15) can also be written as a matrix equation for $\widetilde{\mathbf{E}}$, just like $\mathscr{P}\widetilde{\mathbf{E}} = 0$. The matrix P becomes

$$\mathscr{P} = \begin{pmatrix} n_o^2 + (n_e^2 - n_o^2)d_u^2 + p_u^2 - |\mathbf{p}|^2 & (n_e^2 - n_o^2)d_u d_v + p_u p_v & (n_e^2 - n_o^2)d_u d_w + p_u p_w \\ (n_e^2 - n_o^2)d_v d_u + p_v p_u & n_o^2 + (n_e^2 - n_o^2)d_v^2 + p_v^2 - |\mathbf{p}|^2 & (n_e^2 - n_o^2)d_v d_w + p_v p_w \\ (n_e^2 - n_o^2)d_w d_u + p_w p_u & (n_e^2 - n_o^2)d_w d_v + p_w p_v & n_o^2 + (n_e^2 - n_o^2)d_w^2 + p_w^2 - |\mathbf{p}|^2 \end{pmatrix} \tag{2.20}$$

As before, the equation $\mathscr{P}\widetilde{\mathbf{E}} = 0$ has a solution if the determinant of the matrix is zero. This condition gives[6]

$$\mathcal{H}(\mathbf{u}, \mathbf{p}) = \mathcal{H}_e \mathcal{H}_o = \left(|\mathbf{p}|^2 - n_o^2\right)\left(n_o^2 |\mathbf{p}|^2 + (n_e^2 - n_o^2)\left(\mathbf{p} \cdot \mathbf{d}(\mathbf{u})\right)^2 - n_o^2 n_e^2\right) = 0 \tag{2.21}$$

where $\mathbf{u} = (u, v, w)$. Letting $\mathbf{d} = (0, 0, 1)$ yields the more familiar optical indicatrix in the principal coordinate system

$$\mathcal{H} = \left(\frac{p_u^2 + p_v^2}{n_e^2} + \frac{p_w^2}{n_o^2} - 1\right)\left(\frac{|\mathbf{p}|^2}{n_o^2} - 1\right) = 0$$

see, e.g., [362, Equation (3.36)].

Figure 2.1 illustrates one octant of the ellipsoidal surface $\mathcal{H}_e$ and the spherical surface $\mathcal{H}_0$. In uniaxial materials , these surfaces intersect at just two points.

[6] Really this is the case when $\mathbf{p} = \sqrt{\varepsilon}\mathbf{e}_w$ so that $\mathbf{d} = \mathbf{e}_w$; otherwise the dot product should really be $|\mathbf{p} \cdot \mathbf{e}_w|^2$; see, e.g., equation (4.13) in [384]. In the case when $\mathbf{p} = \sqrt{\varepsilon}\mathbf{e}_w$, then ε is indeed diagonal; in fact $\varepsilon = \mathrm{diag}(n_o^2, n_o^2, n_e^2)$.

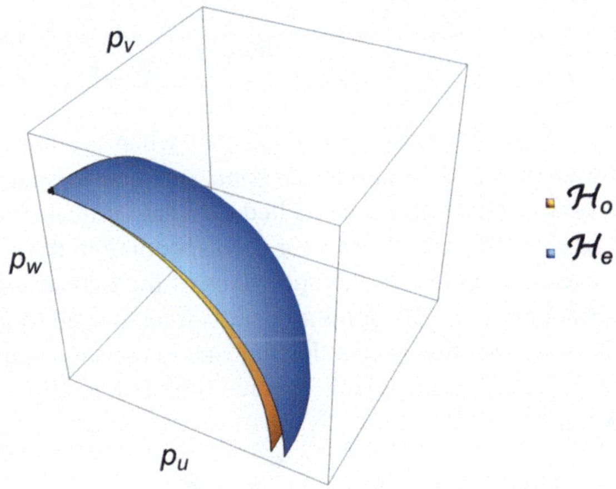

Fig. 2.1 One octant of **p** space with the ellipsoidal surface $\mathcal{H}_e$ and sphere $\mathcal{H}_o$, which for uniaxially symmetric materials intersect at two points

The surface $\mathcal{H}_o = 0$ is a sphere; this is precisely the optical indicatrix for isotropic media. Since the sphere $|\mathbf{p}|^2 = n_o^2$ is rotationally symmetric, we see that there is no directional dependence of optical properties in isotropic media. The same idea applies to ordinary waves in uniaxially anisotropic media, which is why we generally focus on the extraordinary waves.

2.2.2 Biaxial Anisotropy

For biaxial anisotropy, there holds that $\varepsilon_u \neq \varepsilon_v \neq \varepsilon_w$. In the principal coordinate system, the optical indicatrix takes the form

$$\begin{aligned}\mathcal{H} = &\left(\varepsilon_u|p_u|^2 + \varepsilon_v|p_v|^2 + \varepsilon_w|p_w|^2\right)|\mathbf{p}|^2 \\ &- \varepsilon_u|p_u|^2(\varepsilon_v + \varepsilon_w) - \varepsilon_v|p_v|^2(\varepsilon_u + \varepsilon_w) - \varepsilon_w|p_w|^2(\varepsilon_u + \varepsilon_v) + \varepsilon_u\varepsilon_v\varepsilon_w = 0\end{aligned} \tag{2.22}$$

Now, we can write $|\mathbf{p}| = |p_u|\mathbf{e}_u + |p_v|\mathbf{e}_v + |p_w|\mathbf{e}_w$.

In a biaxially anisotropic material, there are two position dependent directions, say $\mathbf{d}_1$ and $\mathbf{d}_2$. The unit vectors $\mathbf{e}_i,\ i = u, v, w$ can be written in terms of these directors (see, e.g., [385]):

$$\mathbf{e}_u(\mathbf{x}) = \frac{\mathbf{d}_1 - \mathbf{d}_2}{|\mathbf{d}_1 - \mathbf{d}_2|}, \qquad \mathbf{e}_v(\mathbf{x}) = \frac{\mathbf{d}_1 \times \mathbf{d}_2}{|\mathbf{d}_1 \times \mathbf{d}_2|}, \qquad \mathbf{e}_w(\mathbf{x}) = \frac{\mathbf{d}_1 + \mathbf{d}_2}{|\mathbf{d}_1 + \mathbf{d}_2|} \tag{2.23}$$

Let us see why this is the case. Since we have assumed $\varepsilon_u < \varepsilon_v < \varepsilon_w$, the optical axis lives in the $u - w$ plane (we will see this below). The angle α between the w-axis and the optical axis is determined by the material properties and is given by

$$\tan(\alpha) = \sqrt{\frac{\varepsilon_w(\varepsilon_v - \varepsilon_u)}{\varepsilon_u(\varepsilon_w - \varepsilon_v)}} \tag{2.24}$$

Indeed, we must understand that when a collimated beam consisting of plane waves having direction inside some small cone is incident on a planar interface with a biaxial medium, the so-called internal conical refraction occurs [398], provided that the refracted wave vector corresponding to the main wave vector of the incident wave is along one of the optic axes of the biaxial medium. During internal conical refraction, an incident light beam is transformed to a hollow cone of light [235]. In this case, the direction of the optic axes is given when the wave vectors on either side of the interface are equal. This requirement, as shown in [398], is fulfilled precisely when

$$\left(\xi_1^2 - \left(k_x^2\frac{\xi_1^2}{\xi_3^2} + k_y^2\right) - \xi_2^2 + \left(k_y^2\frac{\xi_2^2}{\xi_3^2} + k_x^2\right)\right)^2 + 4k_x^2k_y^2\left(1 - \frac{\xi_1^2}{\xi_3^2}\right)\left(1 - \frac{\xi_2^2}{\xi_3^2}\right) = 0 \tag{2.25}$$

where the ξ_j denote the wave numbers associated to the plane waves having their electric field vectors along the appropriate axes, and k_x, k_y denote the components of the transmitted wave vector.

Considering propagation along an optical axis, we see that $|\mathbf{p}| = \sqrt{\varepsilon_v}$ and the corresponding directions of propagation are solutions of (2.25):

$$\mathbf{v} = \pm\sqrt{\frac{\varepsilon_w(\varepsilon_v - \varepsilon_u)}{\varepsilon_v(\varepsilon_w - \varepsilon_u)}}\mathbf{e}_u \pm \sqrt{\frac{\varepsilon_u(\varepsilon_w - \varepsilon_v)}{\varepsilon_v(\varepsilon_w - \varepsilon_u)}}\mathbf{e}_w \tag{2.26}$$

Thus we indeed see that the optic axis lies in the $u - w$ plane and (2.24) follows.

Since $\mathbf{d}_1$ and $\mathbf{d}_2$ are unit vectors parallel to the two local optical axes, from (2.26), we may write

$$\begin{aligned}
\mathbf{d}_1 &= \sqrt{\frac{\varepsilon_w(\varepsilon_v - \varepsilon_u)}{\varepsilon_v(\varepsilon_w - \varepsilon_u)}}\mathbf{e}_u + \sqrt{\frac{\varepsilon_u(\varepsilon_w - \varepsilon_v)}{\varepsilon_v(\varepsilon_w - \varepsilon_u)}}\mathbf{e}_w, \\
\mathbf{d}_2 &= -\sqrt{\frac{\varepsilon_w(\varepsilon_v - \varepsilon_u)}{\varepsilon_v(\varepsilon_w - \varepsilon_u)}}\mathbf{e}_u + \sqrt{\frac{\varepsilon_u(\varepsilon_w - \varepsilon_v)}{\varepsilon_v(\varepsilon_w - \varepsilon_u)}}\mathbf{e}_w
\end{aligned} \tag{2.27}$$

from which we can obtain (2.23). A plot of two such directors when $\varepsilon_u < \varepsilon_v < \varepsilon_w$ can be seen in Fig. 2.2.

Note that all the unit vectors $\mathbf{e}_i$, $i = u, v, w$ depend on position. Then, the surface (2.22) can be written as

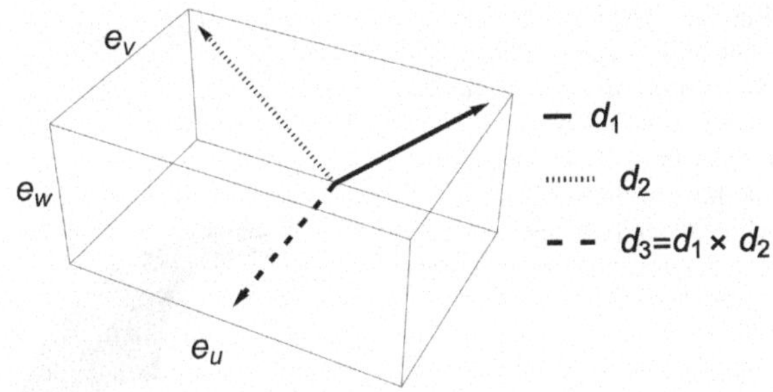

Fig. 2.2 Two directors $\mathbf{d}_1, \mathbf{d}_2$ and their cross product as seen in the principal coordinate system $\{\mathbf{e}_u, \mathbf{e}_v, \mathbf{e}_w\}$. This particular example is as discussed above, when propagation is parallel to $\mathbf{d}_1$ or $\mathbf{d}_2$ and $\mathbf{d}_1, \mathbf{d}_2$ live in the $\mathbf{e}_u - \mathbf{e}_w$ plane. The unit vectors $\{\mathbf{e}_u, \mathbf{e}_v, \mathbf{e}_w\}$ are given by (2.23)

$$\begin{aligned}\mathcal{H}(\mathbf{x}, \mathbf{p}) = &\left(\varepsilon_u(\mathbf{p}\cdot\mathbf{e}_u)^2 + \varepsilon_v(\mathbf{p}\cdot\mathbf{e}_v)^2 + \varepsilon_w(\mathbf{p}\cdot\mathbf{e}_w)^2\right)|\mathbf{p}|^2 \\ &+ \varepsilon_u\varepsilon_v\left[(\mathbf{e}_u\cdot(\mathbf{p}\times\mathbf{e}_v))^2 - |\mathbf{p}|^2\right] + \varepsilon_u\varepsilon_w\left[(\mathbf{e}_u\cdot(\mathbf{p}\times\mathbf{e}_w))^2 - |\mathbf{p}|^2\right] \\ &+ \varepsilon_v\varepsilon_w\left[(\mathbf{e}_v\cdot(\mathbf{p}\times\mathbf{e}_w))^2 - |\mathbf{p}|^2\right] + \varepsilon_u\varepsilon_v\varepsilon_w = 0 \end{aligned} \tag{2.28}$$

Note that taking $\mathbf{e}_u = (1, 0, 0), \mathbf{e}_v = (0, 1, 0), \mathbf{e}_w = (0, 0, 1)$ in (2.28) recovers (2.22)—the indicatrix in the principal coordinate system.

It is convenient to decompose $\mathbf{p}$ into its magnitude and its direction (see also the discussion on effective refractive index in Chap. 3):

$$\mathbf{p} = |\mathbf{p}|\mathbf{v}$$

where $\mathbf{v} = (v_u, v_v, v_w)$. Plugging this into (2.28) yields

$$\begin{aligned}\mathcal{H}(\mathbf{x}, \mathbf{p}) = &\left(\varepsilon_u|v_u|^2 + \varepsilon_v|v_v|^2 + \varepsilon_w|v_w|^2\right)|\mathbf{p}|^4 \\ &- \left(\varepsilon_u|v_u|^2(\varepsilon_v + \varepsilon_w) + \varepsilon_v|v_v|^2(\varepsilon_u + \varepsilon_w) + \varepsilon_w|v_w|^2(\varepsilon_u + \varepsilon_v)\right)|\mathbf{p}|^2 \\ &+ \varepsilon_u\varepsilon_v\varepsilon_w = 0 \end{aligned} \tag{2.29}$$

This implies that for any direction $\mathbf{v}$, there are two solutions $|\mathbf{p}|^2$, which can be found by solving $\mathcal{H} = 0$ for $|\mathbf{p}|^2$:

$$|\mathbf{p}|^2 = \frac{\varepsilon_u|v_u|^2(\varepsilon_v + \varepsilon_w) + \varepsilon_v|v_v|^2(\varepsilon_u + \varepsilon_w) + \varepsilon_w|v_w|^2(\varepsilon_u + \varepsilon_v) \pm \sqrt{f(\varepsilon, \mathbf{v})}}{2\left(\varepsilon_u|v_u|^2 + \varepsilon_v|v_v|^2 + \varepsilon_w|v_w|^2\right)} \tag{2.30}$$

where

$$\begin{aligned} f(\varepsilon, \mathbf{v}) = &\left(\varepsilon_u|v_u|^2(\varepsilon_v + \varepsilon_w) + \varepsilon_v|v_v|^2(\varepsilon_u + \varepsilon_w) + \varepsilon_w|v_w|^2(\varepsilon_u + \varepsilon_v)\right)^2 \\ &- 4\left(\varepsilon_u|v_u|^2 + \varepsilon_v|v_v|^2 + \varepsilon_w|v_w|^2\right)\varepsilon_u\varepsilon_v\varepsilon_w \end{aligned}$$

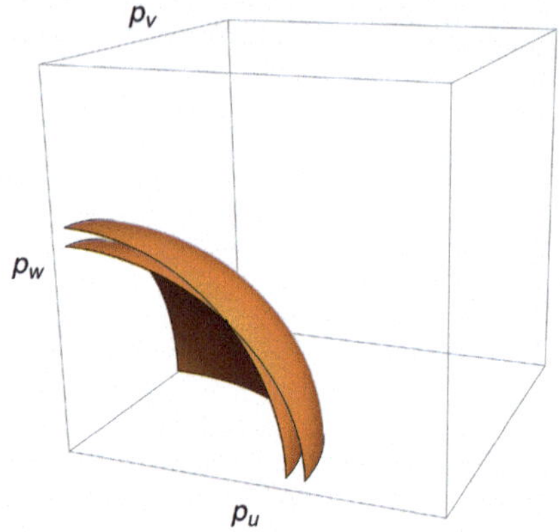

Fig. 2.3 The biaxial indicatrix consists of two overlapping shells, which have four points of intersection. This is one quadrant of the surface, where we have taken $\varepsilon_u = (1.1)^2$, $\varepsilon_v = (1.2)^2$, and $\varepsilon_w = (1.3)^2$

Alternatively, we may write

$$|\mathbf{p}|^2 = \frac{-b \pm \sqrt{b^2 - 4ac}}{2a} \tag{2.31}$$

where

$$\begin{aligned} a &= \varepsilon_u |v_u|^2 + \varepsilon_v |v_v|^2 + \varepsilon_w |v_w|^2, \\ b &= -\varepsilon_u |v_u|^2 (\varepsilon_v + \varepsilon_w) - \varepsilon_v |v_v|^2 (\varepsilon_u + \varepsilon_w) - \varepsilon_w |v_w|^2 (\varepsilon_u + \varepsilon_v), \\ c &= \varepsilon_u \varepsilon_v \varepsilon_w \end{aligned}$$

Figure 2.3 illustrates the two-shell nature of the above surface, arising due to the fact that for each direction $\mathbf{v}$ there are two solutions $|\mathbf{p}|^2$.

From here we can see that for each direction of propagation there are four possible solutions for $n = \pm\sqrt{|\mathbf{p}|^2}$; indeed, we write

$$n = \pm\sqrt{\frac{-b}{2a} \pm \frac{\sqrt{b^2 - 4ac}}{2a}} \tag{2.32}$$

Sometimes the surface associated with the plus sign (in the discriminant part) in (2.32) is called the *b-wave* surface, and the surface associated with the minus sign is call the *a-wave* surface [338]. We thus denote the four solutions in (2.32) by $(n_a, -n_a, n_b, -n_b)$. Now, the two-shelled surface plotted in Fig. 2.3 has four points in common (where the surfaces touch); these points are called the *umbilical points* by Sommerfeld [391]. Note that two solutions for n are identical precisely when $b^2 - 4ac = 0$, for then, at such points, there holds that $n_a = n_b = \sqrt{-b/2a}$.

Finally, we consider a position dependent optical indicatrix with respect to a fixed *position-independent* basis $\{\mathbf{e}_x, \mathbf{e}_y, \mathbf{e}_z\}$; recall that $\{\mathbf{e}_u, \mathbf{e}_v, \mathbf{e}_w\}$ depended on position. As in [384], we start from the constitutive relation $\mathbf{D} = \varepsilon\mathbf{E}$, with $\varepsilon =$

$\varepsilon(\mathbf{e}_u, \mathbf{e}_v, \mathbf{e}_w, \varepsilon_u, \varepsilon_v, \varepsilon_w)$. We now write the elements of ε in terms of the fixed basis $\{\mathbf{e}_x, \mathbf{e}_y, \mathbf{e}_z\}$:

$$\varepsilon = \begin{pmatrix} \varepsilon_{11} & \varepsilon_{12} & \varepsilon_{13} \\ \varepsilon_{21} & \varepsilon_{23} & \varepsilon_{23} \\ \varepsilon_{31} & \varepsilon_{32} & \varepsilon_{33} \end{pmatrix}$$

To find the components of ε, we use the macroscopic material relation [48, 385] (setting ε_0, the free space permittivity, equal to one for simplicity)

$$\mathbf{D} = \varepsilon_u(\mathbf{E}\cdot\mathbf{e}_u)\mathbf{e}_u + \varepsilon_v(\mathbf{E}\cdot\mathbf{e}_v)\mathbf{e}_v + \varepsilon_w(\mathbf{E}\cdot\mathbf{e}_w)\mathbf{e}_w, \qquad \mathbf{D} = \varepsilon\mathbf{E} \tag{2.33}$$

with $\mathbf{e}_u, \mathbf{e}_v, \mathbf{e}_w$ given by (2.23). Then we can write ε in terms of $\varepsilon_u, \varepsilon_v, \varepsilon_w$, which we will denote respectively by n_u^2, n_v^2, n_w^2. Indeed, we find that

$$\varepsilon_{ij} = n_u^2 e_u^i e_u^j + n_v^2 e_v^i e_v^j + n_w^2 e_w^i e_w^j, \qquad i, j = 1, 2, 3 \tag{2.34}$$

where e_u^i denotes the ith component of the vector $\mathbf{e}_u$. In terms of the directors $\mathbf{d}_1, \mathbf{d}_2$, this reads

$$\varepsilon_{ij} - n_u^2 \frac{(\mathbf{d}_1-\mathbf{d}_2)_i(\mathbf{d}_1-\mathbf{d}_2)_j}{|\mathbf{d}_1-\mathbf{d}_2|^2} + n_v^2 \frac{(\mathbf{d}_1\times\mathbf{d}_2)_i(\mathbf{d}_1\times\mathbf{d}_2)_j}{|\mathbf{d}_1\times\mathbf{d}_2|^2} + n_w^2 \frac{(\mathbf{d}_1+\mathbf{d}_2)_i(\mathbf{d}_1+\mathbf{d}_2)_j}{|\mathbf{d}_1+\mathbf{d}_2|^2} \tag{2.35}$$

Note that, if we consider the principal coordinate system by taking $\mathbf{e}_u = (1, 0, 0)$, $\mathbf{e}_v = (0, 1, 0)$, $\mathbf{e}_w = (0, 0, 1)$, and setting $n_u^2 = n_v^2 = n_o^2$ and $n_w^2 = n_e^2$, the permittivity (2.35) reduces to (recall the directors can be written as in equation (2.27)) the uniaxial permittivity (2.15), as expected.

Just as before, the equation $\mathscr{P}\widetilde{\mathbf{E}} = 0$ takes the form

$$\begin{pmatrix} \varepsilon_{11} + |p_x|^2 - |\mathbf{p}|^2 & \varepsilon_{12} + p_x p_y & \varepsilon_{13} + p_x p_z \\ \varepsilon_{21} + p_y p_x & \varepsilon_{22} + |p_y|^2 - |\mathbf{p}|^2 & \varepsilon_{23} + p_y p_z \\ \varepsilon_{31} + p_z p_x & \varepsilon_{32} + p_z p_y & \varepsilon_{33} + |p_z|^2 - |\mathbf{p}|^2 \end{pmatrix} \widetilde{\mathbf{E}} = 0 \tag{2.36}$$

where we have written $\mathbf{p} = (p_x, p_y, p_z)$. This equation has a solution provided that the determinant vanishes, which yields the following condition:

$$\begin{aligned} \mathcal{H}(\mathbf{x}, \mathbf{p}) = &\left(\varepsilon_u|\mathbf{p}\cdot\mathbf{e}_u|^2 + \varepsilon_v|\mathbf{p}\cdot\mathbf{e}_v|^2 + \varepsilon_w|\mathbf{p}\cdot\mathbf{e}_w|^2\right)|\mathbf{p}|^2 \\ &+ \varepsilon_u\varepsilon_v\left(|\mathbf{e}_u\cdot(\mathbf{p}\times\mathbf{e}_v)|^2 - |\mathbf{p}|^2\right) \\ &+ \varepsilon_u\varepsilon_w\left(|\mathbf{e}_u\cdot(\mathbf{p}\times\mathbf{e}_w)|^2 - |\mathbf{p}|^2\right) \end{aligned}$$

$$+\,\varepsilon_v\varepsilon_w\left(|\mathbf{e}_v\cdot(\mathbf{p}\times\mathbf{e}_w)|^2-|\mathbf{p}|^2\right)+\varepsilon_u\varepsilon_v\varepsilon_w=0 \tag{2.37}$$

If $\varepsilon_u = \varepsilon_v = \varepsilon_1$ and $\varepsilon_w = \varepsilon_2$, then (2.37) becomes (2.21) in the case that $\mathbf{d} = \mathbf{e}_w$, i.e., when $\mathbf{p} = \sqrt{\varepsilon_1}\mathbf{e}_w$.

In the biaxial case with directors $\mathbf{d}_1, \mathbf{d}_2$, if $\mathbf{p}$ is parallel to one of these, then in the position dependent basis $\{\mathbf{e}_u, \mathbf{e}_v, \mathbf{e}_w\}$, we can find the components of $\mathbf{v}$ in equation (2.29), but we do not know the phases of the components of $\mathbf{v}$; this would be obtained from the linear transformation of $\mathbf{p}$ in the fixed basis $\{\mathbf{e}_x, \mathbf{e}_y, \mathbf{e}_z\}$ to the basis $\{\mathbf{e}_u, \mathbf{e}_v, \mathbf{e}_w\}$.

2.3 Direction of Propagation and Definition of Light Ray

There is some subtlety in what is meant by propagation direction since in anisotropic media this could mean different things, which is related to how one defines a light ray. In isotropic media, light rays are defined as an integral curve of the wave normal $\mathbf{p}$:

$$\frac{d\mathbf{x}}{dt} = \mathbf{p}(\mathbf{x}(t)) \tag{2.38}$$

That is, the light ray can be considered as the trajectory of a vector which is orthogonal to the *wavefront* ψ = constant [48]. However, there is another important direction in anisotropic media: that of the time averaged Poynting vector $\langle\mathbf{S}\rangle$.

The Poynting vector for the electromagnetic field is defined by

$$\mathbf{S} = \mathbf{E}\times\mathbf{H}$$

and represents the flow of electromagnetic energy. Its time average can be found to be [48, Section 1.4]

$$\langle\mathbf{S}\rangle = \frac{1}{2}\Re\left(\mathbf{E}\times\overline{\mathbf{H}}\right)$$

at least in the high frequency regime. Above, $\Re$ denotes the real part. Indeed, with the ansatz

$$\widetilde{\mathbf{E}}(\mathbf{x})e^{-i\omega t}, \qquad \widetilde{\mathbf{H}}(\mathbf{x})e^{-i\omega t}$$

we define the fields $\mathbf{E}$ and $\mathbf{H}$ to be the real parts of the previous:

$$\mathbf{E}(\mathbf{x},t) = \Re\left(\widetilde{\mathbf{E}}(\mathbf{x})e^{-i\omega t}\right) = \frac{1}{2}\left(\widetilde{\mathbf{E}}(\mathbf{x})e^{-i\omega t} + \overline{\widetilde{\mathbf{E}}(\mathbf{x})}e^{i\omega t}\right)$$

$$\mathbf{H}(\mathbf{x},t)=\Re\left(\widetilde{\mathbf{H}}(\mathbf{x})e^{-i\omega t}\right)=\frac{1}{2}\left(\widetilde{\mathbf{H}}(\mathbf{x})e^{-i\omega t}+\overline{\widetilde{\mathbf{H}}(\mathbf{x})}e^{i\omega t}\right)$$

where overline denotes complex conjugate. We calculate the time averaged Poynting vector:

$$\begin{aligned}\langle\mathbf{S}\rangle&=\frac{1}{2T}\int_{-T}^{T}(\mathbf{E}(\mathbf{x},t)\times\mathbf{H}(\mathbf{x},t))\,dt\\&=\frac{1}{2T}\int_{-T}^{T}\frac{1}{4}\left(\widetilde{\mathbf{E}}\times\widetilde{\mathbf{H}}e^{-2i\omega t}+\widetilde{\mathbf{E}}\times\overline{\widetilde{\mathbf{H}}}+\overline{\widetilde{\mathbf{E}}}\times\widetilde{\mathbf{H}}+\overline{\widetilde{\mathbf{E}}}\times\overline{\widetilde{\mathbf{H}}}e^{2i\omega t}\right)dt\end{aligned}\tag{2.39}$$

Now, notice that

$$\frac{1}{2T}\int_{-T}^{T}e^{-2i\omega t}\,dt=\frac{\sin(2\omega T)}{2\omega T}=\frac{\tau}{4\pi T}\sin(2\omega T),\qquad\tau=2\pi/\omega\tag{2.40}$$

If $T\gg\tau$, i.e., at high frequencies or short wavelengths, then we can neglect the first and last integrals in (2.39), which yields that

$$\langle\mathbf{S}\rangle\approx\frac{1}{2T}\int_{-T}^{T}\frac{1}{4}\left(\widetilde{\mathbf{E}}\times\overline{\widetilde{\mathbf{H}}}+\overline{\widetilde{\mathbf{E}}}\times\widetilde{\mathbf{H}}\right)dt=\frac{1}{2}\Re\left(\widetilde{\mathbf{E}}\times\overline{\widetilde{\mathbf{H}}}\right)$$

as claimed.

Remark 2.2 Notice that the complex amplitude vector $\widetilde{\mathbf{E}}(\mathbf{x})$ can be written as

$$\widetilde{\mathbf{E}}(\mathbf{x})=A(\mathbf{x})e^{i\delta(\mathbf{x})}\widehat{\mathbf{E}}(\mathbf{x})\tag{2.41}$$

where $\widehat{\mathbf{E}}(\mathbf{x})$ is a complex unit vector, $A(\mathbf{x})$ is real and positive, and the phase $\delta(\mathbf{x})$ is real.

By looking at polarization vectors (see Section 4 of [386]), we will show below that the time averaged Poynting vector is parallel to the vector $\nabla_{\mathbf{p}}\mathcal{H}$ defining the optical indicatrix:

$$\langle\mathbf{S}\rangle\propto\nabla_{\mathbf{p}}\mathcal{H}\tag{2.42}$$

This has a number of consequences. First, consider the ordinary rays with $\mathcal{H}_o$ given by (2.18). Then we see that

$$\nabla_{\mathbf{p}}\mathcal{H}_o=2\mathbf{p}$$

which means that the direction of propagation of energy for ordinary rays is in the direction of $\mathbf{p}$, i.e., the normal direction to the wavefronts. Thus there is no distinction between the direction of propagation of the rays and the direction of propagation of energy flux for ordinary rays; this is also the case for isotropic media.

Now, similar to (2.38), we could also define a light ray as an integral curve of the time averaged Poynting vector:

$$\frac{d\mathbf{x}}{dt} = \langle \mathbf{S}(\mathbf{x}(t)) \rangle \tag{2.43}$$

In anisotropic media these directions are typically different since **E** and **D** are not parallel. Indeed, consider the extraordinary rays associated with $\mathcal{H}_e$ in (2.19). From (2.42) we know the direction of energy flux is proportional to $\nabla_{\mathbf{p}}\mathcal{H}_e$, which we can find to be

$$\nabla_p \mathcal{H}_e = 2n_o^2\mathbf{p} + 2(n_e^2 - n_o^2)(\mathbf{p}\cdot\mathbf{d})\mathbf{d} \tag{2.44}$$

in the presence of a director. If the director is perpendicular to the wave vector (or if the medium is isotropic so that $n_e = n_o$), then clearly $\nabla_p \mathcal{H}_e \parallel \mathbf{p}$ and the direction of energy flux coincides with the direction of propagation **p**. Thus, it should be clear that a precise definition of light ray is required in anisotropic media.

Finally, we show that (2.42) holds and follow [384] in the derivation. Without loss of generality we will consider uniaxial anisotropy and consider $\mathcal{H}$ given by (2.17). We start from the Eikonal equation (2.13). From the vector identity

$$\mathbf{A} \times (\mathbf{B} \times \mathbf{C}) = \mathbf{B}(\mathbf{C}\cdot\mathbf{A}) - \mathbf{C}(\mathbf{A}\cdot\mathbf{B})$$

and using the principal basis $\{\mathbf{e}_u, \mathbf{e}_v, \mathbf{e}_w\}$, we can write (2.13) as

$$\begin{aligned}
|\mathbf{p}|^2\widetilde{E}_u - n_o^2\widetilde{E}_u &= (\widetilde{\mathbf{E}}\cdot\mathbf{p})p_u \\
|\mathbf{p}|^2\widetilde{E}_v - n_o^2\widetilde{E}_v &= (\widetilde{\mathbf{E}}\cdot\mathbf{p})p_v \\
|\mathbf{p}|^2\widetilde{E}_w - n_e^2\widetilde{E}_w &= (\widetilde{\mathbf{E}}\cdot\mathbf{p})p_w
\end{aligned} \tag{2.45}$$

where we have used the uniaxial assumption so that $\varepsilon_u = \varepsilon_v = n_o^2$ and $\varepsilon_w = n_e^2$. Thus we see that we may write

$$\widehat{E}_u = C\frac{(\widetilde{\mathbf{E}}\cdot\mathbf{p})p_u}{|\mathbf{p}|^2 - n_o^2}, \qquad \widehat{E}_v = C\frac{(\widetilde{\mathbf{E}}\cdot\mathbf{p})p_v}{|\mathbf{p}|^2 - n_o^2}, \qquad \widehat{E}_w = C\frac{(\widetilde{\mathbf{E}}\cdot\mathbf{p})p_w}{|\mathbf{p}|^2 - n_e^2} \tag{2.46}$$

where C is a normalization factor to make $\widehat{E}$ a unit vector. It is implicitly assumed here that $|\mathbf{p}|^2 \neq \varepsilon_i$. We will show that

$$\widehat{\mathbf{E}}_e \cdot \nabla_{\mathbf{p}}\mathcal{H}_e = 0$$

where $\mathcal{H}_e$ is given by the second term in (2.21) in terms of the director $\mathbf{d}$; this is due to the interest in extraordinary waves. Let us first discuss what happens with the ordinary wave.

For the ordinary wave, we know that $\nabla_{\mathbf{p}}\mathcal{H}_o = 2\mathbf{p}$. Notice that (2.45) implies that $\widetilde{\mathbf{E}}_0 \cdot \mathbf{p} = 0$ since $|\mathbf{p}|^2 = n_o^2$ for ordinary waves; or, that $p_u = p_v = 0$. If $\widetilde{\mathbf{E}}_0 \cdot \mathbf{p} = 0$, then we have $\widetilde{E}_u$, $\widetilde{E}_v$ satisfy $\widetilde{\mathbf{E}} \cdot \mathbf{p} = 0$ with $\widetilde{E}_w = 0$. If the previous product is not zero, and $p_u = p_v = 0$, then $\widetilde{E}_u$, $\widetilde{E}_v$ are arbitrary. Thus $\mathbf{p} = \pm n_o w$ and we obtain from (2.45) that

$$(|\mathbf{p}|^2 - n_e^2)\widetilde{E}_w = (\widetilde{\mathbf{E}}_o \cdot \mathbf{p})p_w = n_o^2 \widetilde{E}_w \tag{2.47}$$

which holds only if $\widetilde{E}_w = 0$. But this then implies that $\widetilde{\mathbf{E}}_0 \cdot \mathbf{p} = 0$.

Thus we conclude that

$$\widehat{\mathbf{E}}_o \cdot \mathbf{p} = 0$$

and since $\nabla_{\mathbf{p}}\mathcal{H}_o = 2\mathbf{p}$, we obtain that

$$\widehat{\mathbf{E}}_o \cdot \nabla_{\mathbf{p}}\mathcal{H}_o = 0$$

Next let us consider the magnetic field. In particular, we will show that

$$\mathbf{H}_o \cdot \nabla_{\mathbf{p}}\mathcal{H}_o = 0$$

Assuming this to be true for the moment, we see that the previous two vanishing inner products imply that the time averaged Poynting vector $\langle\mathbf{S}\rangle$ is perpendicular to the optical indicatrix $\mathcal{H}_o$. But this implies then that

$$\langle\mathbf{S}\rangle \propto \nabla_{\mathbf{p}}\mathcal{H}_o = 2\mathbf{p}$$

as desired. Now let us show vanishing inner product of $\mathbf{H}_o$ with $\nabla_{\mathbf{p}}\mathcal{H}_o$. Indeed, since in the geometrical optics regime the right hand sides of (2.8) vanish; thus from the second equation there we can write

$$\mathbf{H}_o = \frac{1}{c\mu_0}\nabla\psi \times \widetilde{\mathbf{E}}_o$$

Thus

$$\mathbf{H}_o \cdot \nabla_{\mathbf{p}}\mathcal{H} = 2\mathbf{H}_o \cdot \mathbf{p} = \left(\frac{2}{c\mu_o}\mathbf{p} \times \widetilde{\mathbf{E}}_o\right) \cdot \mathbf{p} = \left(\frac{2}{c\mu_0}\mathbf{p} \times A(\mathbf{r})e^{i\delta(\mathbf{r})}\widehat{\mathbf{E}}(\mathbf{r})\right) \cdot \mathbf{p}$$

Now, by using (2.46), we see that

$$(\mathbf{p} \times \widehat{\mathbf{E}}_o) \cdot \mathbf{p} = 0$$

Further, since $\mathcal{H}$ is real, we have that

$$\mathbf{H}_o \cdot \nabla_{\mathbf{p}} \mathcal{H} = \nabla_{\mathbf{p}} \mathcal{H} \cdot \overline{\mathbf{H}}_o$$

Thus we conclude that

$$\mathbf{H}_o \cdot \nabla_{\mathbf{p}} \mathcal{H} = \nabla_{\mathbf{p}} \mathcal{H} \cdot \overline{\mathbf{H}}_o = 0$$

as desired.

Next let us turn our attention to the extraordinary ray and show that $\widehat{\mathbf{E}}_e \cdot \nabla_{\mathbf{p}} \mathcal{H}_e = 0$. We start from the electric flux density as a function of the optical properties (see, e.g., Chapter XI in [237]):

$$\mathbf{D} = \varepsilon_u \left(\mathbf{E}, \mathbf{e}_u\right) \mathbf{e}_u + \varepsilon_v \left(\mathbf{E}, \mathbf{e}_v\right) \mathbf{e}_v + \varepsilon_w \left(\mathbf{E}, \mathbf{e}_w\right) \mathbf{e}_w = \varepsilon \mathbf{E} \tag{2.48}$$

where recall $\varepsilon_u, \varepsilon_v, \varepsilon_w$ are the components of the diagonal permittivity (assuming uniaxial anisotropy) ε. Then, just as in (2.28), using ε in terms of $\mathbf{d}$ from (2.15), we obtain the Fresnel surface (2.21) in terms of $\mathbf{d}$ in the principal coordinate system $\{\mathbf{e}_u, \mathbf{e}_v, \mathbf{e}_w\}$.

Now, by expanding the product $\widehat{\mathbf{E}}_e \cdot \nabla_{\mathbf{p}} \mathcal{H}_e$, using (2.46), we see that

$$\widehat{\mathbf{E}}_e \cdot \nabla_{\mathbf{p}} \mathcal{H}_e = \frac{C(\widetilde{\mathbf{E}} \cdot \mathbf{p})|\mathbf{p}|^2}{(|\mathbf{p}|^2 - n_o^2)(|\mathbf{p}|^2 - n_e^2)} \left(n_o^2 |\mathbf{p}|^2 + (n_e^2 - n_o^2)\left(\mathbf{p} \cdot \mathbf{d}(\mathbf{r})\right)^2 - n_o^2 n_e^2\right)$$

As we have seen above, for extraordinary waves we have $\mathcal{H}_e = 0$, which means that the previous inner product indeed vanishes. Repeating the process above we also see that

$$\mathbf{H}_e \cdot \nabla_{\mathbf{p}} \mathcal{H}_e = 0$$

Thus our conclusion is the same as for the ordinary ray: $\langle \mathbf{S} \rangle$ is perpendicular to $\mathcal{H}_e$ and proportional to $\nabla_{\mathbf{p}} \mathcal{H}_e$, as desired.

Note that we have implicitly assumed that $|\mathbf{p}| \neq n_e$ and $|\mathbf{p}| \neq n_o$ in the calculation of $\widehat{\mathbf{E}}_e \cdot \nabla_{\mathbf{p}} \mathcal{H}_e$.

Finally, we mention how to extend these ideas to absorbing media. As mentioned previously, absorbing media have complex valued permittivity. In particular, we can write [78]

$$\widetilde{\varepsilon}(\mathbf{x}) = \varepsilon(\mathbf{x}) + \frac{i}{\omega} \sigma(\mathbf{x}) \tag{2.49}$$

where as usual ω denotes the angular frequency, ε is the real part of (2.10), and σ is the (real) conductivity matrix. If ε is diagonalizable (e.g., is Hermitian), we can write the principal refractive indices as

$$\widetilde{n}_u = \sqrt{\widetilde{\varepsilon}_u}, \quad \widetilde{n}_v = \sqrt{\widetilde{\varepsilon}_v}, \quad \widetilde{n}_w = \sqrt{\widetilde{\varepsilon}_w}$$

where $\widetilde{\varepsilon}$ and the other terms are the diagonal elements of $\widetilde{\varepsilon}$. Note that each element of the diagonal $\widetilde{\varepsilon}$ could be complex. In particular we can write

$$\widetilde{n}_u = n_u - i\kappa_u, \quad \widetilde{n}_v = n_v - i\kappa_v, \quad \widetilde{n}_w = n_w - i\kappa_w$$

with n_u, n_v, n_w given by (2.12) and the κ's are absorption coefficients. For a biaxial absorbing medium, the three diagonal components of $\widetilde{\varepsilon}$ are distinct, and (2.11) is replaced by the inequalities

$$\Re(\widetilde{\varepsilon}_u) < \Re(\widetilde{\varepsilon}_v) < \Re(\widetilde{\varepsilon}_w)$$

One can obtain a similar formula for the uniaxial optical indicatrix as in equation (2.17), see for instance equation (13) in [428]. This is also true of the biaxial optical indicatrix, see for instance equation (2.14) in [429].

Chapter 3
Liquid Crystals

Before we begin, we mention that we do not attempt to provide an exhaustive overview of the many facets of liquid crystals. One chapter would be an extreme disservice! Rather, we provide some basic background while focusing on the mathematical and optical aspects of liquid crystals.

3.1 Liquid Crystals and Mesophases

There are many references on the basics of liquid crystals for a more detailed description, more than we can possibly list here. We only mention a few of them which we refer to (of course the list is not exhaustive), see in particular [93, 220, 323, 362, 453] as well as [383].

Crystals are one class of materials which exhibit optical anisotropy as discussed previously. Solid crystals such as calcite $CaCO_3$ exhibit optical anisotropy according to three groups [48]: isotropic ($\mathbf{D} = \varepsilon\mathbf{E}$, with ε a scalar multiple of the 3×3 identity matrix), uniaxial anisotropy ($\varepsilon_u = \varepsilon_v = n_o^2, \varepsilon_w = n_e^2$), and biaxial anisotropy ($\varepsilon_u \neq \varepsilon_v \neq \varepsilon_w$). But what about liquid crystals? How do they fit in the picture?

Liquid crystals are anisotropic materials which have properties associated to both solids and liquids. The molecules of a liquid crystal behave like the molecules of a liquid yet retain some order as in a solid crystal.

In fact, it was the French crystallographer G. Friedel in the early twentieth century who suggested liquid crystals are states of matter between isotropic liquids and crystals [135]. However, Friedel was not the first to notice there was a new state of matter between isotropic liquid and solid crystal. This was actually discovered by Otto Lehmann in 1889 [242], and he indeed coined the term "liquid crystal."

Friedel named the new states of matter nematic, cholesteric, and smectic. Let us discuss in detail these phases.

E. Stachura, *Mathematical Methods in Liquid Crystal Optics and Lens Design*,
Springer Tracts in Modern Physics 294,
https://doi.org/10.1007/978-3-031-46614-4_3

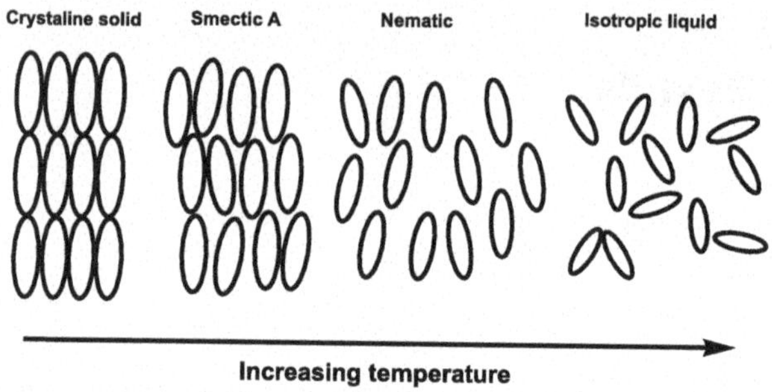

Fig. 3.1 Sequence of phase changes of a liquid crystal with rod-like molecules. Initially, positional order is maintained in the smectic phase but is eventually lost under heating in the nematic phase, where only orientational order remains. Upon enough heating all crystalline order is lost (orientation of the molecules is completely random) and the isotropic liquid phase is reached

Fig. 3.2 The nematic phase, where molecules (here, taken to be rods) exhibit orientational order

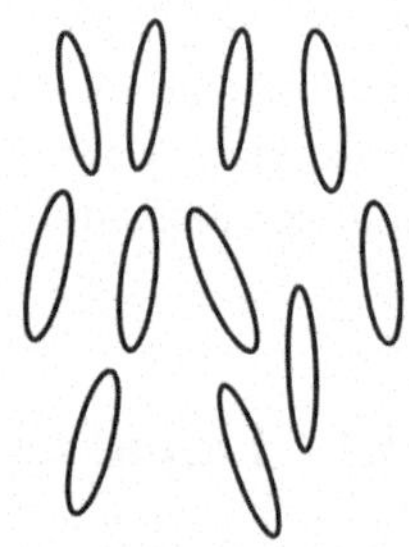

Classically, the differences between solids, liquids, and vapors are to some extent the amount of order in the material; this order is related to the surrounding temperature and pressure (Fig. 3.1). It turns out that some organic substances, which we call liquid crystals, have more condensed phases. The constituent molecules are often called mesogens, and the extra phases are referred to as liquid crystal phases or mesophases. Liquid crystal molecules are inherently anisotropic and often have a rod-like shape; this is the case for instance for the 5CB (4-n-pentyl-4'-cianobiphenyl) liquid crystal [453]. However, molecules may also be disc-like or even banana shaped.[1] All liquid crystals exhibit orientational order; however, the order in the molecule positions is limited. Since the positional arrangement of the molecules in liquid crystal phase varies from complete disorder to order, a substance in liquid crystal phase can actually be seen as a (non-Newtonian) fluid as it assumes the shape of its container. The nematic phase is characterized by orientational order, but there is no positional order (Fig. 3.2). In the nematic phase, molecules can move in any direction. Mostly, the nematic phase is optically uniaxial. Indeed, since in this phase molecules tend to have the same direction, order is orientational. It is also long

[1] There are really three main classes of mesogens: discotics, where molecules are disc shape; calamitics (which we mainly consider here), where molecules are rod-shaped; and nonlinear, bent mesogens, where molecules are banana shaped.

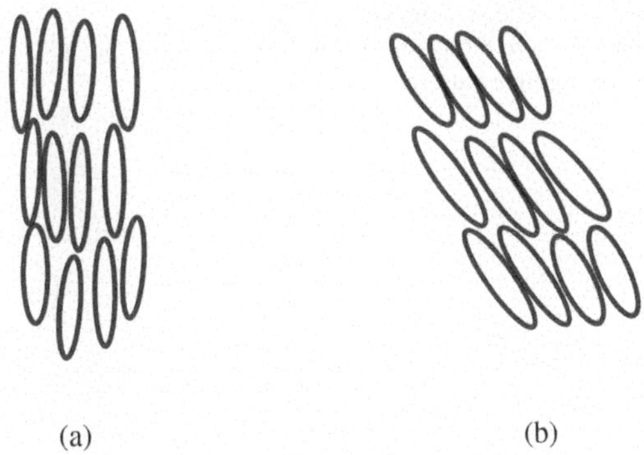

Fig. 3.3 The smectic phase, where orientation of the molecules depends on if the director is tilted with respect to the layer normal or not. (**a**) The smectic A phase. (**b**) The smectic C phase

range since without constraints, all molecules at equilibrium have the same average orientation, leading to the single director for this uniaxially anisotropic material. However, there are biaxial nematics [267, 268], as well; one can often find these in the so-called lyotropic systems: Lyotropic liquid crystals are mesophases essentially in water. Biaxial nematics are actually characterized by three directors: a primary director $\mathbf{d}_1$ and two secondary directors $\mathbf{d}_2$ and $\mathbf{d}_1 \times \mathbf{d}_2$. Here, in 3D, there is partial orientational order without translational order.

In contrast to the nematic phase, Friedel determined that another type of mesophase has molecules which arrange in layers. The rod-like molecules can either be parallel or tilted with respect to the layer's normal axis. The molecules can move within the layers but are less free to move between layers. There are two smectic phases, smectic A and smectic C (Fig. 3.3), which differ in the angle which the director makes with the layer normal as well as the arrangement of molecules within a given layer. In smectic A, the director is parallel to the normal, and there is random positional order of the molecules. If the director is tilted with respect to the layer normal, the phase is called smectic C. The tilt angle usually increases with decreasing temperature [362].

In smectic A, molecules on average are orthogonal to the layer, and so this phase is optically uniaxial. Since the molecules are tilted with respect to the layer normal in smectic C, layers can slide over each other, and so this phase is optically biaxial [323]. There are other smectic phases as well; for instance, the smectic O phase in which molecules form a kind of herringbone pattern and alternate tilting in successive layers; this phase is also optically biaxial (Fig. 3.4).

Finally, the cholesteric phase is a particular case of the nematic phase in which there is no inversion symmetry. In the cholesteric phase (Fig. 3.5), molecules twist around a helix axis which lies orthogonal to the local director field. With every half turn of the helix, the molecules are oriented the same. The full helical turn is called the pitch, p, of the cholesteric liquid crystal.

The cholesteric phase is of particular note since it is one example of a larger class of *chiral* nematic mesogens. The notion of chirality is strictly geometric and is therefore independent of the phase of the mesogen. Informally, chiral materials cannot be superimposed upon their mirror image. Chirality is an important notion in

Fig. 3.4 The smectic O phase, where molecules form a herringbone pattern

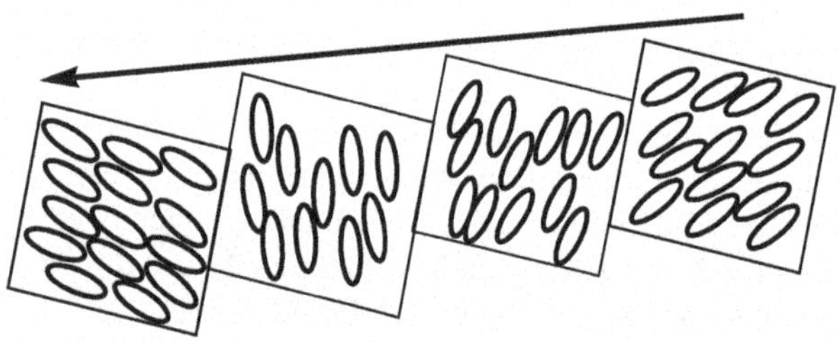

Fig. 3.5 The cholesteric phase, where the average orientation of the molecules turns about the helix axis, denoted in the picture by an arrow

optics as it has a profound impact on the molecular arrangement of the molecules, which is one reason why we devote Chap. 8 to optics of helical media.

In a nematic liquid crystal, the director **d** varies with position. We can distinguish between three types of deformations of the director [220]: splay, twist, and bend. Each of these deformations can be induced by external fields or mechanical distortions. We will show that the director gradient can be decomposed into four parts which represent physically distinct types of spatial variations [266].

We start with a nematic with director **d**. Orientational order can be characterized by local symmetries which preserve **d**. Call this rotation $\mathcal{R}$; under this rotation, the gradient $\nabla\mathbf{d}$ decomposes into irreducible representations as

$$\partial_i d_j = d_i d_k \partial_k d_j + \frac{\nabla \cdot \mathbf{d}}{2}\left(\delta_{ij} - d_i d_j\right) + \frac{\mathbf{d} \cdot \nabla \times \mathbf{d}}{2}\epsilon_{ijk} d_k + \Delta_{ij} \tag{3.1}$$

Let us see in more detail this is the case (here we follow [367]). Since **d** is a unit vector (i.e., so that $\mathbf{d} \cdot \mathbf{d} = 1$), we find that

$$0 = \nabla(\mathbf{d} \cdot \mathbf{d}) = 2(\nabla\mathbf{d})^T\mathbf{d} = 2\mathbf{d} \cdot \nabla\mathbf{d} \tag{3.2}$$

In components this means that $(\partial_i d_j)d_j = 0$. Decomposing $\partial_i d_j$ into terms parallel or perpendicular to **d** and a part living in the plane orthogonal to **d** yields

$$\partial_i d_j = d_i u_j + \alpha_{ij} \tag{3.3}$$

where $\mathbf{u}$ is a vector orthogonal to $\mathbf{d}$ and α_{ij} is a tensor in the plane orthogonal to $\mathbf{d}$. Contracting[2] (3.3) with d_i yields $u_j = d_i \partial_i d_j$; this will be discussed further below and corresponds to the first term in (3.1). Now we handle α_{ij} by decomposing it into an antisymmetric tensor β_{ij} and a symmetric tensor γ_{ij}. Since β is also in the plane orthogonal to $\mathbf{d}$, we may write

$$\beta_{ij} = \frac{1}{2} T \epsilon_{ijk} d_k$$

for some T; this corresponds to the third term in (3.1). Above, ϵ_{ijk} denotes the standard Levi–Civita symbol. Thus (3.3) becomes

$$\partial_i d_j = d_i u_j + \frac{1}{2} \epsilon_{ijk} T d_k + \gamma_{ij} \tag{3.4}$$

Contracting both sides of (3.4) with $\epsilon_{ijm} d_m$ yields that $T = \mathbf{d} \cdot (\nabla \times \mathbf{d})$. Finally we need to handle γ_{ij}. We can break this into its trace[3] S multiplied by half the identity tensor in the plane orthogonal to $\mathbf{d}$, which is $\frac{1}{2}(\delta_{ij} - d_i d_j)$. This provides the second term in (3.1), and at this point we have

$$\partial_i d_j = d_i u_j + \frac{1}{2} \epsilon_{ijk} T d_k + \frac{1}{2} S (\delta_{ij} - d_i d_j) + \Delta_{ij} \tag{3.5}$$

where Δ_{ij} is a traceless symmetric tensor in the plane orthogonal to $\mathbf{d}$. We compute Δ_{ij} by combining (3.5) with its transpose to obtain

$$\begin{aligned}\Delta_{ij} &= \frac{1}{2} \left(\partial_i d_j + \partial_j d_i - d_i u_j - d_j u_i - S(\delta_{ij} - d_i d_j) \right) \\ &= \frac{1}{2} \left(\partial_i d_j + \partial_j d_i + d_i d_k \partial_k d_j + d_j d_k \partial_k d_i - \delta_{ij} \partial_k d_k + d_i d_j \partial_k d_k \right)\end{aligned} \tag{3.6}$$

Let us further explain the terms in (3.1). The first term involves

$$B_j = d_k \partial_k d_j, \text{ i. e. } \mathbf{b} = \mathbf{d} \cdot (\nabla \mathbf{d})^T$$

which is called the bend of $\mathbf{d}$; see Fig. 3.6. A similar definition of bend is used in [313], which defines $\nabla \mathbf{d}$ different from here by a transpose (here we define $(\nabla \mathbf{d})_{ij} = \partial_j d_i$). Notice that $\mathbf{b}$ is orthogonal to $\mathbf{d}$ since $\mathbf{d}$ is a unit vector. Indeed, since $\mathbf{d} \cdot \mathbf{d} = 1$, we have that $0 = \nabla(\mathbf{d} \cdot \mathbf{d}) = 2(\nabla \mathbf{d})^T \mathbf{d}$, and writing in components $\mathbf{b} \cdot \mathbf{d}$, we see this indeed implies that $\mathbf{b} \cdot \mathbf{d} = 0$. Since $\mathbf{d}$ and $-\mathbf{d}$ are equivalent in a nematic , the bend vector $\mathbf{b}$ is invariant under the transformation $\mathbf{d} \mapsto -\mathbf{d}$ and thus does not

[2] A brief review of tensor calculus is provided in Appendix A.1.2.

[3] This is trace of a tensor, not in the sense of restriction to the boundary for a boundary value problem! Explicitly here we have that $S = \partial_k d_k$.

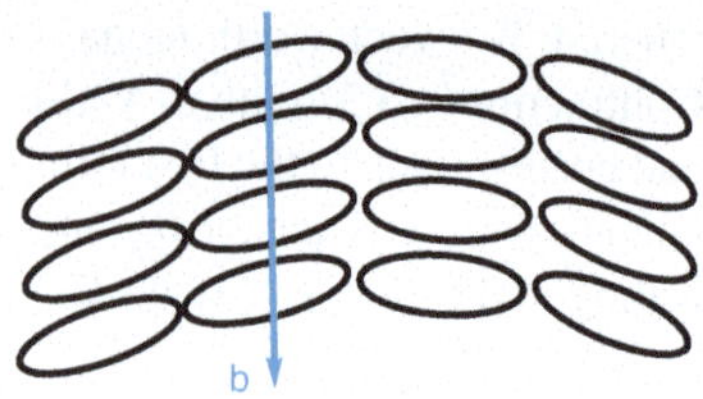

Fig. 3.6 The bend deformation of the director field in a nematic

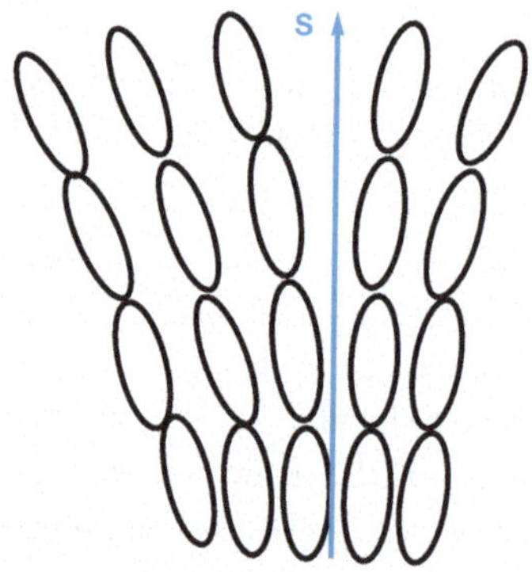

Fig. 3.7 The splay deformation of the director field in a nematic; the splay vector **S** is also pictured

depend on the choice of sign. The vector **b** can be represented by a vector in the plane perpendicular to the local director **d** [367]; see the blue arrow in Fig. 3.6.

The second term in (3.1) contains the scalar

$$S = \nabla \cdot \mathbf{d}$$

called the splay deformation; see Fig. 3.7. Notice that this scalar does change sign under the transformation $\mathbf{d} \mapsto -\mathbf{d}$; if one requires an object that does not depend on this sign choice, it is possible to form the so-called splay vector

$$\mathbf{S} = S\mathbf{d} = \mathbf{d}(\nabla \cdot \mathbf{d})$$

which is a well-known vector in flexoelectricity [295].

The third term in (3.1) contains the term

$$T = \mathbf{d} \cdot \nabla \times \mathbf{d}$$

which is the twist deformation of the director; see Fig. 3.10b. Sometimes this is called "double-twist" since the director changes in both directions orthogonal to the local **d**. The quantity T is a pseudoscalar; indeed, under the transformation $(x, y, z) \mapsto (-x, -y, -z)$, we see that T changes sign (a true scalar would not). The twist T is also invariant under the transformation $\mathbf{d} \mapsto -\mathbf{d}$ and thus does not depend on this choice of sign (Fig. 3.8).

Fig. 3.8 The twist deformation of the director field in a nematic

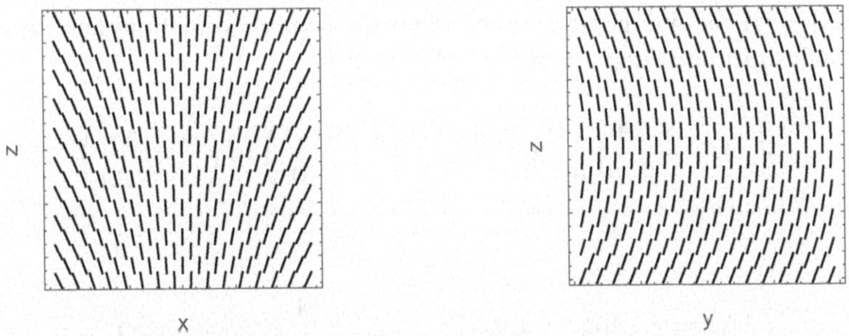

(a) (b)

Fig. 3.9 Slices of a director which exhibits biaxial spray. (**a**) A 2D $(z - x)$ slice of a director $\mathbf{d}$ which exhibits biaxial spray along the z-axis. (**b**) A 2D $(z - y)$ slice of a director $\mathbf{d}$ which exhibits biaxial spray along the z-axis

Finally, we return to Δ_{ij}. It is a symmetric, traceless tensor in the plane orthogonal to $\mathbf{d}$, so that $\Delta_{ij} = \Delta_{ji}$ and $\Delta_{ii} = 0$. This term has been called biaxial splay [367]. Geometrically, in this deformation, $\mathbf{d}$ tips out along one axis orthogonal to the local director and tips in along the other axis orthogonal to $\mathbf{d}$. That is, there is positive splay along one axis and negative splay along the other. The name biaxial spray arose since the symmetry here is similar to a biaxial nematic. Finally, this tensor does change sign under the transformation $\mathbf{d} \mapsto -\mathbf{d}$, so if one wants an object which does not depend on this sign choice, it is necessary to consider the third order tensor $\Delta_{ij} d_k$.

In Fig. 3.9, we illustrate two dimensional slices of a director which exhibits biaxial spray along the z-axis.

Let us compute the above deformations for the classical hedgehog director in 3D given by

$$\mathbf{d}(x, y, z) = \frac{(x, y, z)}{\sqrt{x^2 + y^2 + z^2}}$$

Clearly, since $\nabla \times \mathbf{d} = (0,0,0)$, we see that twist $T = 0$. We further find that $S = \nabla \cdot \mathbf{d} = \frac{2}{\sqrt{x^2+y^2+z^2}}$. Moreover, since $\nabla \mathbf{d} = (\nabla \mathbf{d})^T$, we find that the bend $\mathbf{b} = (0,0,0)$. Finally, we compute the biaxial spray: Since we have shown that bend and spray are both zero, we quickly compute that $\Delta_{ij} = 0$ as well. Thus the hedgehog director only has spray.

Finally, we provide a geometric interpretation of the representation (3.1). Since $\mathbf{d}$ is a unit vector, it defines a splitting of the tangent bundle of $\mathbb{R}^3$

$$T\mathbb{R}^3 \cong \mathbf{L_d} \oplus A$$

where $\mathbf{L_d}$ is a line bundle of vectors which are parallel to $\mathbf{d}$ and A is a rank 2 vector bundle of vectors orthogonal to $\mathbf{d}$ [266]. Gradients of $\mathbf{d}$ take values in $T\mathbb{R}^3 \otimes A \cong \left(\mathbf{L}_\mathbf{d}^* \otimes A\right) \oplus (A^* \otimes A)$; the first part contains the bend deformations and the second part contains orthogonal gradients, locally given by

$$\nabla_\perp \mathbf{d} = \frac{\nabla \cdot \mathbf{d}}{2} \begin{pmatrix} 1 & 0 \\ 0 & 1 \end{pmatrix} + \frac{\mathbf{d} \cdot \nabla \times \mathbf{d}}{2} \begin{pmatrix} 0 & -1 \\ 1 & 0 \end{pmatrix} + \begin{pmatrix} \Delta_1 & \Delta_2 \\ \Delta_2 & -\Delta_1 \end{pmatrix}$$

where

$$\Delta_1 = \frac{1}{2}\left(\partial_1 d_2 - \partial_2 d_1\right), \qquad \Delta_2 = \frac{1}{2}\left(\partial_1 d_2 + \partial_2 d_1\right)$$

The above orthogonal gradients are called the shape operator of the orientational order. Why might this be appropriate? Well, if $\mathbf{d}$ is normal to a surface, then (3.1) is precisely the shape operator [109] of the surface. The last three terms in (3.1), as mentioned above, are gradients orthogonal to $\mathbf{d}$. The $\nabla \cdot \mathbf{d}$ term and $\nabla \cdot \mathbf{d}$ term are the isotropic terms, while the last term Δ_{ij} corresponds to anisotropic orthogonal gradients of $\mathbf{d}$. The splay and twist terms correspond to mean curvature and mean torsion. The eigenvalues of Δ are used to define principal directions of the curvature of the local orientation, that is, the directions of principal curvature (if $\mathbf{d}$ is normal to a surface). Note also that the decomposition (3.1) is compatible with the well-known energy expression

$$|\nabla \mathbf{d}|^2 = \frac{1}{2}(\nabla \cdot \mathbf{d})^2 + \frac{1}{2}(\mathbf{d} \cdot \nabla \times \mathbf{d})^2 + ((\mathbf{d} \cdot \nabla)\mathbf{d})^2 + 2|\Delta|^2$$

from, e.g., [98].

We can write this more generally. Indeed, local rotations about $\mathbf{d}$ provide the bundle A with a $SO(2)$ action under which the product $A^* \otimes A$ decomposes into a direct sum $A^* \otimes A = I \oplus J \oplus K$, where I, J are trivial line bundles and K is a vector bundle of rank two. Thus the shape operator decomposes as

$$\nabla_\perp \mathbf{d} = \frac{\nabla \cdot \mathbf{d}}{2}\mathbf{I} + \frac{\mathbf{d} \cdot \nabla \times \mathbf{d}}{2} j + \Delta$$

where $\mathbf{I}$ is the identity transform, $j = \mathbf{d}\times$ is a complex structure (on A), and Δ is globally [266] given by

$$\Delta(\mathbf{u}) = \frac{1}{2}\left((\mathbf{u}\cdot\nabla)\mathbf{d} + \mathbf{d}\times(\mathbf{d}\times\mathbf{u}\cdot\nabla)\mathbf{d}\right), \qquad \forall\, \mathbf{u}\in A$$

3.2 The Effective Refractive Index

We have seen two refractive index surfaces in an uniaxial medium, corresponding to (2.18) and (2.19), respectively, yielding a sphere and ellipsoid in $\mathbf{p}$ space. We have also seen that there are two points where the surfaces intersect; the axis which connects these points is the optical axis of the uniaxial material. We saw that the ordinary ray (corresponding to $\mathcal{H}_o$) behaves like an isotropic material since $\nabla_{\mathbf{p}}\mathcal{H}_o = \mathbf{p}$. The extraordinary ray (corresponding to $\mathcal{H}_e$), however, has a refractive index which depends on position (inhomogeneity) and direction of propagation (anisotropy). Two *effective* refractive indices can be obtained for the extraordinary ray, depending on which direction is of interest. In both cases we will see that the effective index will depend on the ordinary and extraordinary refractive indices as in (2.16). We see then that there are two rays propagating in an uniaxial material: the ordinary ray and the extraordinary ray. This is the phenomenon of *birefringence* (see, e.g., [362]) and does not appear in isotropic materials. If $n_e > n_o$, we say there is positive birefringence, and if $n_o > n_e$, we say there is negative birefringence.

Below we will derive the effective refractive index for a liquid crystal with uniaxial anisotropy (although a similar formula can be obtained for general anisotropic media; see, e.g., equation (2.56) in [362]).

3.2.1 The Wave Vector Direction

In the spirit of the Eikonal equation (2.14), we write the effective index as

$$n_{\text{eff},p} := |\mathbf{p}| \tag{3.7}$$

In this way, we can decompose the wave normal $\mathbf{p}$ in terms of its magnitude and direction:

$$\mathbf{p} = |\mathbf{p}|\mathbf{v} = n_{\text{eff},p}\mathbf{v} \tag{3.8}$$

where $\mathbf{v} = \mathbf{p}/|\mathbf{p}|$ is the direction of propagation in the liquid crystal.

Using (3.8) in (2.21) yields the effective refractive index

$$n_{\text{eff},p}(\mathbf{x},\mathbf{v}) = \sqrt{\frac{n_o^2 n_e^2}{n_o^2\left(1-(\mathbf{v}\cdot\mathbf{d}(\mathbf{x}))^2\right)+n_e^2(\mathbf{v}\cdot\mathbf{d}(\mathbf{x}))^2}} \tag{3.9}$$

where recall $\mathbf{d}$ is the director.

Note that if $\theta(\mathbf{x},\mathbf{v})$ denotes the angle between the wave vector $\mathbf{p}$ and the director, then (3.9) becomes

$$n_{\text{eff},p}(\mathbf{x},\mathbf{v}) = \frac{n_o n_e}{\sqrt{n_o^2\sin^2(\theta(\mathbf{x},\mathbf{v}))+n_e^2\cos^2(\theta(\mathbf{x},\mathbf{v}))}},$$

which is the more familiar form of the effective refractive index in terms of θ, see, e.g., [362, Equation (3.56)]. Sometimes this effective refractive index is called the effective index associated with the phase velocity (for this reason we index the effective refractive index here with a p).

3.2.2 Poynting Vector Direction

Following Section 3.3 in [220], we let

$$\nabla_{\mathbf{p}}\mathcal{H} = \frac{1}{n_{\text{eff},r}}\mathbf{v}$$

where $\mathbf{v}$ is a vector pointing in the direction of $\nabla_{\mathbf{p}}\mathcal{H}$ (not necessarily a unit vector). Plugging this ansatz into (2.21) and solving for $\mathbf{p}$ yield a new vector $\mathbf{q}$. For example, in 2D $\mathbf{q}$ takes the explicit form

$$\mathbf{q} = \frac{1}{2n_{\text{eff},r}n_o^2n_e^2}\left((n_o^2-n_e^2)d_2(d_1v_2-d_2v_1)+n_o^2v_1,\ (n_o^2-n_e^2)d_1(d_1v_2-v_1d_2)\right)$$

where $\mathbf{d}=(d_1,d_2)$ and $\mathbf{v}=(v_1,v_2)$. Then, we evaluate $\mathcal{H}(\mathbf{r},\mathbf{q})=0$ (the extraordinary term) and solve for the effective index:

$$n_{\text{eff},r}^2 = \frac{n_e^2|\mathbf{v}|^2+(n_o^2-n_e^2)(\mathbf{v}\cdot\mathbf{d})^2}{2n_e^2n_o^2} \tag{3.10}$$

To stay consistent with the literature (e.g., [220], Equation (3.40)), we scale (3.10), which does not affect the extremals (to be discussed in Chap. 4; see also Sect. 5.6), and consider (dropping the index r for convenience)

$$n_{\text{eff}}^2 = n_e^2|\mathbf{v}|^2+(n_o^2-n_e^2)(\mathbf{v}\cdot\mathbf{d})^2 \tag{3.11}$$

Largely the effective index (3.11) is the one we will use, since we are interested in the direction of energy flux. Indeed, $n_{\text{eff},r}$ is sometimes known as the ray index (hence the index r) and is associated with energy velocity. In particular, notice that (3.11) is homogeneous of degree 1, i.e., satisfies

$$n_{\text{eff}}(\mathbf{x}, \lambda\mathbf{v}) = \lambda\, n_{\text{eff}}(\mathbf{x}, \mathbf{v}), \qquad \lambda > 0$$

This turns out to be an important property in the context of Fermat's Principle; this will be discussed more in Chap. 4.

Remark 3.1 As mentioned above, the difference $\Delta n = n_e - n_o$ is called the birefringence. A typical birefringence of a nematic material is roughly $\Delta n \approx 0.1$, see, e.g., [220].

Finally, let us mention the effective refractive index if the nematic is biaxial. Considering the wave vector direction, we again let $\mathbf{p} = n_{\text{eff}}\mathbf{v}$ (so that $n_{\text{eff}} = |\mathbf{p}|$) and plug this into the biaxial optical indicatrix from equation (2.28). We saw in Chap. 2 that this yields the surface (with $n_{\text{eff}}^2 = |\mathbf{p}|^2$)

$$n_{\text{eff}}^2(\mathbf{v}) = \frac{\varepsilon_u|v_u|^2(\varepsilon_v+\varepsilon_w) + \varepsilon_v|v_v|^2(\varepsilon_u+\varepsilon_w) + \varepsilon_w|v_w|^2(\varepsilon_u+\varepsilon_v) \pm \sqrt{f(\varepsilon, \mathbf{v})}}{2\left(\varepsilon_u|v_u|^2 + \varepsilon_v|v_v|^2 + \varepsilon_w|v_w|^2\right)} \tag{3.12}$$

where

$$\begin{aligned} f(\varepsilon, \mathbf{v}) = &\left(\varepsilon_u|v_u|^2(\varepsilon_v+\varepsilon_w) + \varepsilon_v|v_v|^2(\varepsilon_u+\varepsilon_w) + \varepsilon_w|v_w|^2(\varepsilon_u+\varepsilon_v)\right)^2 \\ &- 4\left(\varepsilon_u|v_u|^2 + \varepsilon_v|v_v|^2 + \varepsilon_w|v_w|^2\right)\varepsilon_u\varepsilon_v\varepsilon_w \end{aligned}$$

with $\mathbf{v} = (v_u, v_v, v_w)$. Thus for any direction of propagation $\mathbf{v}$ there are two solutions for n_{eff}^2. If $\mathbf{p}$ is in the direction of $\mathbf{d}_1$ or $\mathbf{d}_2$, then we know from Chap. 2 that $|\mathbf{p}| = \sqrt{\varepsilon_v}$, so that $\mathbf{p} = \sqrt{\varepsilon_v}\mathbf{v}$. We also know that (see equation (2.26))

$$\mathbf{p} = \sqrt{\frac{\varepsilon_w(\varepsilon_v - \varepsilon_u)}{\varepsilon_w - \varepsilon_v}}\mathbf{e}_u + \sqrt{\frac{\varepsilon_u(\varepsilon_w - \varepsilon_v)}{\varepsilon_w - \varepsilon_u}}\mathbf{e}_w$$

which, in light of (2.23), can be written

$$\mathbf{p} = \sqrt{\frac{\varepsilon_w(\varepsilon_v - \varepsilon_u)}{\varepsilon_w - \varepsilon_v}}\frac{\mathbf{d}_1 - \mathbf{d}_2}{|\mathbf{d}_1 - \mathbf{d}_2|} + \sqrt{\frac{\varepsilon_u(\varepsilon_w - \varepsilon_v)}{\varepsilon_w - \varepsilon_u}}\frac{\mathbf{d}_1 + \mathbf{d}_2}{|\mathbf{d}_1 + \mathbf{d}_2|} \tag{3.13}$$

Now, just as in (2.31) (see also (3.12)), setting $\mathbf{p} = n_{\text{eff, bi}}\mathbf{v}$, with $|\mathbf{v}| = \mathbf{p}/|\mathbf{p}|$, in general, we find for the square of the effective refractive index that

$$n^2_{\text{eff, bi}}(\mathbf{x}, \mathbf{v}) = \frac{-B(\mathbf{x}, \mathbf{v}) \pm \sqrt{B(\mathbf{x}, \mathbf{v})^2 - 4A(\mathbf{x}, \mathbf{v})C}}{2A(\mathbf{x}, \mathbf{v})} \tag{3.14}$$

where

$$\begin{aligned}
A(\mathbf{x}, \mathbf{v}) &= \varepsilon_u|\mathbf{v}\cdot\mathbf{e}_u|^2 + \varepsilon_v|\mathbf{v}\cdot\mathbf{e}_v|^2 + \varepsilon_w|\mathbf{v}\cdot\mathbf{e}_w|^2,\\
B(\mathbf{x}, \mathbf{v}) &= \varepsilon_u\varepsilon_v\left(|\mathbf{e}_u\cdot(\mathbf{v}\times\mathbf{e}_v)|^2 - 1\right) + \varepsilon_u\varepsilon_w\left(|\mathbf{e}_u\cdot(\mathbf{v}\times\mathbf{e}_w)|^2 - 1\right)\\
&+ \varepsilon_v\varepsilon_w\left(|\mathbf{e}_v\cdot(\mathbf{v}\times\mathbf{e}_w)|^2 - 1\right)\\
C &= \varepsilon_u\varepsilon_v\varepsilon_w
\end{aligned}$$

In terms of $n_u = \varepsilon_u^2$, $n_v = \varepsilon_v^2$, $n_w = \varepsilon_w^2$ and the directors $\mathbf{d}_1$, $\mathbf{d}_2$, this reads

$$\begin{aligned}
A(\mathbf{x}, \mathbf{v}) &= n_u^2\left|\mathbf{v}\cdot\left(\frac{\mathbf{d}_1-\mathbf{d}_2}{|\mathbf{d}_1-\mathbf{d}_2|}\right)\right|^2 + n_v^2\left|\mathbf{v}\cdot\left(\frac{\mathbf{d}_1\times\mathbf{d}_2}{|\mathbf{d}_1\times\mathbf{d}_2|}\right)\right|^2 + n_w^2\left|\mathbf{v}\cdot\left(\frac{\mathbf{d}_1+\mathbf{d}_2}{|\mathbf{d}_1+\mathbf{d}_2|}\right)\right|^2,\\
B(\mathbf{x}, \mathbf{v}) &= n_u^2n_v^2\left(\left|\frac{\mathbf{d}_1-\mathbf{d}_2}{|\mathbf{d}_1-\mathbf{d}_2|}\cdot\left(\mathbf{v}\times\frac{\mathbf{d}_1\times\mathbf{d}_2}{|\mathbf{d}_1\times\mathbf{d}_2|}\right)\right|^2 - 1\right)\\
&+ n_u^2n_w^2\left(\left|\frac{\mathbf{d}_1-\mathbf{d}_2}{|\mathbf{d}_1-\mathbf{d}_2|}\cdot\left(\mathbf{v}\times\frac{\mathbf{d}_1+\mathbf{d}_2}{|\mathbf{d}_1+\mathbf{d}_2|}\right)\right|^2 - 1\right)\\
&+ n_v^2n_w^2\left(\left|\frac{\mathbf{d}_1\times\mathbf{d}_2}{|\mathbf{d}_1\times\mathbf{d}_2|}\cdot\left(\mathbf{v}\times\frac{\mathbf{d}_1+\mathbf{d}_2}{|\mathbf{d}_1+\mathbf{d}_2|}\right)\right|^2 - 1\right),\\
C &= n_u^2n_v^2n_w^2
\end{aligned} \tag{3.15}$$

Recall that the plus sign in (3.14) stands for the so-called b-wave and the minus sign the a-wave.

Remark 3.2 Note that the above effective refractive index is not homogeneous in $\mathbf{v}$ (either of degree one or two). As such, one cannot consider the functional

$$\mathcal{Q}_{\text{bi}}(\mathbf{x}, \mathbf{v}) = \int_0^1 n^2_{\text{eff,bi}}(\mathbf{x}, \mathbf{v})dt$$

as a parametric variational integral in the usual sense [72, 139] since one usually requires homogeneity of the integrand. In particular, homogeneity of the integrand implies that the functional would be invariant under reparameterizations (e.g., by arclength).

Next we move on to some basic mathematics used to study liquid crystals.

3.3 Order Parameters: The Q-Tensor and Director d

It is useful to have an order parameter to somehow define a degree of order (or rather, disorder); this was developed by Landau to describe phase transitions. The idea is to provide a continuum-level description of nematic liquid crystals, and various liquid crystal theories (to be described in the following sections) describe the material as functions taking values in the so-called order parameter spaces. Somehow these spaces encode the local order of the molecules [457]. We have already seen one such space for the director: the unit sphere (recall **d** is a unit vector). But let us think a bit more on this. If we have a uniaxial nematic, then **d** is equivalent to $-\mathbf{d}$, so we have to factor the unit sphere S^2 by a set of opposite points on the sphere [220]. Thus the order parameter space for a uniaxial nematic is

$$\mathcal{OP}_{UN} = S^2/Z_2$$

where Z_2 is the group $\{0, 1\}$ consisting of just two numbers. But S^2/Z_2 is also denoted RP^2, the real projective plane. Note that the notation G/H means the coset space of the subgroup H in the group G.

For a biaxial nematic, the order is specified by three directors actually: $\mathbf{d}_1$, $\mathbf{d}_2$ and $\mathbf{d}_1 \times \mathbf{d}_2$, such that $\mathbf{d}_1$ is equivalent to $-\mathbf{d}_1$, $\mathbf{d}_2$ is equivalent to $-\mathbf{d}_2$, and $\mathbf{d}_1 \times \mathbf{d}_2$ is equivalent to $-\mathbf{d}_1 \times \mathbf{d}_2$. The order parameter space is the group $SO(3)$ of rotations of $\mathbf{d}_1, \mathbf{d}_2, \mathbf{d}_1 \times \mathbf{d}_2$ factored by the four element point group D_2 of π-rotations about the directions $\mathbf{d}_1, \mathbf{d}_2, \mathbf{d}_1 \times \mathbf{d}_2$:

$$\mathcal{OP}_{BN} = SO(3)/D_2$$

It is possible to push further and use these spaces to topologically classify the so-called defects (see the next section). This can be done by defining homotopy groups of the order parameter spaces. Each element of these homotopy groups corresponds to a class of stable defects, but each element of the homotopy groups is also topological invariants (topological charge of the defects). See, for example, [220] for more in this direction.

To derive an order parameter besides the director, we start by thinking of the director **d** for a nematic as a result of averaging over the axes carried by the individual molecules. Denote by **a** a unit vector parallel to a given molecule, which we will treat as a rod. Both **a** and $-\mathbf{a}$ should be taken with equal probability since the nematic does not recognize different extremities of the molecules. We can write $\mathbf{a}$ in terms of the polar angle θ and azimuthal angle ϕ:

$$\mathbf{a} = (\sin(\theta)\cos(\phi), \sin(\theta)\sin(\phi), \cos(\theta))$$

This is an "ellipsoidal" coordinate system—notice θ in the last coordinate in $\mathbf{a}$. The state of alignment of the molecule is described by a distribution function $f(\theta, \phi)d\Omega$, which gives the probability of finding the molecule in a small solid angle $d\Omega = \sin(\theta)d\theta d\phi$. Again, since $\mathbf{a}$ and $-\mathbf{a}$ are the same in the nematic, we must have $f(\theta, \phi) = f(\pi - \theta, \phi)$. Since f does not actually depend on ϕ, we find that $\frac{\partial f}{\partial \phi} = 0$, and the function f is normalized so that

$$2\pi \int_0^\pi f(\theta)\sin(\theta)d\theta = 1$$

In the isotropic phase, f is actually constant, and this normalization condition implies that

$$f(\theta) = \frac{1}{4\pi}$$

Let us get to a tensor which describes the average molecular orientation along $\mathbf{d}$ which vanishes in the isotropic phase. Define the scalar

$$s = 2\pi \int_0^\pi \frac{(3\cos^2(\theta) - 1)}{2} f(\theta)\sin(\theta)d\theta = 2\pi \int_0^\pi P_2(\cos(\theta)) f(\theta)\sin(\theta)d\theta$$

where $P_2(\cos(\theta))$ is the Legendre polynomial of order two. The quantity s measures the degree of orientational order. Why might this work? Well, in an isotropic phase, all orientations have equal probability, so $f(\theta) = \frac{1}{4\pi}$ as seen before, and this yields $s = 0$ (one can simply do the integral by u substitution). On the other hand, when there is the most order possible, the only orientation is $\theta = 0$, and so $f(\theta) = \frac{1}{4\pi}\delta(\theta)$ where δ denotes the Dirac Delta function. This yields $s = 1$; indeed, since the δ function pulls out the value of P_2 at $\theta = 0$, we find

$$s = 2\pi \int f(\theta)\sin(\theta)d\theta = 1$$

Thus, the parameter s can be used to describe the nematic–isotropic liquid phase transition [220].

Let us work a little more carefully now. Let S^2 be the unit sphere and write our probability distribution $f = f(\mathbf{x}, \mathbf{a})$ with $\mathbf{a} \in S^2$. We compute the first moment:

$$\int_{S^2} \mathbf{a} f(\mathbf{x}, \mathbf{a})d\mathbf{a} = -\int_{S^2} \mathbf{a} f(\mathbf{x}, -\mathbf{a})d\mathbf{a} = 0$$

where $d\mathbf{a}$ denotes the area element on the sphere. Let $\mathbf{M}(\mathbf{x})$ denote the second moment:

$$\mathbf{M}(\mathbf{x}) = \int_{S^2} \mathbf{a} \otimes \mathbf{a} f(\mathbf{x}, \mathbf{a}) d\mathbf{a}$$

We recall that the tensor product $\mathbf{a} \otimes \mathbf{a}$ stands for the matrix $\mathbf{a}\mathbf{a}^T$. $\mathbf{M}$ is a symmetric, second order tensor such that $\mathbf{M}(\mathbf{x})\mathbf{e} \cdot \mathbf{e} > 0$ for all $\mathbf{e} \in S^2$; i.e., $\mathbf{M}$ is positive definite [26]. This is since

$$\mathbf{M}(\mathbf{x})\mathbf{e} \cdot \mathbf{e} = \int_{S^2} (\mathbf{a} \cdot \mathbf{e})^2 f(\mathbf{x}, \mathbf{a}) d\mathbf{a} \geq 0$$

and equality would imply that $f(\mathbf{x}, \mathbf{a}) = 0$ for $\mathbf{a} \cdot \mathbf{e} \neq 0$, which contradicts the fact that f is a probability distribution. Now, in the isotropic phase, when (as we have seen above) $f = \frac{1}{4\pi}$, we find that $\mathbf{M}(\mathbf{x}) = \frac{1}{3}\mathbf{I}$, where $\mathbf{I}$ denotes the identity. Finally, we define the de Gennes $\mathbf{Q}$ tensor by

$$\mathbf{Q}(\mathbf{x}) = \mathbf{M}(\mathbf{x}) - \frac{1}{3}\mathbf{I} = \int_{S^2} \left(\mathbf{a} \otimes \mathbf{a} - \frac{1}{3}\mathbf{I} \right) f(\mathbf{x}, \mathbf{a}) d\mathbf{a} \tag{3.16}$$

This tensor measures how far $\mathbf{M}(\mathbf{x})$ is from its isotropic value. From the properties of $\mathbf{M}$ and f, we find that $\mathbf{Q}$ satisfies

$$\mathbf{Q}(\mathbf{x}) = \mathbf{Q}^T(\mathbf{x}), \quad \mathrm{Tr}(\mathbf{Q}(\mathbf{x})) = 0, \quad \lambda_{\min}(\mathbf{Q}(\mathbf{x})) > -\frac{1}{3} \tag{3.17}$$

where $\lambda_{\min}(\mathbf{Q}(\mathbf{x}))$ denotes the smallest eigenvalue of $\mathbf{Q}(\mathbf{x})$. Thus, as will be discussed below, in the Landau–de Gennes theory of liquid crystals, one takes the Q-tensor as the order parameter from the space of Q-tensors

$$\mathcal{S}_0 = \left\{ \mathbf{Q} \in \mathbb{R}^{3\times 3} : \mathbf{Q} = \mathbf{Q}^T, \ \mathrm{Tr}(\mathbf{Q}) = 0 \right\} \tag{3.18}$$

For a biaxial nematic, the tensor $\mathbf{Q}$ takes the form

$$\mathbf{Q}(\mathbf{x}) = s_1 \left(\mathbf{d}_1 \otimes \mathbf{d}_1 - \frac{1}{3}\mathbf{I} \right) + s_2 \left(\mathbf{d}_2 \otimes \mathbf{d}_2 - \frac{1}{3}\mathbf{I} \right) \tag{3.19}$$

where s_1, s_2 are scalar order parameters. In the isotropic phase, we have $s_1 = s_2 = 0$, and in the uniaxial phase either $s_1 = 0, s_2 \neq 0, s_1 \neq 0, s_2 = 0$, or $s_1 = s_2$, which recovers (3.16). We can see this more explicitly by considering a perfectly aligned system where the molecules align along a pair of unit vectors $(\mathbf{d}, -\mathbf{d})$ (named this way on purpose!); in this case, the distribution function f is given by [271]

$$f(\mathbf{a}) = \frac{1}{8\pi} (\delta(\mathbf{d}, \mathbf{a}) + \delta(-\mathbf{d}, \mathbf{a}))$$

where δ again denotes the Dirac Delta on S^2. In this case the Q-tensor reduces to

$$\mathbf{Q}(\mathbf{x}) = \left(\mathbf{d} \otimes \mathbf{d} - \frac{1}{3}\mathbf{I}\right)$$

Finally, we can talk about liquid crystal phases in terms of eigenvalues of the Q-tensor. The phase is isotropic when $\mathbf{Q}$ has three equal eigenvalues (in fact then $\mathbf{Q}$ vanishes, as we can see above). The phase is uniaxial when $\mathbf{Q}$ has two nonzero equal eigenvalues, and the phase is biaxial when $\mathbf{Q}$ has three distinct eigenvalues. This Q-tensor plays a big role in the Landau–de Gennes theory which we briefly describe below.

3.4 Defects

We have seen a few examples of order parameters; here for simplicity we will just focus on the director $\mathbf{d}$, and for a nematic at first. There can be distortions of $\mathbf{d}$, and this can be when $\mathbf{d}$ has singularities or not [220, 239, 246]. In $\mathbb{R}^3$, singular regions could be zero dimensional; this corresponds to what are called *point defects* [221]. They could also be one dimensional; this corresponds to *line defects*. Line defects along which a symmetry of rotation is broken are called a *disclination*. Essentially, disclinations arise when the director $\mathbf{d}$ makes a continuous π-rotation along a closed loop (which thus changes $\mathbf{d}$ to $-\mathbf{d}$). For example, in $\mathbb{R}^3$, let ϕ denote the angle between the director $\mathbf{d}$ and the z-axis, and θ the polar angle between the projection of $\mathbf{d}$ onto the $x - y$ plane and the x-axis; we may then write

$$\mathbf{d} = (\cos(\theta)\sin(\phi), \sin(\theta)\sin(\phi), \cos(\phi)) \tag{3.20}$$

The director around a wedge disclination along y can be described by

$$\phi = s \arctan(z/x) + \phi_0, \qquad \theta = 0$$

and a twist disclination along y can be described by

$$\phi = \frac{\pi}{2}, \qquad \theta = s \arctan(z/x) + \theta_0$$

where s is the strength of the disclination and θ_0, ϕ_0 are constant angles. Now, the strength of a disclination is the ratio $\alpha/2\pi$, where α is the angle the director rotates after traversing a certain curve (the so-called Burgers circuit [220, 222]). The strength s takes the values $\pm\frac{1}{2}, \pm 1, \pm\frac{3}{2}, \ldots$. Most often disclinations of strength $\pm\frac{1}{2}$ are observed in nematics. There are different types of disclinations in nematics: wedge and twist [238]. In a wedge disclination, the axis of rotation is along the

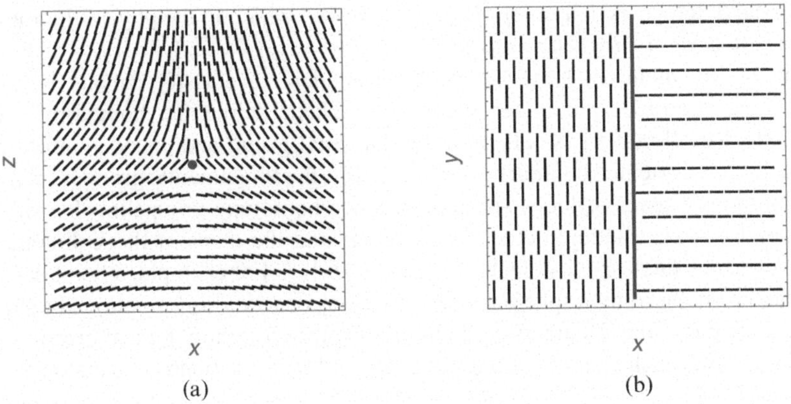

Fig. 3.10 Example wedge and twist disclinations. The left plot is in the $x - z$ plane, while the right plot is in the $x - y$ plane. (**a**) A wedge disclination. (**b**) A twist disclination in the $z = 0$ plane

disclination line, while in the twist disclination, the axis of rotation is perpendicular to this line. In Fig. 3.10 we illustrate a wedge and twist disclination. Figure 3.10a illustrates a wedge disclination in the $x - z$ plane (recall how we defined **d** in (3.20); the disclination line is denoted by the blue dot and comes out of the page (along y). We have taken this wedge to be $s = -\frac{1}{2}$ strength. Figure 3.10b is a $s = \frac{1}{2}$ strength twist in the $x - y$ plane (we have plotted the plane which contains the disclination). The blue line denotes the disclination line; the director **d** rotates around z.

Predicting defect patterns is useful in many areas, such as to the self-assembly of biomolecules and colloidal suspensions [187], but it is quite difficult in general to do this. We will go into this more here, but merely refer the interested reader to the following references [121, 127, 213, 259, 299, 451].

Since we are typically interested in interfaces in this text, it is worth noting that a disclination is usually observed near curved surface interfaces separating a nematic from another material (such as air or glass) [90]. This is due to the boundary condition imposed on the director **d** at the surface interface, and the curved interface itself causes the director to turn. In fact, the line defect can even have a loop shape [408].

Finally, the singular region could be three dimensional; this corresponds to *wall defects*. All of these defects are rather striking optical features of a nematic, as they arise when polarized light is passed through a thin liquid crystal layer. The defects manifest as localized changes in intensity and color of light.

The main mathematical techniques used to study liquid crystal defects involve homotopy theory from algebraic geometry. Recall that, given the interval [0, 1], a family of maps $T_t : X \to Y$ is called a homotopy from the map T_0 to T_1 if $T : I \times X \to Y,\ (t, x) \mapsto T_t(x)$ is a continuous function. One of the key notions to understand defects is that of *topological invariant*, which in the physics literature is often called *topological charge*.

As briefly mentioned before, the general ideal to classify defects is to first determine the order parameter space $\mathcal{OP}$. Then, define the homotopy groups $\pi_i(\mathcal{OP})$. Recall that π_1 is called the fundamental group and encodes information about loops or oriented contours (more precisely, the fundamental group π_1 is the set of homotopy classes of loops at a point). Since homotopy groups involve spheres, the key for understanding defects is to understand the maps of the k dimensional spheres which enclose the defects, where $k = 0, 1, 2$. For example, a point defect in $\mathbb{R}^3$ can be enclosed by a sphere so $k = 2$; i.e., π_2 encodes[4] information about point defects. In particular, elements of the homotopy group are topological invariants. At the same time, each element of the homotopy group corresponds to a class of stable defects [294].

Consider the two dimensional hedgehog $\mathbf{d}(x, y) = \frac{(x,y)}{\sqrt{x^2+y^2}}$. We can realize this as the projection of the full three dimensional hedgehog into the $x - y$ plane. The order parameter space $\mathcal{OP}$ depends on the angle ϕ made between the three dimensional hedgehog and the normal vector to the $x - y$ plane, i.e., $(0, 0, 1)$. Suppose this angle is $0 < \phi < \pi/2$; in this case, $\mathcal{OP} = S^1$, the entire circle [220]. Note that S^1 is not simply connected (here we explicitly use the dimension of the sphere since the unit sphere *is* simply connected in higher dimensions), and thus not every curve is contractible to a point. Consider a closed contour around the "interesting point" $(0, 0)$. Then, the image of the contour under the map $\mathbf{d}$ produces another loop on S^1, which actually runs the whole circle S^1, which we know cannot be contracted continuously. To "remove" the defect, a substantial amount of energy is required, e.g., by cutting S^1, and this energy would be much larger than any 2D distortions. Thus this defect we say is topologically stable, i.e., the vector field cannot be continuously deformed into a uniform state $\mathbf{d} =$ constant. What we see in action here is that the fundamental group $\pi_1(S^1) = \mathbb{Z}$, whose elements k (called the topological charge) count the number of times the path winds around, i.e., the winding number. Note also that $\pi_2(S^2) = \mathbb{Z}$ (we write "=" but really it should be understood that these groups are isomorphic).

In particular, we saw above that $\pi_1(S^1)$ is abelian, but this is not always the case for defects: It depends on the order parameter space $\mathcal{OP}$. In non-abelian groups, things get trickier, and non-commutativity of the fundamental group can lead to interesting physical effects. For instance, in a non-abelian space, a point defect does not correspond to one element of the fundamental group π_1, but rather an entire class of conjugated elements. Recall that, in a group G, two elements p and q are called conjugate to each other if there exists an element $g \in G$ such that $q = g^{-1}pg$. Physically, this results in the fact that the merging of two defects is not well-defined.

---

[4] Recall that π_2 is the set of homotopy classes of the so-called 2-loop, which can be defined as follows. Let I_2 denote the unit cube $\{(x, y) : 0 \leq x, y \leq 1\}$ and ∂I_2 its geometrical boundary, i.e., $\partial I_2 = \{(x, y) \in I_2 : x \text{ or } y = 0 \text{ or } 1\}$. Then $I_2/\partial I_2 \cong S^2$, and a 2-loop is the map α which maps the boundary ∂I_2 to a single point.

Finally, all of the above discussion has been for (uniaxial) nematics, so we briefly discuss now defects in smectics , cholesterics, and biaxial nematics.

We start with biaxial nematics. As we saw previously, the order parameter space for biaxial nematics is

$$\mathcal{OP}_{BN} = SO(3)/D_2$$

where D_2 is the four element group of π rotations about the directors $\mathbf{d}_1, \mathbf{d}_2$ as well as $\mathbf{d}_1 \times \mathbf{d}_2$. First, note that biaxial nematics cannot have point defects since[5] $\pi_2(SO(3)/D_2) = 0$, but disclinations are possible, and they differ from the uniaxial case.

Since π-rotations around axes oriented in opposite directions produce the same thing, we can represent $SO(3)$ as RP^3, the real projective space. It can be shown [220] that the fundamental group of this group is the non-abelian group Q

$$\pi_1(S^3/D_2) = Q$$

where Q denotes the eight element group of *quaternions*. Note that we have considered S^3/D_2 rather than $SO(3)/D_2$. This is because S^3 "is" the universal cover of $SO(3)$. Recall that the universal cover of a connected topological space X is a simply connected space Y with a map $f : X \to Y$ which is a covering map, i.e., a map which is surjective, open, and locally a homeomorphism [179]. The point of introducing the universal cover is that any property of X can be lifted to its universal cover.

We have previously written "is" in quotations because this is slightly imprecise. This is because the universal cover of $SO(3)$ is actually the Lie group Spin(3) [175], which is diffeomorphic to the unit sphere S^3. Indeed, since the group $SU(2) \cong SO(3)$, the quotient map $f : SU(2) \to \mathbb{Z}_2\ SU(2) \cong SO(3)$ is the universal cover of $SO(3)$.

Now, each of the directors $\mathbf{d}_1, \mathbf{d}_2, \mathbf{d}_1 \times \mathbf{d}_2$ can carry disclinations of (uniaxial) nematic type. In particular, a singular distortion can affect only part of the "full" order parameter [219]. For instance, $\mathbf{d}_1, \mathbf{d}_2$ could be singular, but $\mathbf{d}_1 \times \mathbf{d}_2$ is not. For biaxial nematics, there are five (rather than one, as is the case for uniaxial nematics) classes of topologically stable lines; these correspond to the five classes of conjugated elements of Q.

Topological classification of defects for biaxial nematics can also be applied to cholesterics , at least in the case when the pitch is much smaller than the scale of deformations , i.e., the size of the liquid crystal sample. Perhaps this is unsurprising, since at large scale, orientational order of cholesterics is also prescribed by three directors: λ, along the local direction of the molecules; χ, along the helix axis; and $\tau = \lambda \times \chi$. It is interesting to note, though, that there is only really "one" director λ, since the cost of removing singularities due to the other directors would cost more

[5] This is also true for cholesterics.

energy than removing singularities due to λ. We also mention that χ disclinations and the so-called *dislocations* are essentially the same since symmetry of $n\pi$ rotations about the χ axis is actually equivalent to the symmetry of translations $n(p/2)\chi$, where p is the pitch of the cholesteric.

We have introduced now the notion of dislocation, which, in this setting, is essentially just a χ disclination. But what about more generally? While disclinations, as we have seen, are topologically related to rotational symmetry, dislocations are topologically related to *translational* symmetry. Dislocations are plastic deformations of solids and are the main kind of defect in solid crystals. However, this is not true for liquid crystals as we have seen: there are also disclinations and point defects. A thorough discussion of the elastic properties of dislocations is beyond the scope of this book, so we refer the interested reader in particular to Chapter 8 in [220]. We merely mention that there are different types of dislocations (which are of particular relevance in smectics, which we discuss below), including edge dislocations and screw dislocations, which are the simplest defects in translational order.

Consider smectic A, where the nematic director aligns with the layer normal. Recalling that smectics have a layered structure, in 2D, defects include point defects (just like 2D nematics) as well as dislocations (where a layer begins or ends). In 3D, again there are disclinations and a hedgehog-type point defect from the 3D nematic, as well as edge and screw dislocations.

There is much more to be said about defects in liquid crystals (see for instance the recent review [136]), but since we are interested in optics here, we end this section by discussing the relevance of defects in optics. It turns out that optical waveguiding can be accomplished along disclination lines, both singular and non-singular (defects can involve a director which is not necessarily singular). For non-singular disclinations, this can be realized through liquid crystal optical fibers. In particular, a liquid crystal slab can act as a converging lens when there is negative birefringence, i.e., $n_o > n_e$. Positive birefringence leads to diverging lens behavior, see [54] and section I.5 in [323]. Waveguiding can also occur for singular disclination lines; this has been observed, for instance, using filaments of smectic A grown in water [332]. This waveguiding occurs when light passes through a liquid crystal, but the liquid crystal can also be viewed as a light source when doped with laser dye and pumped by pulse light. In this case, one can obtain lasing from point or line defects [54].

Speaking of lasing, the bandgap structure of cholesteric liquid crystals allows for lasing [68, 330, 403, 444], even without external mirrors [129] (which usually form a laser cavity). It is also known that lasing can be achieved in dye-doped cholesteric mixtures [153]. Moreover, tunable lasing can be achieved with cholesteric elastomers by imposing uniaxial strain along the helical axis [419]. We have not yet mentioned elastomers, but elastomers in general are rubbery polymers[6] which can be easily stretched and quickly recover their original size when the applied stress

---

[6] A polymer is a compound composed of molecules which have long sequences of atoms or groups of atoms linked to each other by certain bonds [362].

is released. A liquid crystal elastomer is then a network of liquid crystal polymers, which can be thought of as polymers with liquid crystal properties, e.g., containing mesogens.

More recently, it has been experimentally shown how to control the flow of random laser light (i.e., lasing via recurrent scattering) by using spatial optical solitons (the so-called nematicons [23]) in dye-doped nematic liquid crystals [336]. A spatial optical soliton is a self-trapped optical beam which propagates in a nonlinear medium without diffraction [85]. Nematicons also present interesting mathematical problems, as the following example illustrates. In dimensionless form, the dynamics of nematicons in liquid crystals is governed by the system

$$
\begin{aligned}
&i\frac{\partial q}{\partial t} + a\frac{\partial^2 q}{\partial x^2} + b\theta q = 0,\\
&c\frac{\partial^2 \theta}{\partial x^2} + \lambda\theta + \alpha F(|q|)^2 = 0
\end{aligned}
\tag{3.21}
$$

where $q(x, t)$ represents the wave profile and $\theta(x, t)$ denotes the tilt angle of the liquid crystal molecule. The first term in the first equation in (3.21) represents the temporal evolution of nematicons, while the second term represents group velocity dispersion. Finally, $a, b, c, \lambda, \alpha \in \mathbb{R}$ are constant and F represents various nonlinearities, such as $F(s) = s$ for the Kerr law of nonlinear optics (cubic nonlinearity). Different kinds of soliton solutions of (3.21) can be found, e.g., the so-called bright and dark solitons. Given an ansatz of the form

$$
q(x, t) = P(x, t)e^{i\phi}, \qquad \theta(x, t) = Q(x, t)
$$

with phase $\phi = -\kappa x + \omega t_\sigma$ where κ is the soliton frequency, ω the wave number of the soliton, and σ a phase constant, bright solitons correspond to

$$
P(x, t) = A_1 \operatorname{sech}^{p_1}(B(x - vt)), \qquad Q(x, t) = A_2 \operatorname{sech}^{p_2}(B(x - vt))
$$

for some p_1, p_2, B. Dark solitons correspond to

$$
P(x, t) = A_1 + B_1 \tanh^{p_1}(B(x - vt)), \qquad Q(x, t) = A_2 + B_2 \tanh^{p_2}(B(x - vt))
$$

for some A_1, A_2, B_1, B_2. See for instance [361] for an analysis of bright and dark soliton solutions in liquid crystals and [20] for more general soliton solutions.

Colloids[7] in a dye-doped nematic liquid crystal can focus light and make a stable laser cavity [331, Section 9]. In particular, the dye-doped liquid crystal above and below the colloid can provide enough gain to enable lasing [189].

[7] A colloidal system is such that one of the components (the solute) is made of small enough particles so that surface effects dominate bulk effects. Familiar examples are milk and detergent.

Actually, it is possible to generate defects within initially homogeneous anisotropic media. This can be used to generate light beams having polarization-dependent helical wavefronts, i.e., the so-called optical vortices [52]. This can be accomplished by using Gaussian light beams to create reorientation defects. The optical torque of such a Gaussian beam (at normal incidence) on a nematic generates elastic distortions of the liquid crystal [53].

These optical beams are quite relevant, so let us briefly expand on them. The concept of optical vortex dates back to the work of Coullet et al. in [94], where they found vortex solutions of the Maxwell–Bloch equations and created the notion of optical vortex as inspired by hydrodynamic vortices [373]. Essentially an optical vortex is a special type of optical field which exists in a laser cavity, which is somehow analogous to a superfluid vortex. Indeed, like a fluid vortex, an optical vortex beam has a phase singularity with a particular topological charge. Here we will define the topological charge: If θ denotes the tilt angle of the molecules, then for point defects , the topological charge of θ (and more generally, the topological charge of a singularity for a phase distribution) is given by

$$k = \frac{1}{2\pi} \oint_{\gamma} \nabla\theta \cdot dl = 0, \pm 1, \pm 2, \dots$$

where γ is a closed loop around the point defect. Recall in our discussion of point defects, we found that $\pi_1(S^1) \cong \mathbb{Z}$, so each element of the fundamental group of S^1 can be labeled by an integer k as above. However, we should note that non-integer topological charge was investigated in optical vortices [41, 240]. Overall, a phase singularity with certain topological charge is a representation of a simple vortex soliton. Moreover, optical vortices have also been introduced (by symmetry considerations) as topological defects which arise above the laser transition.

Related to optical vortices are the so-called vortex beams, which are paraxial light beams carrying optical vortices along the propagation axis. There are also cylindrical *vector* beams, which, mathematically, are solutions of the Maxwell equations with axial symmetry in both amplitude and phase. These can be generated by either active or passive methods [458]. Active methods involve the use of laser intracavity devices which force the laser to oscillate in cylindrical vortex modes. Passive methods convert spatially homogeneous polarizations into spatially inhomogeneous cylindrical vortex polarizations (this can be done using optical fibers).

More precisely, in free space, paraxial solutions are obtained from the scalar Helmholtz equation [458]

$$(\Delta + k^2)E = 0$$

where Δ is the Laplace operator and $k = 2\pi/\lambda$ is the wave number. Beamlike paraxial solutions in free space take the form

$$E(\mathbf{x}, t) = u(\mathbf{x})e^{i(kz-\omega t)}, \qquad \mathbf{x} = (x, y, z)$$

Under the so-called slowly varying envelope approximation, one has

$$u_{zz} \ll k^2 u, \qquad u_{zz} \ll k u_z$$

From here, one can obtain the so-called Gauss–Hermite solution $HG_{m,n}$ modes simply by separation of variables. This leads to

$$u(\mathbf{x}) = E_0 H_m\left(\sqrt{2}\frac{x}{w(z)}\right) H_n\left(\sqrt{2}\frac{y}{w(z)}\right)\frac{w_0}{w(z)}\times$$
$$\exp\left(-i\phi_{mn}(z)\right)\exp\left(i\frac{k}{2q(z)}r^2\right)$$

where H_m denote the Hermite polynomials, E_0 is a constant, $w(z)$ is the beam size, $q(z) = z - iz_0$ is a complex beam parameter, and $\phi_{mn}(z) = (m + n + 1)\arctan(z/z_0)$ is the so-called Gouy phase shift. The case $m = n = 0$ corresponds to what is known as the Gaussian beam solution. One can do a similar analysis in cylindrical coordinates (under the same slowly varying approximation); this leads, via separation of variables, to the famous Laguerre–Gauss solution modes

$$u(r, \theta, z)$$
$$= E_0\left(\sqrt{2}\frac{r}{\omega}\right)^{\ell} L_p^{\ell}\left(2\frac{r^2}{\omega^2}\right)\frac{w_0}{w(z)}\exp\left(-i\phi_{p\ell}(z)\right)\exp\left(i\frac{k}{2q(z)}r^2\right)\exp\left(i\ell\phi\right)$$

where $\phi_{p\ell}(z) = (2p+\ell+1)\arctan(z/z_0)$ and L_p^{ℓ} denote the Laguerre polynomials. We mention these last modes because these are the earliest reported vortex beams carrying the so-called orbital angular momentum [5]. Finally we mention that the full vector Helmholtz equation can also be considered:

$$\nabla \times \nabla \times \mathbf{E} - k^2\mathbf{E} = 0$$

and similar solutions can be derived, and they involve Bessel functions [458].

Vortex [373, 427] and vector [352, 458] beams can be generated directly in a laser cavity [130]. But, a simple and easy to manufacture laser microcavity which generates structured light on demand is not readily available. But in [331] it was shown that tunable laser vector beams can be generated from self-assembled liquid crystal microstructures with topological defects inside a Fabry–Pérot microcavity.

The literature on defects is ever-growing, even from the mathematical perspective; see for instance [9, 14, 176, 177, 190–192, 215, 236, 377]. It is clear that the study of liquid crystal defects is quite rich, even if we have only alluded to a few aspects relevant to optics in particular.

Remark 3.3 We have been a bit cavalier in our distinctions between wedge disclinations in 2D and 3D, but one should really be more careful, especially as it relates to the famous "escape into the third dimension" [89, 296]. Roughly, the idea here is that the singularity of the director can be removed if the strength s of the disclination is an integer. However, if s is not an integer, then the singularity cannot be removed. To see this, we look at the fundamental homotopy group for a uniaxial nematic . Recall that the order parameter space here is $S^2/\mathbb{Z}_2$ (or RP^2), whose fundamental group [179] is $\mathbb{Z}_2$. Since the fundamental group, by definition, is the group of equivalence classes (under homotopy) of loops, we need to understand loops in the space [220]. A moment's thought reveals two types of loops on RP^2 (which we think of as being obtained by identification of antipodal points on the sphere S^2). In particular, there are closed contours (corresponding to disclinations of strength $s \in \mathbb{Z}$) and curves which run between antipodal points (corresponding to disclinations of strength $s \in \frac{1}{2}\mathbb{Z}$). The former curves are contractible (i.e., can be shrunk to a point), while the later curves cannot be contracted to a point while keeping the endpoints fixed. The first homotopy class (of loops contractible to a point) corresponds to a no defect system (the singularity has been removed), while the second homotopy class corresponds to a system with defect that cannot be removed (a so-called stable defect). However, in 2D, we have $\pi_1(RP^1) = \mathbb{Z}$, so there are an infinite number of defects (labeled by $k \in \mathbb{Z}$). Since RP^1 is isomorphic to S^1, one can think of identifying loops in S^1 with the number of times it wraps around the circle. Loops here cannot be contracted to a point; thus the disclination in this case is stable.

3.5 Competing (Nematic) Liquid Crystal Theories

There are three main competing theories [15, 457] to describe nematics which will *briefly* discuss below. Since we emphasize the power of the calculus of variations throughout this text, we will also mention typically variational problems considered in each theory. There is, though, a quite expansive literature, so we refer especially to [26, 399, 457] for many more results and references.

First, there is the Oseen–Frank theory, which uses the director $\mathbf{d} : \Omega \to S^2$ as order parameter (recall S^2 is the unit sphere in $\mathbb{R}^3$). The domain $\Omega \subset \mathbb{R}^3$ is viewed as a container completely filled with a liquid crystal. In this theory, defects are defined as discontinuities of the director $\mathbf{d}$. This model describes just the nematic phase [422]. The fundamental problem in this theory [26] is to find the director $\mathbf{d}$ which minimizes the Oseen–Frank energy functional

$$I[\mathbf{d}] = \int_\Omega W(\mathbf{d}, \nabla\mathbf{d})d\mathbf{x} \tag{3.22}$$

subject to the constraint that $\mathbf{d} \in S^2$ (i.e., $\mathbf{d}$ is a unit vector). Typically, a boundary condition is also given (e.g., prescribing $\mathbf{d}$ on $\partial\Omega$). The function W is given by

$$\begin{aligned} W(\mathbf{d}, \nabla\mathbf{d}) = K_1(\nabla\cdot\mathbf{d})^2 + K_2(\mathbf{d}\cdot\nabla\times\mathbf{d})^2 + K_3|\mathbf{d}\times\nabla\times\mathbf{d}|^2 \\ + (K_2+K_4)\left(\mathrm{Tr}(\nabla\mathbf{d})^2 - (\nabla\cdot\mathbf{d})^2\right) \end{aligned} \tag{3.23}$$

where the K_j are called the Frank constants, and typically [122] are assumed to satisfy the so-called Eriksen inequalities:

$$K_1 > 0,\ K_2 > 0,\ K_3 > 0,\ K_2 > |K_4|,\ 2K_1 > K_2 + K_4$$

Also, Tr denotes the tensor trace. Now, since we can write $\mathbf{d} \times (\nabla \times \mathbf{d}) = -(\mathbf{d} \cdot \nabla)\mathbf{d}$, then by comparison with (3.1) it is clear why the constants K_1, K_2, K_3 describe the splay, twist, and bend of $\mathbf{d}$.

The Euler–Lagrange equation for (3.22) can be found to be

$$(\mathbf{I} - \mathbf{d} \otimes \mathbf{d}) \left(\nabla \cdot \frac{\partial W}{\partial \nabla \mathbf{d}} - \frac{\partial W}{\partial \mathbf{d}} \right) = 0$$

which is a system of nonlinear partial differential equations with the constraint that $\mathbf{d} \in S^2$. A typical use of the direct method in the calculus of variations provides a minimizer $\mathbf{d}^*$ of (3.22) over all $\mathbf{d} \in H^1(\Omega; S^2)$, assuming $K_1 > 0$, $K_2 > 0$, $K_3 > 0$.

Why the Sobolev space H^1? Consider for instance the hedgehog director: $\mathbf{d}(\mathbf{x}) = \mathbf{x}/|\mathbf{x}|$; then $\mathbf{d} \in H^1(\Omega; S^2)$. It is important to be able to capture singular director fields (recall our previous discussion on defects), so $\nabla \mathbf{d}$ should exist in at least a weak sense.

A simplifying case occurs when $K_1 = K_2 = K_3, K_4 = 0$; in this case the Oseen–Frank functional reduces to

$$I[\mathbf{d}] = K_1 \int_\Omega |\nabla \mathbf{d}|^2 d\mathbf{x}$$

which is the energy functional for harmonic maps. The Euler–Lagrange equation in this case becomes

$$\Delta \mathbf{d} + |\nabla \mathbf{d}|^2 \mathbf{d} = 0 \tag{3.24}$$

which is the so-called harmonic map equation. In this simplifying case, more can be said about minimizers, such as that in the one-constant approximation, and minimizers are smooth except for a finite number of point defects , see [56, 365].

Then there is the Ericksen theory with order parameters $(\mathbf{d}, s) : S^2 \rightarrow \times[-1/2, 1]$ and additionally can account for local isotropic melting [123]. The well-known issue in the previous Oseen–Frank theory is the inability to predict certain liquid crystal configurations; the Ericksen model in part was introduced to address this issue. Defects in this theory correspond to regions where the order parameter s vanishes.

This Eriksen model is described mathematically for instance in [65], and we describe the essential model here. The evolution for s and $\mathbf{d}$ is governed by the following system [123]:

$$\beta(s)\dot{s} = \nabla \cdot \left(\frac{\partial W_1}{\partial \nabla s}\right) - \frac{\partial W_1}{\partial s} - W'(s)$$

$$\gamma(s)\dot{\mathbf{d}} \times \mathbf{d} = \left(\nabla \cdot \left(\frac{\partial W_1}{\partial \nabla \mathbf{d}}\right) - \frac{\partial W_1}{\partial \mathbf{d}}\right) \times \mathbf{d} \tag{3.25}$$

where $\mathcal{W} = W(s) + W_1(s, \mathbf{d}, \nabla s, \nabla \mathbf{d})$ denotes the so-called Helmholtz free energy. As in the Oseen–Frank theory, W_1 is quadratic in the gradients of $\mathbf{d}$ (and also in s). The coefficients β, γ are given based on molecular predictions and typically take the form

$$\beta(s) \equiv 1, \qquad \gamma(s) = cs^2$$

for some constant c. It is interesting to note that in the stationary case, adding s into the energy allows one to model new kinds of defects [247]. Indeed, consider for instance the Oseen–Frank energy for $\mathbf{d}(x)$ where $x \in \mathbb{R}$, i.e., 1D. Then the Euler–Lagrange equation is just

$$\mathbf{d}'' = |\mathbf{d}'|^2\mathbf{d}$$

Multiply this equation by $\mathbf{d}'$ and use the fact that $\mathbf{d}$ is a unit vector to conclude that

$$(|\mathbf{d}'|^2)' = 0, \quad \text{and} \quad -\mathbf{d}'' = c\mathbf{d}$$

for some constant c. This means that $\mathbf{d}$ is a smooth vector field with no defects. However, when s is introduced into the model, defects can exist in 1D, see, e.g., [255]. A sample result is the following: In [65], it was shown that the system (3.25) admits a global weak solution as well as a local classical solution.

Finally there is the Landau–de Gennes theory using Q-tensors $\mathbf{Q} : \Omega \to \mathcal{S}_0$ with $\mathcal{S}_0$ given by (3.18), the space of Q-tensors. This theory allows for the description of phase transitions as well as biaxial phases [98, 184, 272, 273, 308]. Here, there is not yet a universally accepted definition of defect [457], but regions with high gradients (for example) are seen as good candidates for the definition of defect [328, 456]. For example, if $\mathbf{Q}$ is continuous everywhere, defects can be described in this framework as discontinuities of eigenvalues which can only occur at the biaxial–uniaxial, uniaxial–isotropic, or biaxial–isotropic interfaces [207]. One can use the so-called biaxial coefficient

$$\beta(\mathbf{Q}) = 1 - 6\frac{\mathrm{Tr}(\mathbf{Q}^3)^2}{\mathrm{Tr}(\mathbf{Q}^2)^3}.$$

It can be shown that $0 \le \beta \le 1$, $\beta \ne 0$ for a biaxial region, and $\beta = 0$ for uniaxial region [274].

In the absence of external fields and surface energies, the Landau–de Gennes energy functional is given by

$$I_{LG}[\mathbf{Q}] = \int_\Omega f_B(\mathbf{Q}) + L|\nabla \mathbf{Q}|^2 d\mathbf{x}$$

where f_B is called the bulk energy density and $L > 0$ is a constant. The simplest form of the bulk energy density takes the form [27, 271]

$$f_B(\mathbf{Q}) = \frac{a}{2}\mathrm{Tr}(\mathbf{Q})^2 - \frac{b}{3}\mathrm{Tr}(\mathbf{Q})^3 + \frac{c}{4}\left(\mathrm{Tr}(\mathbf{Q}^2)\right)^2$$

where a, b, c are material-dependent constants. This particular form allows for a (first order) nematic–isotropic phase transition. The direct method again provides the existence of a global minimizer $\mathbf{Q}^*$ of I_{LG} in the class $H^1(\Omega; \mathcal{S}_0)$. This minimizer is a weak solution of the corresponding Euler–Lagrange equation, which here takes the form

$$2L\Delta \mathbf{Q}_{\alpha\beta} = a\mathbf{Q}_{\alpha\beta} + b\left(\frac{1}{3}\mathrm{Tr}\mathbf{Q}^2\delta_{\alpha\beta} - \mathbf{Q}_{\alpha p}\mathbf{Q}_{p\beta}\right) + c(\mathbf{Q}_{pq})^2\mathbf{Q}_{\alpha\beta}, \quad \alpha, \beta = 1, 2, 3,$$

and it can be shown that $\mathbf{Q}^*$ is also a classical solution of the above system. In fact, $\mathbf{Q}^*$ is smooth and analytic everywhere in Ω [140].

There are a number of results regarding passing from Landau de–Gennes theory to Oseen–Frank theory, say in the one-constant approximation. For instance, in [274] it was shown that for any sequence $L_k \to 0$ there is a subsequence $L_{k'}$ such that minimizers $\mathbf{Q}_{k'}$ of $I_{LG_{k'}}$ converge in $H^1(\Omega; \mathbb{R}^3)$ to a minimizer $\mathbf{Q}^*$ of $\int_\Omega |\nabla \mathbf{Q}|^2 dx$ subject to a certain constraint. Moreover, the convergence rate was improved in [311].

Remark 3.4 In the Oseen–Frank theory, only point defects in 3D are allowed (having finite energy), but line and wall defects have infinite energy [56, 457]. This is in contrast to the Ericksen theory, where defects are described where the parameter s vanishes and includes wall and line defects [8]. Moreover, a deficiency in the Oseen–Frank theory is that in 2D, point defects are predicted to have infinite energy and in fact does not allow for half-integer strengths [28]. But, the Landau–de Gennes theory addresses this issue in particular [98] and has been studied mathematically rigorously [105].

Finally, we will mention some related research for smectics (recall all the above was for nematics; moreover, in Chap. 8 we discuss cholesterics). A variational model which accounts for phase transitions between nematic and smectic A or C phases was developed in [80], extending earlier work from de Gennes in [97] which only modeled nematic to smectic A transitions. This model, with the so-called Chen–Lubensky energy, was shown to admit minimizers in [32]. Let us briefly recount this model here. The Chen–Lubensky functional takes the form

$$I_{CL} = \int_\Omega f_{CL}(\psi, \mathbf{d}) + f_N(\mathbf{d}) d\mathbf{x}$$

where $\mathbf{d}$ is the director as usual, and the new parameter $\psi : \Omega \to \mathbb{C}$ is a complex valued parameter, which, upon writing $\psi(\mathbf{x}) = \rho(\mathbf{x})e^{i\omega(\mathbf{x})}$, we have $\rho(\mathbf{x})$ the amplitude of smectic layering and measures how well the smectic phase has developed near the point $\mathbf{x}$. In particular, the material is in the nematic phase at $\mathbf{x} \in \mathbb{R}^3$ if $\psi(\mathbf{x}) = 0$. To define the Chen–Lubensky energy density, we consider the covariant derivative of ψ, written as $D\psi = \nabla\psi - i\overline{q}\mathbf{d}\psi$, where $\overline{q} = 2\pi/d$ is the wave number with d the layer spacing. We consider the parts of $D\psi$ which are parallel to $\mathbf{d}$ and perpendicular to $\mathbf{d}$ and write

$$D_{\|}\psi = (\mathbf{d} \cdot \nabla\psi - i\overline{q}\psi)\mathbf{d}, \qquad D_{\perp}\psi = D\psi - D_{\|}\psi$$

Then the Chen–Lubensky energy density can be written as

$$\begin{aligned} f_{CL} &= a_{\perp}|D \cdot D_{\perp}\psi|^2 + a_{\|}|D \cdot D_{\|}\psi|^2 - c_{\perp}|D_{\perp}\psi|^2 \\ &\quad + c_{\|}|D_{\|}\psi|^2 - r'|\psi|^2 + g|\psi|^4 \end{aligned}$$

where $a_{\|}, a_{\perp}, c_{\|}, g > 0$ and $c_{\perp} > 0$ for the smectic C phase and $c_{\perp} < 0$ for smectic A. Moreover, r' denotes the difference between the temperature of the material and the nematic–smectic transition at $\tau = 0$, where τ is the chiral pitch of the material. Finally, f_N denotes the Oseen–Frank energy density (3.23).

We again wish to reiterate the wealth of literature on related variational problems, and we have by no means been exhaustive. Rather, we have just given a flavor of the types of problems that have been considered.

Chapter 4
Fermat Principle and Snell's Law of Refraction

Fermat's principle is a classical principle, established by Pierre de Fermat in the 1600s, originally developed to explain the refraction of light [133]. This principle essentially says that the actual trajectory of a light ray is along the path which can be traveled in the least amount of time. For this reason Fermat's principle is sometimes known as the principle of least time. We will frequently use the equivalent formulation of Fermat's principle, which says that the light rays passing between two points choose the shortest *optical* path [244]. This can be formulated in the language of the calculus of variations by considering the optical path length

$$s = \int n\,d\ell$$

defined in terms of the refractive index of the material, where $d\ell = \sqrt{dx^2 + dy^2 + dz^2}$ is the path length element in $\mathbb{R}^3$. Thus, Fermat's principle reduces to minimizing the functional s.

Snell's Law is a classical law which describes the behavior of light rays at an interface between two materials. Even though the law is attributed to Willebrord Snellius, the ideas date all the way back to the work of Ibn Sahl in the study of burning mirrors and lenses [345]. Classically, Snell's Law says the following [155]: Suppose a surface separates two homogeneous and isotropic media in $\mathbb{R}^3$ with refractive indices $n_1, n_2 > 0$, recall Fig. 1.1. Then, the angle θ_1 between the incident ray $\mathbf{a}$ and the normal $\boldsymbol{\nu}$ is related to the angle between the refracted ray $\boldsymbol{m}$ and the normal ν by the formula

$$n_1 \sin(\theta_1) = n_2 \sin(\theta_2).$$

We shall see that it is often more useful to work with a vector form of this law, which says that

E. Stachura, *Mathematical Methods in Liquid Crystal Optics and Lens Design*, Springer Tracts in Modern Physics 294,
https://doi.org/10.1007/978-3-031-46614-4_4

$$n_1(\mathbf{a} \times \boldsymbol{\nu}) = n_2(\boldsymbol{m} \times \boldsymbol{\nu})$$

where $\times$ denotes cross product. Note that the scalar law is recovered by taking magnitude of the vectors in the vector law.

In this section we will discuss Fermat's Principle and Snell's Law more precisely and in the context of liquid crystals. We will see that (the classical) Snell's Law is particularly useful given *linear* light trajectories, as this law relates the directions of the incoming and refracted rays at an interface.

4.1 Fermat's Principle

Fermat's principle is one of the cornerstones of geometric optics and entails minimizing optical path length. Fermat's principle is a variational principle which provides the trajectory of a light ray and in anisotropic media (where the refractive index depends on both position and direction) was introduced by F. Grandjean [149].

Theorem 4.1 *Consider a material whose refractive index $n(\mathbf{x}, \mathbf{v})$, depending on both position and direction of propagation, only varies over distances much larger than the wavelength of light used. The trajectory of the light ray between two points A and B is an extremum of the optical path*

$$\int_A^B n(\mathbf{x}, \mathbf{v}) \cdot \mathbf{dl} \tag{4.1}$$

which can be expressed as a zero variation condition:

$$\delta \int_A^B n(\mathbf{x}, \mathbf{v}) \cdot \mathbf{dl} = 0 \tag{4.2}$$

The variation δ will be made precise shortly. It is convenient to work with parametric integrals of the form

$$\mathcal{F}(\mathbf{x}) = \int_a^b F(\mathbf{x}(t), \dot{\mathbf{x}}(t))dt \tag{4.3}$$

whose integrand is homogeneous of degree one with respect to $\mathbf{v} = \dot{\mathbf{x}}$, i.e., $F(\mathbf{x}, \lambda\dot{\mathbf{x}}) = \lambda F(\mathbf{x}, \dot{\mathbf{x}})$ for $\lambda > 0$. Thus we will formulate Fermat's principle for parametric integrals. Indeed, the Fermat principle states that light rays follow an extremum of the optical path length

$$\int_a^b n(\mathbf{x}(s), \dot{\mathbf{x}}(s))ds$$

subject to taking s to be the arclength parameterization of $\mathbf{x}$, that is, $|\dot{\mathbf{x}}(s)| = 1$. It should be noted that the 1-homogeneity condition makes this a parametric integral, that is, the value of the integral is independent of the parameterization of the curve. Indeed, if $\sigma : [t_1, t_2] \to [t_1, t_2]$ is a C^1 diffeomorphism,[1] with $\sigma'(t) > 0$, then, if $\mathbf{z} = \mathbf{x}(\sigma(t))$, then by 1-homogeneity it follows that

$$
\begin{aligned}
&\int_a^b n(\mathbf{x}(s), \dot{\mathbf{x}}(s))ds \\
&\quad = \int_{t_1}^{t_2} n(\mathbf{x}(\sigma(t)), \dot{\mathbf{x}}(\sigma(t)))\sigma'(t)dt = \int_{t_1}^{t_2} n\left(\mathbf{x}(\sigma(t)), (\dot{\mathbf{x}}(\sigma))\frac{d\sigma}{dt}\right)dt \\
&\quad = \int_{t_1}^{t_2} n(\mathbf{z}(t), \dot{\mathbf{z}}(t))dt
\end{aligned}
$$

Remark 4.1 Parameterization by arclength can be done in a much more general setting, even that of a metric space. Indeed, let (X, d) be a metric space. A curve γ in X is a continuous map $\gamma : [a, b] \to X$. The image of the curve is $\gamma([a, b])$, and the length of γ is defined as

$$
\ell(\gamma) = \sup \left\{ \sum_{j=0}^{n-1} d(\gamma(t_j), \gamma(t_{j+1})) \right\},
$$

where the supremum is taken over all partitions $a = t_0 < t_1 < \cdots < t_n = b$. We say the curve is *rectifiable* if $\ell(\gamma) < \infty$. If γ is rectifiable, then it is possible to find [174, Theorem 3.2] a unique curve $\widetilde{\gamma} : [0, \ell(\gamma)] \to X$ such that $\gamma = \widetilde{\gamma} \circ \ell(\gamma|_{[a,t]})$, and such that $\ell(\widetilde{\gamma}|_{[0,t]}) = t$ for all $t \in [0, \ell(\gamma)]$. For this last reason, we call $\widetilde{\gamma}$ parameterized by arclength. In particular, the arclength parameterization is Lipschitz and is such that $|\dot{\widetilde{\gamma}}(t)| = 1$ for a.e. $t \in [0, \ell(\gamma)]$.

If γ is not continuous, then one can define the length (total variation) of $\mathbf{x}$ in a similar way. Here we require at most countably many points of discontinuity. Thus we see that this is intimately related to functions of bounded variation: γ is rectifiable if and only if the coordinate functions are continuous and of bounded variation. Recall that a function $u \in L^1(0, T)$ is of bounded variation if its derivative $\dot{u}$ belongs to the Banach space of real, regular Borel measures $\mathcal{M}(0, T)$. In fact, u is of bounded variation if and only if

$$
||Du|| := \sup \left\{ \left\{ \int_0^T u(t)\dot{\phi}(t)dt : \phi \in C_0^1(0, T),\ |\phi(t)| \le 1\ \forall\, t \in [0, T] \right\} < \infty
$$

See, e.g., [6] for more details on BV functions.

---

[1] That is, σ is continuously differentiable, it is bijective, and its inverse σ^{-1} is also continuously differentiable.

In what follows, we will freely move between an optical path length parametric integral and a time parametric variational integral by setting the speed of light $c = 1$. Indeed, we may consider the time $t(s)$ that it takes for light to move from $\mathbf{x}(0)$ to $\mathbf{x}(s)$ as

$$t(s) = \int_0^s n(\mathbf{x}(s), \dot{\mathbf{x}}(s))ds$$

where $n(\mathbf{x}(s), \dot{\mathbf{x}}(s))$ is the refractive index of the material. Then

$$\frac{dt}{ds} = n(\mathbf{x}(s), \dot{\mathbf{x}}(s))$$

so that $1/n(\mathbf{x}(s), \dot{\mathbf{x}}(s))$ can be understood as the velocity of light at $\mathbf{x}(s)$ in the direction $\mathbf{v}(s) = \dot{\mathbf{x}}(s)$.

Similarly, we consider the optical path length from $\mathbf{x}(0) = A$ to $\mathbf{x}(1) = B$ as the integral of the refractive index

$$\text{OPL}_{A\to B} = \int_0^1 n(\mathbf{x}(s), \dot{\mathbf{x}}(s))ds$$

In the context of liquid crystals we will consider the following optical path length integral for the extraordinary ray

$$\int_a^b n_{\text{eff}}(\mathbf{x}(s), \dot{\mathbf{x}}(s))\, ds \tag{4.4}$$

with s being the arclength parameterization of $\mathbf{x}$. Thus we are considering the parametric integral (4.3) with

$$F(\mathbf{x}, \dot{\mathbf{x}}) = n_{\text{eff}}(\mathbf{x}, \dot{\mathbf{x}}) \tag{4.5}$$

where the effective index is given by (3.11).

For the ordinary ray (or more generally, rays in homogeneous, isotropic media), Fermat's principle says

$$\delta \int_a^b nds = 0 \tag{4.6}$$

where $n = n_o$ is the ordinary refractive index. But (4.6) implies that the trajectory of the ordinary ray is just a straight line. Indeed, we may write

$$\int_a^b nds = \int n\sqrt{\dot{x}^2 + \dot{y}^2 + \dot{z}^2}dt$$

in 3D. Recall that arclength of a curve $\mathbf{r}(t) = [x(t), y(t), z(t)]$ in 3D is given by

$$s = \int_a^b \sqrt{\dot{x}^2 + \dot{y}^2 + \dot{z}^2} dt$$

Then, by taking variation (4.6), we obtain the system of Euler–Lagrange equations

$$\frac{\partial n}{\partial x} - \frac{d}{ds}\left(n\frac{dx}{ds}\right) = 0,$$
$$\frac{\partial n}{\partial y} - \frac{d}{ds}\left(n\frac{dy}{ds}\right) = 0,$$
$$\frac{\partial n}{\partial z} - \frac{d}{ds}\left(n\frac{dz}{ds}\right) = 0$$

or, in more compact notation:

$$\frac{d}{ds}\left(n(\mathbf{r}(s))\frac{d\mathbf{r}}{ds}\right) = \nabla n(\mathbf{r}(s)) \tag{4.7}$$

which is known as the equation of the light rays (in isotropic media) [48]. Now, if n is constant, clearly this yields

$$\frac{d^2\mathbf{r}}{ds^2} = 0 \implies \mathbf{r}(s) = as + b, \quad a, b \in \mathbb{R}^3$$

so that indeed, light rays are straight lines. In fact this also holds for homogeneous, *anisotropic* media, see, e.g., [48, 159]. That is, in general homogeneous media, light rays trace out straight lines.

Before continuing, let us spell out carefully the variation δ that we took above. For the parametric integral (4.3), the first variation $\delta\mathcal{F}$ is given by

$$\delta\mathcal{F}[\mathbf{x}, \phi] = \int_a^b \left(\nabla_{\mathbf{x}} F(\mathbf{x}, \dot{\mathbf{x}}) \cdot \phi + \nabla_{\mathbf{v}} F(\mathbf{x}, \dot{\mathbf{x}}) \cdot \dot{\phi}\right) dt$$

for each $\phi \in C_0^1\left([a, b]; \mathbb{R}^3\right)$; that is, for each ϕ continuously differentiable with compact support on (a, b). Roughly speaking (this will be made more precise in Chap. 5), if $\delta\mathcal{F}[\mathbf{x}, \phi] = 0$ for all $\phi \in C_0^\infty((a, b); \mathbb{R}^3)$, and $\mathbf{x}$ is sufficiently smooth (i.e., twice continuously differentiable), then $\mathbf{x}$ satisfies the Euler–Lagrange equations

$$\nabla_{\mathbf{x}} F(\mathbf{x}, \dot{\mathbf{x}}) - \frac{d}{dt}\nabla_{\mathbf{v}} F(\mathbf{x}, \dot{\mathbf{x}}) = 0$$

which is how we obtained (4.7). Such $\mathbf{x}$ as above are called extremals of $\mathcal{F}$, and in Chap. 5 we will distinguish between various types of extremals, which depend on how smooth, i.e., how differentiable, they are.

Now, Fermat's principle can be used to obtain Snell's Law of Refraction at an interface between isotropic and anisotropic media.

To this end, we consider the setup from [139, Chapter 8].

Let Ω denote a domain in space separated by a surface Σ into two nonempty subdomains Ω_1 and Ω_2, such that $\Omega = \Omega_1 \cup \Omega_2 \cup \Sigma$. Consider a jump across Σ, so that $F = F_1$ in Σ_1 and $F = F_2$ in Σ_2, i.e.,

$$F(\mathbf{x}, \mathbf{v}) = \begin{cases} F_1(\mathbf{x}, \mathbf{v}), & \text{for } x \in \Omega_1 \cup \Sigma \\ F_2(\mathbf{x}, \mathbf{v}), & \text{for } x \in \Omega_2 \end{cases}$$

In particular, F may not be continuous along the surface Σ. Take a piecewise differentiable curve $\mathbf{x}(s)$ on some time interval $s \in [s_1, s_2]$ parameterized with arclength, such that $\mathbf{x}(s)$ moves from Ω_1 to Ω_2 and crosses the surface Σ at s_0. In particular, suppose that the restrictions $\mathbf{x}(s)|_{\Omega_1}$ and $\mathbf{x}(s)|_{\Omega_2}$ are sufficiently smooth (say twice continuously differentiable) and satisfy the Euler–Lagrange equations there. *Thus, we are considering curves* $\mathbf{x}(s)$ *such that the restrictions in each* Ω_j *are extremals of the functionals*

$$\mathcal{F}_j = \int F_j(\mathbf{x}, \mathbf{v}) ds, \qquad j = 1, 2$$

We use the following definition of light ray:

Definition 4.1 The curve $\mathbf{x}(s)$ for $s_1 \leq s \leq s_2$, continuous on the interval $[s_1, s_2]$ such that $\mathbf{x}([s_1, s_0)) \subset \Omega_1$, and $\mathbf{x}((s_0, s_2]) \subset \Omega_2$, such that $\mathbf{x}$ is twice continuously differentiable on the intervals $[s_1, s_0]$ and $[s_0, s_2]$ is called a light ray in Ω if

$$\delta\mathcal{F}(\mathbf{x}, \phi) := \int_{s_1}^{s_2} \left(\nabla_{\mathbf{x}} F(\mathbf{x}, \dot{\mathbf{x}}) \cdot \phi + \nabla_{\mathbf{v}} F(\mathbf{x}, \dot{\mathbf{x}}) \cdot \frac{d\phi}{dt} \right) dt = 0$$

for all $\phi \in C_0^\infty((s_1, s_2); \mathbb{R}^3)$ such that $\phi(s_0) \in T_{\mathbf{x}_0}\Sigma$ for $\mathbf{x}_0 := \mathbf{x}(s_0)$; that is, for all variations ϕ with compact support such that $\phi(s_0)$ is tangent to the surface Σ at $\mathbf{x}_0$.

Note that Definition 4.1 essentially says that a light ray is a weak extremal of $\mathcal{F}$, since the first variation of $\mathcal{F}$ is precisely

$$\delta\mathcal{F}(\mathbf{x}, \phi) = \int_{s_1}^{s_2} \left(\nabla_{\mathbf{x}} F(\mathbf{x}, \dot{\mathbf{x}}) \cdot \phi + \nabla_{\mathbf{v}} F(\mathbf{x}, \dot{\mathbf{x}}) \cdot \frac{d\phi}{dt} \right) dt$$

which is well-defined for Lipschitz curves $\mathbf{x}(t)$ which are regular, i.e., such that $\dot{\mathbf{x}}(t) \neq 0$ for a.e. $t \in [0, 1]$.

We use this definition of light ray here since in general, for an arbitrary variation $\xi(s, \epsilon)$ of the curve $\mathbf{x}(s)$, the function $\mathcal{F}(\xi(\cdot, \epsilon))$ will not be differentiable due to the discontinuities of F and ξ; for more on this, see [139, Chapter 8.1.3]. To obtain

Snell's Law of Refraction, we need the following Weierstrass–Erdmann corner condition [139, Chapter 8.1.3, Cor. 2]:

If a light ray $\mathbf{x}(s)$ *crosses a surface discontinuity* Σ *at* $\mathbf{x}_0 = \mathbf{x}(s_0)$, *then at* $\mathbf{x}_0$, *it satisfies*

$$\nabla_{\mathbf{v}} F\left(\mathbf{x}_0, \dot{\mathbf{x}}(s_0^+)\right) - \nabla_{\mathbf{v}} F\left(\mathbf{x}_0, \dot{\mathbf{x}}(s_0^-)\right) \perp \text{ to } T_{\mathbf{x}_0}\Sigma \tag{4.8}$$

Above, $\dot{\mathbf{x}}(s_0^{\pm})$ denote the one-sided limits $\lim_{s\to s_0^{\pm}} \dot{\mathbf{x}}(s)$.

To see how this is obtained, we can write the general variation ξ of a curve $\mathbf{x}$ in the direction ϕ as $\xi(s,\epsilon) = \mathbf{x}(s) + \epsilon\phi(s)$ plus a lower order term in ϵ via Taylor expansion of ξ. More precisely, by general variation ξ of a curve $\mathbf{x}(s)$ in the direction $\phi(s)$, we mean a C^1 function $\xi : [s_1, s_2] \times (-\epsilon_0, \epsilon_0) \to \mathbb{R}^2$ such that $\xi(s, 0) = \mathbf{x}(s)$ and

$$\left.\frac{\partial \xi}{\partial \epsilon}(s,\epsilon)\right|_{\epsilon=0} = \phi(s).$$

Note that such ϕ belongs to the tangent space[2] to Σ at $\mathbf{x}(s_0)$.

Thus, for any ray as in Definition 4.1, we have

$$\begin{aligned}
&\int_{s_1}^{s_2} \left(\nabla_{\mathbf{x}} F(\mathbf{x}, \dot{\mathbf{x}}) \cdot \phi + \nabla_{\mathbf{v}} F(\mathbf{x}, \dot{\mathbf{x}}) \cdot \frac{d\phi}{ds}\right) ds = \\
&\int_{s_1}^{s_0} \left(\nabla_{\mathbf{x}} F(\mathbf{x}, \dot{\mathbf{x}}) \cdot \phi + \nabla_{\mathbf{v}} F(\mathbf{x}, \dot{\mathbf{x}}) \cdot \frac{d\phi}{ds}\right) ds \\
&\qquad + \int_{s_0}^{s_2} \left(\nabla_{\mathbf{x}} F(\mathbf{x}, \dot{\mathbf{x}}) \cdot \phi + \nabla_{\mathbf{v}} F(\mathbf{x}, \dot{\mathbf{x}}) \cdot \frac{d\phi}{ds}\right) ds \\
&= \int_{s_1}^{s_0} \left(\nabla_{\mathbf{x}} F(\mathbf{x}, \dot{\mathbf{x}}) - \frac{d}{ds}\nabla_{\mathbf{v}} F(\mathbf{x}, \dot{\mathbf{x}})\right) \cdot \phi ds \\
&\qquad + \int_{s_0}^{s_2} \left(\nabla_{\mathbf{x}} F(\mathbf{x}, \dot{\mathbf{x}}) - \frac{d}{ds}\nabla_{\mathbf{v}} F(\mathbf{x}, \dot{\mathbf{x}})\right) \cdot \phi ds \\
&\qquad + \lim_{s\to s_0^-} \left(\nabla_{\mathbf{v}} F(\mathbf{x}, \dot{\mathbf{x}}) \cdot \phi\right) - \nabla_{\mathbf{v}} F(\mathbf{x}(s_1), \dot{\mathbf{x}}(s_1)) \cdot \phi(s_1) \\
&\qquad + \nabla_{\mathbf{v}} F(\mathbf{x}(s_2), \dot{\mathbf{x}}(s_2)) \cdot \phi(s_2) - \lim_{s\to s_0^+} \left(\nabla_{\mathbf{v}} F(\mathbf{x}, \dot{\mathbf{x}}) \cdot \phi\right) \\
&= \lim_{s\to s_0^-} \left(\nabla_{\mathbf{v}} F(\mathbf{x}, \dot{\mathbf{x}}) \cdot \phi\right) - \lim_{s\to s_0^+} \left(\nabla_{\mathbf{v}} F(\mathbf{x}, \dot{\mathbf{x}}) \cdot \phi\right)
\end{aligned} \tag{4.9}$$

[2] If M is a submanifold of $\mathbb{R}^n$, we say $\mathbf{v}$ is a tangent vector of M at p if there exists a C^1 curve $c : (-\epsilon_0, \epsilon_0) \to \Sigma$ such that $\left.\frac{dc}{d\epsilon}\right|_{\epsilon=0} = \mathbf{v}$.

where we have used the fact that for the ray splits at $\mathbf{x}(s_0)$ from being in Ω_1 for $s \in [s_1, s_0)$ and being in Ω_2 for $s \in (s_0, s_2]$. We have also integrated by parts, which is justified assuming all quantities are sufficiently smooth, and used the fact that $\phi(s_1) = \phi(s_2) = 0$ since it is of compact support.

Since $\mathbf{x}(s)$ is a light ray, all of (4.9) vanishes, yielding the corner condition (4.8) since $\phi(s_0) \in T_{\mathbf{x}_0}\Sigma$. Finally, notice that the integral terms

$$\int_{s_1}^{s_0} \left(\nabla_{\mathbf{x}} F(\mathbf{x}, \dot{\mathbf{x}}) - \frac{d}{ds} \nabla_{\mathbf{v}} F(\mathbf{x}, \dot{\mathbf{x}}) \right) \cdot \phi ds, \quad \int_{s_0}^{s_2} \left(\nabla_{\mathbf{x}} F(\mathbf{x}, \dot{\mathbf{x}}) - \frac{d}{ds} \nabla_{\mathbf{v}} F(\mathbf{x}, \dot{\mathbf{x}}) \right) \cdot \phi ds$$

vanish since we have assumed that the restrictions of $\mathbf{x}$ to $[s_1, s_0]$ and to $[s_0, s_2]$ are smooth and satisfy the Euler–Lagrange equations there.

4.2 Isotropic, Inhomogeneous Snell's Law

Consider an isotropic, inhomogeneous material in Ω_2; then we let $F_2(\mathbf{x}, \mathbf{v}) = n_2(\mathbf{x})|\mathbf{v}|$. Suppose Ω_1 is also isotropic and inhomogeneous, with refractive index $n_1(\mathbf{x})$, so that $F_1(\mathbf{x}, \mathbf{v}) = n_1(\mathbf{x})|\mathbf{v}|$. Suppose the light ray travels from Ω_1 to Ω_2 and crosses the surface at $\mathbf{x}_0$. Further define $\mathbf{a} = \dot{\mathbf{x}}(s_0^-)$ and $\mathbf{m} = \dot{\mathbf{x}}(s_0^+)$. Notice that

$$\nabla_{\mathbf{v}} F_j(\mathbf{x}, \mathbf{v}) = n_j(\mathbf{x}) \frac{\mathbf{v}}{|\mathbf{v}|}, \qquad j = 1, 2$$

Recalling that light rays are such that $|\mathbf{v}| = 1$, the condition (4.8) can be written as

$$n_2(\mathbf{x}_0)\mathbf{m} - n_1(\mathbf{x}_0)\mathbf{a} = \lambda \boldsymbol{\nu}(\mathbf{x}_0), \qquad \lambda \in \mathbb{R} \tag{4.10}$$

where $\boldsymbol{\nu}$ is a normal vector to Σ at $\mathbf{x}_0$, pointing from Ω_1 to Ω_2.

If n_1, n_2 were constant, i.e., Ω_1, Ω_2 are homogeneous, then (4.10) agrees with the vector form of Snell's Law for isotropic media [157, Equation (2.1)] and implies the usual scalar law

$$n_1 \sin(\theta_1) = n_2 \sin(\theta_2) \tag{4.11}$$

where θ_1 denotes the angle between the vectors $\mathbf{a}$ and the normal $\boldsymbol{\nu}$ and θ_2 denotes the angle between $\mathbf{m}$ and $\boldsymbol{\nu}$. Indeed, to obtain (4.11), take cross product of (4.10) with $\boldsymbol{\nu}$:

$$n_2 \mathbf{m} \times \boldsymbol{\nu} - n_1 \mathbf{a} \times \boldsymbol{\nu} = 0$$

Then taking magnitude yields (4.11).

4.3 Anisotropic, Inhomogeneous Snell's Law

Let us generalize the previous results to inhomogeneous, anisotropic media, where the refractive index depends on the direction of the light ray. This may not be as useful if the trajectory of the ray after striking an interface is nonlinear (which is generally the case in inhomogeneous media). But, if once forces the trajectory to be linear, e.g., by choosing a particular director in a liquid crystal (see Sect. 6.3), then these results will be just as useful as in the homogeneous case. Otherwise, the results below just provide the direction of the tangent to the ray trajectory at the interface.

Remark 4.2 There are other versions of Snell's Law for other types of media. For instance, when the refractive index of a medium is only a function of the distance $\mathbf{r}$ from the origin, then there is a generalized Snell's Law

$$n(\mathbf{r})\sin(\phi) = \kappa$$

where ϕ is the angle between the positive direction along the ray to the direction of increasing $\mathbf{r}$ and κ is a constant for a particular ray; see, e.g., [48, 307]. The above formulation is equivalent to the differential equation

$$\frac{n(r)r^2\dfrac{d\theta}{dr}}{\sqrt{1+\left(r\dfrac{d\theta}{dr}\right)^2}} = -\kappa$$

where (r, θ) are polar coordinates of points along the ray.

Suppose that each Ω_j, $j = 1, 2$, has refractive index $n_j = n_j(\mathbf{x}, \mathbf{v})$, where again $\mathbf{v}(s) = \dot{\mathbf{x}}(s)$. Then $F_j(\mathbf{x}, \mathbf{v}) = n_j(\mathbf{x}, \mathbf{v})|\mathbf{v}|$.

We compute

$$\nabla_{\mathbf{v}} F_j(\mathbf{x}, \mathbf{v}) = \left(\nabla_{\mathbf{v}} n_j(\mathbf{x}, \mathbf{v})\right)|\mathbf{v}| + n_j(\mathbf{x}, \mathbf{v})\frac{\mathbf{v}}{|\mathbf{v}|}, \qquad j = 1, 2.$$

Again we define

$$\mathbf{a} = \dot{\mathbf{x}}(s_0^-), \qquad \mathbf{m} = \dot{\mathbf{x}}(s_0^+)$$

Thus the corner condition (4.8) becomes

$$\nabla_{\mathbf{v}} n_2(\mathbf{x}_0, \mathbf{m}) + n_2(\mathbf{x}_0, \mathbf{m})\mathbf{m} - (\nabla_{\mathbf{v}} n_1(\mathbf{x}_0, \mathbf{a}) + n_1(\mathbf{x}_0, \mathbf{a})\mathbf{a}) = \lambda \boldsymbol{\nu}(\mathbf{x}_0), \qquad \lambda \in \mathbb{R} \tag{4.12}$$

where again $\boldsymbol{\nu}$ denotes a normal vector to Σ at $\mathbf{x}_0$. Equation (4.12) is a general Snell's Law at the interface of anisotropic, inhomogeneous media. Note that if each

Ω_j were isotropic (and inhomogeneous), then $n_j = n_j(\mathbf{x})$ only, and the derivatives in (4.12) vanish, so we recover (4.10), as expected.

Remark 4.3 If each Ω_j were homogeneous and anisotropic, then $n_j = n_j(\mathbf{v})$ only. We show that in this case (4.12) becomes the anisotropic Snell's Law from [159]. The result from [159] requires to provide each Ω_j with a norm N_j associated with it, such that the surfaces for the wavefronts in each Ω_j are given by N_j. Indeed, in each Ω_j one can impose the norm

$$N_j(\mathbf{v}) = n_j(\mathbf{v})|\mathbf{v}|, \qquad j = 1, 2 \tag{4.13}$$

Suppose that Ω_1 is air and Ω_2 is a uniaxial nematic liquid crystal with uniform director $\mathbf{d}$ (i.e., one which is constant in $\mathbf{x}$). In the latter case, n_j could stand for n_o for the ordinary ray, or the effective refractive index (3.9) for the extraordinary ray. The ordinary ray is straightforward: We know from (2.18) that the optical indicatrix is given by

$$|\mathbf{p}|^2 = n_o^2, \text{ i.e. } \frac{|\mathbf{p}|^2}{n_o^2} = 1$$

This induces a dual norm

$$N^*(\mathbf{p}) = \left|\frac{\mathbf{p}}{n_o}\right|$$

which is dual to the norm given by

$$N(\mathbf{m}) = \sup_{N^*(\mathbf{p})=1} |\mathbf{m} \cdot \mathbf{p}| = n_o|\mathbf{m}|$$

which is the norm we associate to the liquid crystal for the ordinary ray. Equation (4.12) above becomes

$$n_o\mathbf{m} - n_1\mathbf{a} = \lambda\boldsymbol{\nu}$$

for the ordinary ray. Snell's Law via equation 3.2 in [159] gives in this case

$$(\nabla N(\mathbf{m}) - n_1\mathbf{a}) \parallel \boldsymbol{\nu}$$

Computing the gradient of $N(\mathbf{m})$, we find that the last two equations agree (recall $\mathbf{m}$ is a unit vector). Now let us see the agreement for the extraordinary ray.

From the uniaxial indicatrix (2.21), using that $|\mathbf{p}|^2 = n_{\text{eff},p}^2$ as we saw in (3.9), one can impose the extraordinary ray norm

$$N(\mathbf{v}) = n_{\text{eff},p}(\mathbf{v})|\mathbf{v}| = \sqrt{\frac{n_o^2 n_e^2}{n_o^2(1-(\mathbf{v}\cdot\mathbf{d})^2) + n_e^2(\mathbf{v}\cdot\mathbf{d})^2}}|\mathbf{v}|$$

as claimed in (4.13). With such N, equation (3.2) in [159] becomes

$$(\nabla_{\mathbf{v}} N(\mathbf{v}) - n_1 \mathbf{a}) \parallel \boldsymbol{\nu}$$

which agrees with (4.12) above.

Now, let us determine the value of λ in (4.12). For ease let us define the vector

$$\mathbf{g} = \mathbf{g}(\mathbf{x}_0, \mathbf{m}, \mathbf{a}) := \nabla_{\mathbf{v}} n_2(\mathbf{x}_0, \mathbf{m}) - \nabla_{\mathbf{v}} n_1(\mathbf{x}_0, \mathbf{a})$$

Starting from (4.12) take dot product with the vector $\mathbf{a}$:

$$\mathbf{g} \cdot \mathbf{a} + n_2(\mathbf{x}_0, \mathbf{m})\mathbf{m} \cdot \mathbf{a} - n_1(\mathbf{x}_0, \mathbf{a}) = \lambda \mathbf{a} \cdot \boldsymbol{\nu}(\mathbf{x}_0)$$

which implies

$$\mathbf{m} \cdot \mathbf{a} = \frac{n_1(\mathbf{x}_0, \mathbf{a}) + \lambda \mathbf{a} \cdot \boldsymbol{\nu}(\mathbf{x}_0) - \mathbf{g} \cdot \mathbf{a}}{n_2(\mathbf{x}_0, \mathbf{m})} \tag{4.14}$$

Now, take dot product of (4.12) with itself and simplify, using (4.14), results in the following quadratic equation for λ:

$$\lambda^2 - 2\left[(n_1(\mathbf{x}_0, \mathbf{a})\mathbf{a} - \mathbf{g}) \cdot \boldsymbol{\nu}(\mathbf{x}_0)\right] \lambda + |n_1(\mathbf{x}_0, \mathbf{a})\mathbf{a} - \mathbf{g}|^2 - n_2(\mathbf{x}_0, \mathbf{m})^2 = 0 \tag{4.15}$$

Solving for λ yields

$$\begin{aligned} \lambda &= (n_1(\mathbf{x}_0, \mathbf{a})\mathbf{a} - \mathbf{g}) \cdot \boldsymbol{\nu}(\mathbf{x}_0) \\ &\pm \sqrt{n_2(\mathbf{x}_0, \mathbf{m})^2 - \left[|n_1(\mathbf{x}_0, \mathbf{a})\mathbf{a} - \mathbf{g}|^2 - ((n_1(\mathbf{x}_0, \mathbf{a})\mathbf{a} - \mathbf{g}) \cdot \boldsymbol{\nu}(\mathbf{x}_0))^2\right]} \end{aligned} \tag{4.16}$$

Now, which sign ($\pm$) in (4.16) should be taken? Taking dot product of (4.12) with $\boldsymbol{\nu}$ and using (4.16), we obtain

$$\begin{aligned} &- n_2(\mathbf{x}_0, \mathbf{m})\mathbf{m} \cdot \boldsymbol{\nu}(\mathbf{x}_0) \\ &\quad = \pm\sqrt{n_2(\mathbf{x}_0, \mathbf{m})^2 - \left[|n_1(\mathbf{x}_0, \mathbf{a})\mathbf{a} - \mathbf{g}|^2 - ((n_1(\mathbf{x}_0, \mathbf{a})\mathbf{a} - \mathbf{g}) \cdot \boldsymbol{\nu}(\mathbf{x}_0))^2\right]} \end{aligned}$$

Since $\mathbf{m} \cdot \boldsymbol{\nu}(\mathbf{x}_0) \geq 0$, and if we assume $n_2(\mathbf{x}_0, \mathbf{m}) > 0$, we obtain

$$\begin{aligned} \lambda &= (n_1(\mathbf{x}_0, \mathbf{a})\mathbf{a} - \mathbf{g}) \cdot \boldsymbol{\nu}(\mathbf{x}_0) \\ &- \sqrt{n_2(\mathbf{x}_0, \mathbf{m})^2 - \left[|n_1(\mathbf{x}_0, \mathbf{a})\mathbf{a} - \mathbf{g}|^2 - ((n_1(\mathbf{x}_0, \mathbf{a})\mathbf{a} - \mathbf{g}) \cdot \boldsymbol{\nu}(\mathbf{x}_0))^2\right]} \end{aligned} \tag{4.17}$$

4.4 Reflection

Let us start by reviewing the Law of Reflection in the homogeneous, isotropic case. This is a special case of Snell's Law

$$n_1(\mathbf{a} \times \boldsymbol{\nu}) = n_2(\mathbf{m} \times \boldsymbol{\nu}) \tag{4.18}$$

when $n_1 = n_2$. Recall that **a** denotes the incident ray vector and **m** the refracted ray. Indeed, Snell's Law (4.18) implies

$$(\mathbf{a} - \mathbf{m}) \times \boldsymbol{\nu} = 0$$

so there exists some constant λ such that $\mathbf{a} - \mathbf{m} = \lambda\boldsymbol{\nu}$. Taking dot product with **a** and then with **m** yields $\mathbf{a} \cdot \boldsymbol{\nu} = -\mathbf{m} \cdot \boldsymbol{\nu}$. Also taking dot product in the equation $\mathbf{a} - \mathbf{m} = \lambda\boldsymbol{\nu}$ with $\boldsymbol{\nu}$ yields $\lambda = 2\mathbf{a} \cdot \boldsymbol{\nu}$, so

$$\mathbf{m} = \mathbf{a} - 2(\mathbf{a} \cdot \boldsymbol{\nu})\boldsymbol{\nu} \tag{4.19}$$

which gives the direction of the reflected ray. The classical Law of Reflection can be obtained directly from Snell's Law (4.18) by taking absolute values when $n_1 = n_2$:

$$\sin(\theta_1) = \sin(\theta_2) \implies \theta_1 = \theta_2$$

That is, the angle of reflection is equal to the angle of incidence; see Fig. 4.1.

Now let us see how the above analysis applies to the general case of anisotropic and homogeneous media (we again restrict to this case to ensure linear trajectories). Since $\mathbf{m} \cdot \boldsymbol{\nu}(\mathbf{x}_0) \leq 0$ for reflection (the reflected ray must be on the same side of the surface interface), we see that (4.12) becomes, writing $n_1 = n_2 = n$:

$$\nabla_{\mathbf{v}} n(\mathbf{x}_0, \mathbf{m}) + n(\mathbf{x}_0, \mathbf{m})\mathbf{m} - (\nabla_{\mathbf{v}} n(\mathbf{x}_0, \mathbf{a}) + n(\mathbf{x}_0, \mathbf{a})\mathbf{a}) = \lambda\boldsymbol{\nu}(\mathbf{x}_0), \qquad \lambda \in \mathbb{R} \tag{4.20}$$

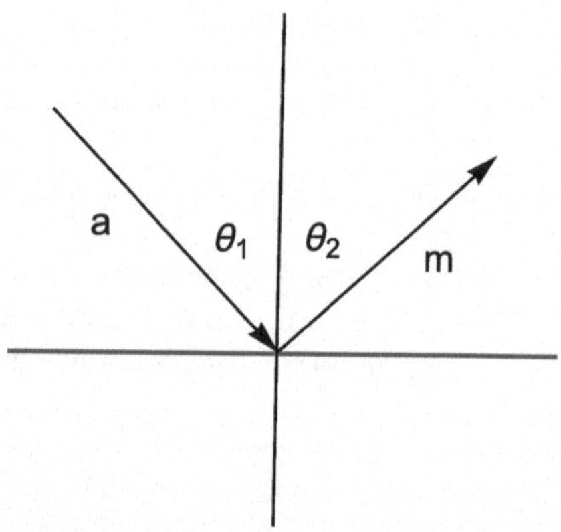

Fig. 4.1 The law of reflection

with λ now given by

$$\lambda = (n(\mathbf{x}_0, \mathbf{a})\mathbf{a} - \mathbf{g}) \cdot \boldsymbol{\nu}(\mathbf{x}_0) + \sqrt{n(\mathbf{x}_0, \mathbf{m})^2 - \left[|n(\mathbf{x}_0, \mathbf{a})\mathbf{a} - \mathbf{g}|^2 - ((n(\mathbf{x}_0, \mathbf{a})\mathbf{a} - \mathbf{g}) \cdot \boldsymbol{\nu}(\mathbf{x}_0))^2\right]} \tag{4.21}$$

The plus sign in front of the square root appears now since, as mentioned above, $\mathbf{m} \cdot \boldsymbol{\nu}(\mathbf{x}_0) \leq 0$. In the case when n is independent of direction, this indeed recovers the classical law of reflection [155].

4.5 Application to Liquid Crystals

4.5.1 Poynting Vector Direction

We apply now the above Snell's Law to the case of an interface of a uniaxial nematic liquid crystal with effective refractive index given by (3.11). That is, we formulate a Snell's Law in the direction of the Poynting vector . In the notation above, we consider

$$F(\mathbf{x}(s), \mathbf{v}(s)) = n_{\text{eff}}(\mathbf{x}(s), \mathbf{v}(s)) = \sqrt{n_e^2|\mathbf{v}|^2 + (n_o^2 - n_e^2)(\mathbf{v} \cdot \mathbf{d}(\mathbf{x}))^2}$$

We compute

$$\nabla_{\mathbf{v}} F = \frac{n_e^2\mathbf{v} + (n_o^2 - n_e^2)(\mathbf{v} \cdot \mathbf{d})\mathbf{d}}{\sqrt{n_e^2|\mathbf{v}|^2 + (n_o^2 - n_e^2)(\mathbf{v} \cdot \mathbf{d})^2}}$$

If we consider the interface to separate the liquid crystal from air (so that $n_1 \equiv 1$), from (4.12) we obtain the following Snell's Law:

$$\sqrt{n_e^2|\mathbf{m}|^2 + (n_o^2 - n_e^2)(\mathbf{m} \cdot \mathbf{d}(\mathbf{x}_0))^2}\mathbf{m} - \mathbf{a} = \lambda\boldsymbol{\nu}(\mathbf{x}_0) - \frac{n_e^2\mathbf{m} + (n_o^2 - n_e^2)(\mathbf{m} \cdot \mathbf{d}(\mathbf{x}_0))\mathbf{d}(\mathbf{x}_0)}{\sqrt{n_e^2|\mathbf{m}|^2 + (n_o^2 - n_e^2)(\mathbf{m} \cdot \mathbf{d}(\mathbf{x}_0))^2}} \tag{4.22}$$

which becomes (if $\mathbf{m}$ is a unit vector)

$$\sqrt{n_e^2 + (n_o^2 - n_e^2)(\mathbf{m} \cdot \mathbf{d}(\mathbf{x}_0))^2}\mathbf{m} - \mathbf{a} = \lambda\boldsymbol{\nu}(\mathbf{x}_0) - \frac{n_e^2\mathbf{m} + (n_o^2 - n_e^2)(\mathbf{m} \cdot \mathbf{d}(\mathbf{x}_0))\mathbf{d}(\mathbf{x}_0)}{\sqrt{n_e^2 + (n_o^2 - n_e^2)(\mathbf{m} \cdot \mathbf{d}(\mathbf{x}_0))^2}} \tag{4.23}$$

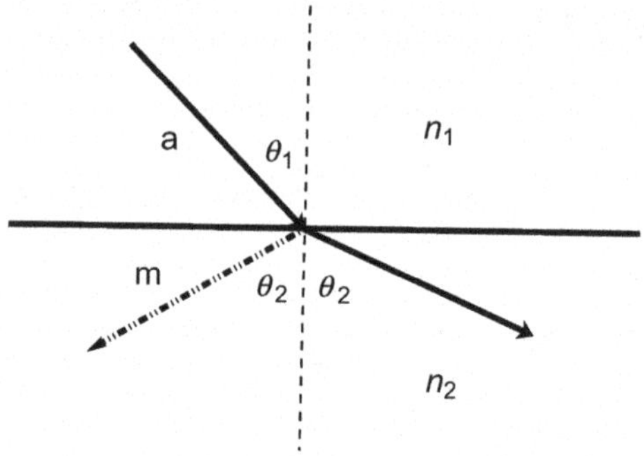

Fig. 4.2 Positive (solid line) and negative refraction (dashed line)

As an application of (4.23), we will obtain the critical angle for amphoteric refraction as in [263]. The phenomenon of amphoteric refraction is when refraction can be positive or negative. For left-handed materials (e.g., those with negative refractive index), when a light beam that is incident on it from air, the refracted beam lies on the same side of the normal to the interface as the incident beam, see Fig. 4.2. This is in contrast to the classical case (solid line in Fig. 4.2). It turns out that the refracted ray can be refracted positively or negatively depending on the incident direction and the orientation of the optic axis.

The critical angle for negative refraction occurs precisely when $\mathbf{m} \parallel \boldsymbol{\nu}$, so taking cross product in (4.23) with $\boldsymbol{\nu}$ yields

$$\mathbf{a} \times \boldsymbol{\nu} = \frac{(n_o^2 - n_e^2)(\mathbf{m} \cdot \mathbf{d}(\mathbf{x}_0))}{\sqrt{n_o^2(\mathbf{m} \cdot \mathbf{d}(\mathbf{x}_0))^2 + n_e^2(1 - (\mathbf{m} \cdot \mathbf{d}(\mathbf{x}_0))^2)}} \mathbf{d}(\mathbf{x}_0) \times \boldsymbol{\nu}(\mathbf{x}_0)$$

Let γ denote the angle between $\boldsymbol{\nu}(\mathbf{x}_0)$ and $\mathbf{d}(\mathbf{x}_0)$ (and thus the angle between $\mathbf{m}$ and $\mathbf{d}(\mathbf{x}_0)$). Taking absolute values in the previous equation yields

$$\sin(\theta_{c,a}) = \frac{(n_o^2 - n_e^2)\cos(\gamma)\sin(\gamma)}{\sqrt{n_o^2 \cos^2(\gamma) + n_e^2 \sin^2(\gamma)}} \tag{4.24}$$

where $\theta_{c,a}$ denotes the critical angle for amphoteric refraction. Equation (4.24) is exactly the critical angle obtained in [263], see also [340] and [256].

Remark 4.4 In the determination of λ in (4.17), we have assumed that $n_2(\mathbf{x}_0, \mathbf{m}) > 0$. However, this need not be the case. In standard liquid crystals, there holds that $n_o > 0$. However, in metamaterials, it could be that $n_o < 0$, see, e.g., [55]. Similarly, we could have $n_e > 0$ or $n_e < 0$. Thus we see that the sign of n_2 is determined by the product $n_o n_e$. To make this more precise, we include magnetic effects and define

$$n_o = \pm\sqrt{\mu\varepsilon_1}, \qquad n_e = \pm\sqrt{\mu\varepsilon_2} \tag{4.25}$$

Thus the sign of n_2 is dictated by the signs of μ, ε_1 and ε_2. To summarize, if $n_o n_e > 0$, then λ is given by (4.17), whereas if $n_o n_e < 0$, then $n_2 < 0$ and so λ is given by

$$\lambda = (n_1(\mathbf{x}_0, \mathbf{a})\mathbf{a} - \mathbf{g}) \cdot \boldsymbol{\nu}(\mathbf{x}_0) + \sqrt{n_2(\mathbf{x}_0, \mathbf{m})^2 - \left[|n_1(\mathbf{x}_0, \mathbf{a})\mathbf{a} - \mathbf{g}|^2 - ((n_1(\mathbf{x}_0, \mathbf{a})\mathbf{a} - \mathbf{g}) \cdot \boldsymbol{\nu}(\mathbf{x}_0))^2\right]} \tag{4.26}$$

This can be compared with what happens in isotropic media. Indeed, consider an interface between two isotropic media with refractive indices $n_1, n_2 > 0$. Then Snell's Law at this interface is given by

$$\mathbf{a} - \frac{n_2}{n_1}\mathbf{m} = \lambda_0 \boldsymbol{\nu} \tag{4.27}$$

where **a** is the incident vector and **m** is the refracted vector; see, e.g., [155]. In this case λ_0 can be found to be

$$\lambda_0 = \mathbf{a} \cdot \boldsymbol{\nu} - \frac{n_2}{n_1}\sqrt{1 - (n_1/n_2)^2(1 - (\mathbf{a} \cdot \boldsymbol{\nu})^2)}$$

Now, if $n_2 < 0$, then Snell's Law takes the form

$$\mathbf{a} - \kappa \mathbf{m} = \lambda_1 \boldsymbol{\nu} \tag{4.28}$$

where $\kappa = n_2/n_1 < 0$, and in this case

$$\lambda_1 = \mathbf{a} \cdot \boldsymbol{\nu} + \sqrt{(\mathbf{a} \cdot \boldsymbol{\nu})^2 - (1 - \kappa)^2}$$

see [168], Section 2.B.

4.5.2 Wave Vector Direction

Similar to the previous section, we can formulate a vector form of Snell's Law for the refracted wave vector direction. Consider the case when Ω_1 is isotropic and homogeneous (e.g., air) and Ω_2 is filled with a liquid crystal. In this case, $n_1(\mathbf{x}_0, \mathbf{a}) = n_1$ is just constant. We are mainly interested in the extraordinary ray, so the index $n_2(\mathbf{x}_0, \mathbf{m})$ will be given by (3.9), i.e.,

$$n_2(\mathbf{x}_0, \mathbf{m}) = \frac{|n_o n_e|}{\sqrt{n_o^2\left(1 - (\mathbf{m} \cdot \mathbf{d}(\mathbf{x}_0))^2\right) + n_e^2(\mathbf{m} \cdot \mathbf{d}(\mathbf{x}_0))^2}} \tag{4.29}$$

This yields the following vector form of Snell's Law (after computing the $\mathbf{v}$ gradient):

$$\frac{|n_o n_e|}{\sqrt{n_o^2\left(1-(\mathbf{m}\cdot\mathbf{d}(\mathbf{x}_0))^2\right)+n_e^2(\mathbf{m}\cdot\mathbf{d}(\mathbf{x}_0))^2}}\mathbf{m}-n_1\mathbf{a}=$$
$$\tilde{\lambda}\boldsymbol{\nu}(\mathbf{x}_0)-\frac{n_o n_e(\mathbf{m}\cdot\mathbf{d}(\mathbf{x}_0))(n_o^2-n_e^2)}{\left(n_o^2\left(1-(\mathbf{m}\cdot\mathbf{d}(\mathbf{x}_0))^2\right)+n_e^2(\mathbf{m}\cdot\mathbf{d}(\mathbf{x}_0))^2\right)^{3/2}}\mathbf{d} \tag{4.30}$$

From (4.30) we can recover a scalar version of Snell's Law if we assume the following homeotropic boundary condition holds[3]

$$\mathbf{d}(\mathbf{x}_0)\parallel\boldsymbol{\nu}(\mathbf{x}_0) \tag{4.31}$$

Then by taking cross product in (4.30) with $\boldsymbol{\nu}$, we obtain

$$\frac{|n_o n_e|}{\sqrt{n_o^2\left(1-(\mathbf{m}\cdot\mathbf{d}(\mathbf{x}_0))^2\right)+n_e^2(\mathbf{m}\cdot\mathbf{d}(\mathbf{x}_0))^2}}\mathbf{m}\times\boldsymbol{\nu}(\mathbf{x}_0)=n_1\mathbf{a}\times\boldsymbol{\nu}(\mathbf{x}_0) \tag{4.32}$$

Taking absolute value in (4.32) yields

$$n_1\sin(\alpha)=\frac{|n_o n_e|}{\sqrt{n_o^2\left(1-(\mathbf{m}\cdot\mathbf{d}(\mathbf{x}_0))^2\right)+n_e^2(\mathbf{m}\cdot\mathbf{d}(\mathbf{x}_0))^2}}\sin(\beta) \tag{4.33}$$

where α is the angle between $\mathbf{a}$ and $\boldsymbol{\nu}$, and β is the refracted angle. Then, (4.33) is precisely the scalar law obtained in [55], see their equation (8). Note that (4.33) depends on the angle between the refracted vector $\mathbf{m}$ and the director $\mathbf{d}$, which corresponds to the angle $\gamma-\beta$ in [55].

There can be two critical angles for total internal reflection of the extraordinary ray at the interface of an isotropic and anisotropic material, see, e.g., [384, Section 3.4.2]. We will determine what the critical angles are for the wave vector direction by using (4.12).

Since $\lambda\in\mathbb{R}$, the quantity under the square root should be nonnegative:

$$n_2(\mathbf{x}_0,\mathbf{m})^2\geq|n_1(\mathbf{x}_0,\mathbf{a})\mathbf{a}-\mathbf{g}|^2-((n_1(\mathbf{x}_0,\mathbf{a})\mathbf{a}-\mathbf{g})\cdot\boldsymbol{\nu}(\mathbf{x}_0))^2 \tag{4.34}$$

Notice this is equivalent to

[3] That is, the director is perpendicular to the interface. This is opposed to a planar boundary condition, where we would impose $\mathbf{d}(\mathbf{x}_0)\cdot\boldsymbol{\nu}(\mathbf{x}_0)=0$.

$$\left[\left(\mathbf{a} - \frac{\mathbf{g}}{n_1(\mathbf{x}_0, \mathbf{a})}\right) \cdot \boldsymbol{\nu}(\mathbf{x}_0)\right]^2 \geq \left|\mathbf{a} - \frac{\mathbf{g}}{n_1(\mathbf{x}_0, \mathbf{a})}\right|^2 - \left(\frac{n_2(\mathbf{x}_0, \mathbf{m})}{n_1(\mathbf{x}_0, \mathbf{a})}\right)^2$$

So if $\mathbf{a}$ is such that

$$\left|\mathbf{a} - \frac{\mathbf{g}}{n_1(\mathbf{x}_0, \mathbf{a})}\right| \leq \frac{n_2(\mathbf{x}_0, \mathbf{m})}{n_1(\mathbf{x}_0, \mathbf{a})}$$

then (4.34) holds. Now, if

$$\left|\mathbf{a} - \frac{\mathbf{g}}{n_1(\mathbf{x}_0, \mathbf{a})}\right| > \frac{n_2(\mathbf{x}_0, \mathbf{m})}{n_1(\mathbf{x}_0, \mathbf{a})}$$

then (4.34) holds when either

$$\mathbf{a} \cdot \boldsymbol{\nu}(\mathbf{x}_0) \geq \frac{\mathbf{g}}{n_1(\mathbf{x}_0, \mathbf{a})} \cdot \boldsymbol{\nu}(\mathbf{x}_0) + \sqrt{\left|\mathbf{a} - \frac{\mathbf{g}}{n_1(\mathbf{x}_0, \mathbf{a})}\right|^2 - \left(\frac{n_2(\mathbf{x}_0, \mathbf{m})}{n_1(\mathbf{x}_0, \mathbf{a})}\right)^2}$$

or

$$\mathbf{a} \cdot \boldsymbol{\nu}(\mathbf{x}_0) \leq \frac{\mathbf{g}}{n_1(\mathbf{x}_0, \mathbf{a})} \cdot \boldsymbol{\nu}(\mathbf{x}_0) - \sqrt{\left|\mathbf{a} - \frac{\mathbf{g}}{n_1(\mathbf{x}_0, \mathbf{a})}\right|^2 - \left(\frac{n_2(\mathbf{x}_0, \mathbf{m})}{n_1(\mathbf{x}_0, \mathbf{a})}\right)^2}$$

Thus the critical angles are the angles between $\mathbf{a}$ and $\boldsymbol{\nu}$, denoted θ_c, such that

$$\mathbf{a} \cdot \boldsymbol{\nu}(\mathbf{x}_0) = \cos(\theta_c) = \frac{\mathbf{g} \cdot \boldsymbol{\nu}(\mathbf{x}_0)}{n_1(\mathbf{x}_0, \mathbf{a})} + \sqrt{\left|\mathbf{a} - \frac{\mathbf{g}}{n_1(\mathbf{x}_0, \mathbf{a})}\right|^2 - \left(\frac{n_2(\mathbf{x}_0, \mathbf{m})}{n_1(\mathbf{x}_0, \mathbf{a})}\right)^2}, \quad \text{or} \tag{4.35}$$

$$\mathbf{a} \cdot \boldsymbol{\nu}(\mathbf{x}_0) = \cos(\theta_c) = \frac{\mathbf{g} \cdot \boldsymbol{\nu}(\mathbf{x}_0)}{n_1(\mathbf{x}_0, \mathbf{a})} - \sqrt{\left|\mathbf{a} - \frac{\mathbf{g}}{n_1(\mathbf{x}_0, \mathbf{a})}\right|^2 - \left(\frac{n_2(\mathbf{x}_0, \mathbf{m})}{n_1(\mathbf{x}_0, \mathbf{a})}\right)^2} \tag{4.36}$$

Now, what do these calculations imply? Again, under the strong anchoring condition (4.31), the critical angle occurs when

$$\frac{n_o n_e}{\sqrt{n_o^2\left(1 - (\mathbf{m} \cdot \mathbf{d}(\mathbf{x}_0))^2\right) + n_e^2(\mathbf{m} \cdot \mathbf{d}(\mathbf{x}_0))^2}} \mathbf{m} \cdot \boldsymbol{\nu}(\mathbf{x}_0) = 0$$

Since $\mathbf{d}(\mathbf{x}_0) \parallel \boldsymbol{\nu}(\mathbf{x}_0)$, this means that the effective refractive index becomes just n_e. We further find that $\mathbf{g} = \mathbf{0}$. Thus the previous critical angle calculations give

$$\cos(\theta_c) = \pm\sqrt{1 - \frac{n_e^2}{n_1^2}}$$

if Ω_1 is homogeneous and isotropic with refractive index n_1. Squaring both sides of the last equation yields $\sin^2(\theta_c) = n_e^2/n_1^2$, which gives the critical angle for refraction

$$\theta_c = \arcsin\left(\frac{n_e}{n_1}\right) \tag{4.37}$$

This is precisely the critical angle calculated in [449] in the case when the optic axis is parallel to the normal $\boldsymbol{\nu}$, and there is no twist or tilt of the optic axis; see their equation (5). For the ordinary ray, $n_2 = n_o$, so the above calculation yields the critical angle for the ordinary ray

$$\theta_c^o = \arcsin\left(\frac{n_o}{n_1}\right)$$

which agrees with [449, Equation (7)].

Now let us understand the critical angles under the planar boundary condition

$$\mathbf{d}(\mathbf{x}_0) \cdot \boldsymbol{\nu}(\mathbf{x}_0) = 0 \tag{4.38}$$

Denote the angle between $\mathbf{m}$ and $\mathbf{d}(\mathbf{x}_0)$ by $\gamma - \beta$, where γ is the angle between $\mathbf{d}(\mathbf{x}_0)$ and $\boldsymbol{\nu}(\mathbf{x}_0)$, and β is the angle between $\mathbf{m}$ and $\boldsymbol{\nu}(\mathbf{x}_0)$ (thus $\gamma = \pi/2$).

Then from (4.35)–(4.36), since $\mathbf{g} \cdot \boldsymbol{\nu}(\mathbf{x}_0) = 0$ in this case (recall $\mathbf{g} \parallel \mathbf{d}(x_0)$), we obtain by using the definition of $\mathbf{g}$

$$\cos^2(\theta_c) = \left|\mathbf{a} - \frac{\mathbf{g}}{n_1}\right|^2 - \frac{n_o^2 n_e^2}{n_1^2\left[n_o^2\left(\sin^2(\gamma-\beta)\right) + n_e^2\cos^2(\gamma-\beta)\right]}$$

Expanding the square on the right hand side yields, using the fact that $\mathbf{a} \cdot \mathbf{g} = \cos(\pi/2 - \theta_c)|\mathbf{g}|$:

$$\begin{aligned}\cos^2(\theta_c) = 1 &- \frac{2}{n_1}\frac{n_o n_e(n_o^2 - n_e^2)\cos(\gamma-\beta)\cos(\pi/2-\theta_c)}{\left[n_o^2\left(\sin^2(\gamma-\beta)\right) + n_e^2\cos^2(\gamma-\beta)\right]^{3/2}} \\ &+ \frac{n_o^2 n_e^2 (n_o^2 - n_e^2)^2\cos^2(\gamma-\beta)}{n_1^2\left[n_o^2\left(\sin^2(\gamma-\beta)\right) + n_e^2\cos^2(\gamma-\beta)\right]^3} \\ &- \frac{n_o^2 n_e^2}{n_1^2\left[n_o^2\left(\sin^2(\gamma-\beta)\right) + n_e^2\cos^2(\gamma-\beta)\right]}\end{aligned}$$

Since $\cos(\pi/2 - \theta_c) = \sin(\theta_c)$, we obtain the following equation satisfied by θ_c:

$$
\begin{aligned}
&\sin^2(\theta_c) - \frac{2}{n_1}\frac{n_o n_e (n_o^2 - n_e^2)\cos(\gamma-\beta)}{\left[n_o^2\left(\sin^2(\gamma-\beta)\right) + n_e^2\cos^2(\gamma-\beta)\right]^{3/2}}\sin(\theta_c) \\
&+ \frac{n_o^2 n_e^2 (n_o^2 - n_e^2)^2\cos^2(\gamma-\beta)}{n_1^2\left[n_o^2\left(\sin^2(\gamma-\beta)\right) + n_e^2\cos^2(\gamma-\beta)\right]^3} \\
&\quad - \frac{n_o^2 n_e^2}{n_1^2\left[n_o^2\left(\sin^2(\gamma-\beta)\right) + n_e^2\cos^2(\gamma-\beta)\right]} \\
&= 0
\end{aligned}
\tag{4.39}
$$

Letting $\gamma = \pi/2$ and solving this quadratic equation yield

$$
\theta_c = \pm\arcsin\left(\frac{n_e n_o\left(\sin(\beta)\left(n_e^2\sin(\beta) - n_e^2 + n_o^2\right) + n_e^2\cos^2(\beta)\right)}{n_1\left(n_e^2\sin^2(\beta) + n_o^2\cos^2(\beta)\right)^{3/2}}\right)
\tag{4.40}
$$

If we further impose that $\beta = 0$ (which in this context means there is no twist of the optical axis), then (4.40) implies that

$$
\theta_c = \pm\arcsin\left(\frac{n_e}{n_1}\right)
\tag{4.41}
$$

which agrees with (4.37) and hence the special case in [449].

4.5.3 *Biaxial Nematics*

In the previous section we implicitly assumed the liquid crystal was a uniaxial nematic. However, we can also consider a biaxial nematic, and to do so, we start from Snell's Law using wave normals $\mathbf{p}$, as considered in [48]

$$
\mathbf{p}_i \times \boldsymbol{\nu} = \mathbf{p}_r \times \boldsymbol{\nu}
\tag{4.42}
$$

where $\mathbf{p}_r$ denotes the reflected or refracted wave normal, $\boldsymbol{\nu}$ as before is the unit normal pointing from an isotropic medium (say) to the second material (here we will take this to be a biaxial nematic), and $\mathbf{p}_i$ is the incident wave normal. In an isotropic medium, the wave normal is given by

$$
\mathbf{p}_i = n\langle\mathbf{S}\rangle
$$

where n is the index of refraction of the material and $\langle\mathbf{S}\rangle$ is the (normalized) incident Poynting vector [384]. When the material is anisotropic, though, the wave normal $\mathbf{p}_i$ satisfies $\mathcal{H}(\mathbf{x}, \mathbf{p}) = 0$, where $\mathcal{H}$ is the optical indicatrix.

The refracted wave normal can be determined by substitution of

$$\mathbf{p} = \mathbf{p}_r := \mathbf{p}_t + \lambda \boldsymbol{\nu}, \quad \lambda \geq 0$$

into $\mathcal{H}(\mathbf{x}, \mathbf{p}) = 0$ and solving for λ. Here, $\mathbf{p}_t$ denotes the tangential wave normal, which can be computed as

$$\mathbf{p}_t = \mathbf{p}_i - (\mathbf{p}_i \cdot \boldsymbol{\nu})\boldsymbol{\nu}$$

We merely subtracted the normal component of $\mathbf{p}_i$ from $\mathbf{p}_i$.

Without loss of generality, we assume that the principal coordinate system $\{\mathbf{e}_u, \mathbf{e}_v, \mathbf{e}_w\}$ coincides with the coordinate system in which $\mathbf{p}_t$ and $\boldsymbol{\nu}$ are defined. If not, one can consider a linear transformation to change bases. For example, if ε is not diagonal, one can apply two rotations: If ε is diagonal in a double primed coordinate system $\mathbf{x}'' = (x'', y'', z'')$, then one can transform to an unprimed (i.e., laboratory) frame as follows. First, transform to a primed frame $\mathbf{x}' = (x', y', z')$ by rotating about the x'' axis by an angle ψ_1, i.e., via the rotation matrix

$$R_1 = \begin{pmatrix} 1 & 0 & 0 \\ 0 & \cos(\psi_1) & \sin(\psi_1) \\ 0 & -\sin(\psi_1) & \cos(\psi_1) \end{pmatrix}$$

Then rotate the new ε about the z' axis by an angle ψ_2, i.e., by applying the rotation matrix

$$R_2 = \begin{pmatrix} \cos(\psi_2) & \sin(\psi_2) & 0 \\ -\sin(\psi_2) & \cos(\psi_2) & 0 \\ 0 & 0 & 1 \end{pmatrix}$$

This results in the final ε having the form, denoting the components of ε by ε_{ij}:

$$\begin{aligned}
\varepsilon_{11} &= \varepsilon_u \cos^2(\psi_2) + (\varepsilon_v \cos^2(\psi_1) + \varepsilon_w \sin^2(\psi_1)) \sin^2(\psi_2), \\
\varepsilon_{12} &= \left(-\varepsilon_u + \varepsilon_v \cos^2(\psi_1) + \varepsilon_w \sin^2(\psi_1)\right) \sin(\psi_2) \cos(\psi_2), \\
\varepsilon_{13} &= (\varepsilon_w - \varepsilon_v) \sin(\psi_1) \cos(\psi_1) \sin(\psi_2), \\
\varepsilon_{22} &= \varepsilon_u \sin^2(\psi_2) + (\varepsilon_v \cos^2(\psi_1) + \varepsilon_w \sin^2(\psi_1)) \cos^2(\psi_2), \\
\varepsilon_{23} &= (\varepsilon_w - \varepsilon_v) \sin(\psi_1) \cos(\psi_1) \cos(\psi_2), \\
\varepsilon_{33} &= \varepsilon_v \sin^2(\psi_1) + \varepsilon_w \cos^2(\psi_1), \\
\varepsilon_{21} &= \varepsilon_{12}, \ \varepsilon_{31} = \varepsilon_{13}, \ \varepsilon_{32} = \varepsilon_{23}
\end{aligned}$$

Now, from (2.42), we know that

$$\langle \mathbf{S} \rangle = \frac{\nabla_{\mathbf{p}} \mathcal{H}}{|\nabla_{\mathbf{p}} \mathcal{H}|}$$

where $\mathcal{H}$ satisfies (2.22). This allows one to solve for $\mathbf{p}_i$ since $\mathbf{p}_i = |p_{i,u}|\mathbf{e}_u + |p_{i,v}|\mathbf{e}_v + |p_{i,w}|\mathbf{e}_w$, so from here one can compute the tangential wave normal $\mathbf{p}_t$.

Again, an incident wave upon a biaxial medium (assuming a planar interface) will split into two characteristic waves, which we saw in Chap. 2 are called the *a*-wave and *b*-wave.

Let us compute the refracted wave normals. To start, a quick calculation shows that this method for the isotropic case (or, for the ordinary ray in uniaxial media), i.e., plugging $\mathbf{p} = \mathbf{p}_i - (\mathbf{p}_i \cdot \boldsymbol{\nu})\boldsymbol{\nu} + \lambda \boldsymbol{\nu}$ into $\mathcal{H}(\mathbf{x}, \mathbf{p}) = |\mathbf{p}|^2 - n_o^2 = 0$, yields the same λ_0 as in (4.27), assuming $\mathbf{p}_i$ to be a unit vector. What about the extraordinary ray in uniaxial media? In this case, recall that

$$\mathcal{H}_e(\mathbf{x}, p) = (|p_u|^2 + |p_v|^2)n_o^2 + |p_w|^2 n_e^2 - n_e^2 n_o^2$$

The plugging $\mathbf{p} = \mathbf{p}_i - (\mathbf{p}_i \cdot \boldsymbol{\nu})\boldsymbol{\nu} + \lambda\boldsymbol{\nu}$ into $\mathcal{H}_e = 0$ yields

$$\begin{aligned}
&n_o^2 \left(|p_u^i|^2 - 2(\mathbf{p}_i \cdot \boldsymbol{\nu})(p_u^i \nu_u) + (\mathbf{p}_i \cdot \boldsymbol{\nu})^2 \nu_u^2 - 2\lambda(\mathbf{p}_i \cdot \boldsymbol{\nu})\nu_u^2 + \lambda^2 \nu_u^2 + \right.\\
&\quad \left. + |p_v^i|^2 - 2(\mathbf{p}_i \cdot \boldsymbol{\nu})(p_v^i \nu_v) + (\mathbf{p}_i \cdot \boldsymbol{\nu})^2 \nu_v^2 - 2\lambda(\mathbf{p}_i \cdot \boldsymbol{\nu})\nu_v^2 + \lambda^2 \nu_v^2\right)\\
&\quad + n_e^2 \left(|p_w^i|^2 - 2(\mathbf{p}_i \cdot \boldsymbol{\nu})(p_w^i \nu_w) + (\mathbf{p}_i \cdot \boldsymbol{\nu})^2 \nu_w^2 - 2\lambda(\mathbf{p}_i \cdot \boldsymbol{\nu})\nu_w^2 + \lambda^2 \nu_w^2\right)\\
&\quad - n_o^2 n_e^2 = 0
\end{aligned}$$

where $\boldsymbol{\nu} = (\nu_u, \nu_v, \nu_w)$ and $\mathbf{p}_i = (p_u^i, p_v^i, p_w^i)$. This is a quadratic equation in λ:

$$\begin{aligned}
&\lambda^2 \left(n_o^2 \left(\nu_u^2 + \nu_v^2\right) + n_e^2 \nu_w^2\right) - 2(\mathbf{p}_i \cdot \boldsymbol{\nu})\lambda \left(n_o^2 \left(\nu_u^2 + \nu_v^2\right) + n_e^2 \nu_w^2\right)\\
&\quad + \left(n_o^2 \left(|p_u^i|^2 + |p_v^i|^2\right) + n_e^2 |p_w^i|^2 - 2n_o^2 (\mathbf{p}_i \cdot \boldsymbol{\nu}) \left(p_u^i \nu_u + p_v^i \nu_v\right)\right.\\
&-2n_e^2(\mathbf{p}_i \cdot \boldsymbol{\nu})(p_w^i \nu_w)\\
&\quad \left. + n_o^2 (\mathbf{p}_i \cdot \boldsymbol{\nu})^2 \left(\nu_u^2 + \nu_v^2\right) + n_e^2 (\mathbf{p}_i \cdot \boldsymbol{\nu})^2 \nu_w^2 - n_o^2 n_e^2\right) = 0
\end{aligned} \tag{4.43}$$

Defining the quantities

$$\begin{aligned}
A_1 &= n_o^2 \left(\nu_u^2 + \nu_v^2\right) + n_e^2 \nu_w^2,\\
A_2 &= -2(\mathbf{p}_i \cdot \boldsymbol{\nu}) \left(n_o^2 \left(\nu_u^2 + \nu_v^2\right) + n_e^2 \nu_w^2\right),\\
A_3 &= n_o^2 \left(|p_u^i|^2 + |p_v^i|^2\right) + n_e^2 |p_w^i|^2 - 2n_o^2 (\mathbf{p}_i \cdot \boldsymbol{\nu}) \left(p_u^i \nu_u + p_v^i \nu_v\right)
\end{aligned}$$

$$
\begin{aligned}
&- 2n_e^2(\mathbf{p}_i \cdot \boldsymbol{\nu})(p_w^i \nu_w) \\
&+ n_o^2(\mathbf{p}_i \cdot \boldsymbol{\nu})^2 \left(\nu_u^2 + \nu_v^2\right) + n_e^2(\mathbf{p}_i \cdot \boldsymbol{\nu})^2 \nu_w^2 - n_o^2 n_e^2
\end{aligned}
$$

we write (4.43) as

$$
A_1\lambda^2 + A_2\lambda + A_3 = 0
$$

whose solution is

$$
\lambda = \frac{-A_2 \pm \sqrt{A_2^2 - 4A_1A_3}}{2A_1}
$$

The plus or minus above applies to refracted rays and the minus sign to reflected rays.

Finally we apply this method for the biaxial case to find the refracted wave vectors for both the a- and b-waves. So we use (2.22) and plug in $\mathbf{p} = \mathbf{p}_i - (\mathbf{p}_i \cdot \boldsymbol{\nu})\boldsymbol{\nu} + \lambda\boldsymbol{\nu}$ as usual to find

$$
\begin{aligned}
&\left(|\mathbf{p}_i|^2 - (\mathbf{p}_i \cdot \boldsymbol{\nu})^2 + \lambda^2\right) \\
&\quad \Big\{\varepsilon_u \left(|p_u^i|^2 - 2(\mathbf{p}_i \cdot \boldsymbol{\nu})(p_u^i \nu_u) + (\mathbf{p}_i \cdot \boldsymbol{\nu})^2 \nu_u^2 - 2\lambda(\mathbf{p}_i \cdot \boldsymbol{\nu})\nu_u^2 + \lambda^2 \nu_u^2\right) \\
&\quad + \varepsilon_v \left(|p_v^i|^2 - 2(\mathbf{p}_i \cdot \boldsymbol{\nu})(p_v^i \nu_v) + (\mathbf{p}_i \cdot \boldsymbol{\nu})^2 \nu_v^2 - 2\lambda(\mathbf{p}_i \cdot \boldsymbol{\nu})\nu_v^2 + \lambda^2 \nu_v^2\right) \\
&\quad + \varepsilon_w \left(|p_w^i|^2 - 2(\mathbf{p}_i \cdot \boldsymbol{\nu})(p_w^i \nu_w) + (\mathbf{p}_i \cdot \boldsymbol{\nu})^2 \nu_w^2 - 2\lambda(\mathbf{p}_i \cdot \boldsymbol{\nu})\nu_w^2 + \lambda^2 \nu_w^2\right)\Big\} \\
&- \varepsilon_u(\varepsilon_v + \varepsilon_w) \left(|p_u^i|^2 - 2(\mathbf{p}_i \cdot \boldsymbol{\nu})(p_u^i \nu_u) + (\mathbf{p}_i \cdot \boldsymbol{\nu})^2 \nu_u^2 - 2\lambda(\mathbf{p}_i \cdot \boldsymbol{\nu})\nu_u^2 + \lambda^2 \nu_u^2\right) \\
&- \varepsilon_v(\varepsilon_u + \varepsilon_w) \left(|p_v^i|^2 - 2(\mathbf{p}_i \cdot \boldsymbol{\nu})(p_v^i \nu_v) + (\mathbf{p}_i \cdot \boldsymbol{\nu})^2 \nu_v^2 - 2\lambda(\mathbf{p}_i \cdot \boldsymbol{\nu})\nu_v^2 + \lambda^2 \nu_v^2\right) \\
&- \varepsilon_w(\varepsilon_u + \varepsilon_v) \left(|p_w^i|^2 - 2(\mathbf{p}_i \cdot \boldsymbol{\nu})(p_w^i \nu_w) + (\mathbf{p}_i \cdot \boldsymbol{\nu})^2 \nu_w^2 - 2\lambda(\mathbf{p}_i \cdot \boldsymbol{\nu})\nu_w^2 + \lambda^2 \nu_w^2\right) \\
&\quad + \varepsilon_u\varepsilon_v\varepsilon_w = 0
\end{aligned}
$$

This is now a fourth order equation in λ. We get

$$
B_1\lambda^4 + B_2\lambda^3 + B_3\lambda^2 + B_4\lambda + B_5 = 0 \tag{4.44}
$$

where

$$
B_1 = \varepsilon_u \nu_u^2 + \varepsilon_v \nu_v^2 + \varepsilon_w \nu_w^2,
$$

$$
\begin{aligned}
B_2 &= -2(\mathbf{p}_i \cdot \boldsymbol{\nu}) \left(\varepsilon_u \nu_u^2 + \varepsilon_v \nu_v^2 + \varepsilon_w \nu_w^2 \right), \\
B_3 &= \left(|\mathbf{p}_i|^2 - (\mathbf{p}_i \cdot \boldsymbol{\nu})^2 \right) \left(\varepsilon_u \nu_u^2 + \varepsilon_v \nu_v^2 + \varepsilon_w \nu_w^2 \right) \\
&\quad + \varepsilon_u \left(|p_u^i|^2 - 2(\mathbf{p}_i \cdot \boldsymbol{\nu})(p_u^i \nu_u) + (\mathbf{p}_i \cdot \boldsymbol{\nu})^2 \nu_u^2 \right) \\
&\quad + \varepsilon_v \left(|p_v^i|^2 - 2(\mathbf{p}_i \cdot \boldsymbol{\nu})(p_v^i \nu_v) + (\mathbf{p}_i \cdot \boldsymbol{\nu})^2 \nu_v^2 \right) \\
&\quad + \varepsilon_w \left(|p_w^i|^2 - 2(\mathbf{p}_i \cdot \boldsymbol{\nu})(p_w^i \nu_w) + (\mathbf{p}_i \cdot \boldsymbol{\nu})^2 \nu_w^2 \right) \\
&\quad - \varepsilon_u(\varepsilon_v + \varepsilon_w)\nu_u^2 - \varepsilon_v(\varepsilon_u + \varepsilon_w)\nu_v^2 - \varepsilon_w(\varepsilon_u + \varepsilon_v)\nu_w^2, \\
B_4 &= -2(\mathbf{p}_i \cdot \boldsymbol{\nu}) \left(|\mathbf{p}_i|^2 - (\mathbf{p}_i \cdot \boldsymbol{\nu})^2 \right) \left(\varepsilon_u \nu_u^2 + \varepsilon_v \nu_v^2 + \varepsilon_w \nu_w^2 \right) \\
&\quad + 2(\mathbf{p}_i \cdot \boldsymbol{\nu}) \left(\varepsilon_u(\varepsilon_v + \varepsilon_w)\nu_u^2 + \varepsilon_v(\varepsilon_u + \varepsilon_w)\nu_v^2 + \varepsilon_w(\varepsilon_u + \varepsilon_v)\nu_w^2 \right), \\
B_5 &= \varepsilon_u \varepsilon_v \varepsilon_w
\end{aligned}
$$

The four roots of the above quartic equation for λ (two corresponding to the a-wave, two to the b-wave) are given by the formulas

$$
\begin{aligned}
\lambda_{1,2} &= -\frac{B_2}{4B_1} - S \pm \frac{1}{2}\sqrt{-4S^2 - 2\beta + \frac{\gamma}{S}}, \\
\lambda_{3,4} &= -\frac{B_2}{4B_1} + S \pm \frac{1}{2}\sqrt{-4S^2 - 2\beta - \frac{\gamma}{S}}
\end{aligned}
\tag{4.45}
$$

where

$$
\beta = \frac{8B_1B_3 - 3B_2^2}{8B_1^2}, \qquad \gamma = \frac{B_2^3 - 4B_1B_2B_3 + 8B_1^2B_4}{8B_1^3}
$$

and

$$
S = \frac{1}{2}\sqrt{-\frac{2}{3}\beta + \frac{1}{3B_1}\left(Q + \frac{\Delta_0}{Q} \right)}
$$

where

$$
Q = \left(\frac{\Delta_1 + \sqrt{\Delta_1^2 - 4\Delta_0^3}}{2} \right)^{1/3}
$$

where

$$\Delta_0 = B_3^2 - 3B_2B_4 + 12B_1B_5,$$

$$\Delta_1 = 2B_3^3 - 9B_2B_3B_4 + 27B_2^2B_5 + 27B_1B_4^2 - 72B_1B_3B_5$$

See, for instance, [445] for more details on obtaining general solutions to a quartic.

Let us compare this method with the notation of previous sections. Suppose a light ray originating in an isotropic medium with refractive index n_1 strikes a surface separating the isotropic medium from a biaxial nematic. We know that two rays will be refracted: the so-called a-wave and b-wave. Let $\mathbf{x}$ denote the incident ray vector and let $\mathbf{m}_a, \mathbf{m}_b$ denote the refracted a and b rays. Then, with $\mathbf{p}_i = n_1\mathbf{x}$, $\mathbf{p}_a = n_a(\mathbf{m}_a)\mathbf{m}_a$, and $\mathbf{p}_b = n_b(\mathbf{m}_b)\mathbf{m}_b$, Snell's Law takes the form

$$n_a(\mathbf{m}_a)\mathbf{m}_a = n_i\mathbf{x} - (n_i\mathbf{x}\cdot\boldsymbol{\nu})\boldsymbol{\nu} + \lambda\boldsymbol{\nu}$$

for the a ray and

$$n_b(\mathbf{m}_b)\mathbf{m}_b = n_i\mathbf{x} - (n_i\mathbf{x}\cdot\boldsymbol{\nu})\boldsymbol{\nu} + \lambda\boldsymbol{\nu}$$

for the b ray. We have emphasized that the refractive indices $n_a(\mathbf{m}_a)$ and $n_b(\mathbf{m}_b)$ depend on the direction of propagation, as already seen in (2.31) and (2.32). Thus one cannot simply "solve" for the refracted direction $\mathbf{m}_a$ or $\mathbf{m}_b$ since the associated refractive indices also depend on the directions $\mathbf{m}_a$ and $\mathbf{m}_b$.

If the incoming ray was instead perhaps another biaxial material, then an incident a ray (say) would refract another a-wave according to

$$n_a(\mathbf{m}_a)\mathbf{m}_a = n_{i,a}(\mathbf{x}_a)\mathbf{x}_a - (n_{i,a}(\mathbf{x}_a)\mathbf{x}_a\cdot\boldsymbol{\nu})\boldsymbol{\nu} + \lambda\boldsymbol{\nu}$$

The difficulty here, as seen above, is that the unknown λ now solves a fourth order equation.

4.5.4 *Snell's Law with Absorption*

We can also consider absorbing media (recall the end of Chap. 2) where a complex version of Snell's Law holds. We will first recall this for isotropic absorbing media and then focus on uniaxial anisotropic media. We recall the basic idea from [78] and provide more details in the derivation. We know that an absorbing medium can be characterized by a complex refractive index $\widetilde{n} = n + i\kappa$. Seeking plane wave solutions of the isotropic Maxwell equations, in particular taking the form $\mathbf{E}(\mathbf{x}, t) = \mathbf{E}e^{i\mathbf{k}\cdot\mathbf{x}-i\omega t}$ and similar for $\mathbf{H}$, we obtain

$$
\begin{aligned}
\mathbf{k}\times\mathbf{E} &= \omega\mathbf{H},\\
\mathbf{k}\times\mathbf{H} &= -\widetilde{\varepsilon}\omega\mathbf{E},\\
\mathbf{k}\cdot\mathbf{E} &= 0,\\
\mathbf{k}\cdot\mathbf{H} &= 0
\end{aligned}
\tag{4.46}
$$

where the last two equations above come from the divergence free conditions in the isotropic Maxwell equations $\nabla\cdot\mathbf{E}(\mathbf{x},t)=0$ and $\nabla\cdot\mathbf{H}(\mathbf{x},t)=0$. Here, just as at the end of Chap. 2, we have

$$
\widetilde{\varepsilon}=\varepsilon+i\frac{\sigma}{\omega}:=\widetilde{n}^2
$$

where σ is the conductivity of the medium. The wave vector $\mathbf{k}$ is complex, and we write it as $\mathbf{k}=k_0(N\mathbf{e}+iK\mathbf{f})$ where $k_0=\omega/c$ with c as usual denoting the speed of light in vacuum (we set this equal to one for simplicity in what follows). In a non-absorbing medium, $\mathbf{e}\parallel\mathbf{f}$, while in general they are unit normals to the planes of constant amplitude and phase, respectively. With $\mathbf{k}$ as above, we compute

$$
\mathbf{k}\cdot\mathbf{k}=\omega^2\left(N^2+i2NK\mathbf{e}\cdot\mathbf{f}-K^2\right)
\tag{4.47}
$$

At the same time, from (4.46), we have

$$
\begin{aligned}
\mathbf{k}\times\mathbf{H} &= \mathbf{k}\times\left(\frac{1}{\omega}\mathbf{k}\times\mathbf{E}\right)\\
&= \frac{1}{\omega}\left((\mathbf{k}\cdot\mathbf{E})\mathbf{k}-(\mathbf{k}\cdot\mathbf{k})\mathbf{E}\right)\\
&= -\widetilde{\varepsilon}\omega\mathbf{E}
\end{aligned}
\tag{4.48}
$$

Here we have used the vector identity

$$
\mathbf{a}\times(\mathbf{b}\times\mathbf{c})=(\mathbf{a}\cdot\mathbf{c})\mathbf{b}-(\mathbf{a}\cdot\mathbf{b})\mathbf{c}
$$

From (4.48) we see that

$$
\frac{1}{\omega}(\mathbf{k}\cdot\mathbf{k})\mathbf{E}=\widetilde{\varepsilon}\omega\mathbf{E}
$$

so that $\mathbf{k}\cdot\mathbf{k}=\omega^2\widetilde{\varepsilon}$. From (4.47) then we find

$$
\omega^2\left(N^2+i2NK\mathbf{e}\cdot\mathbf{f}-K^2\right)=\omega^2\widetilde{n}^2
$$

Equating real and imaginary parts of the last equality yields

$$N^2 - K^2 = n^2 - \kappa^2 \quad \text{and} \quad NK\mathbf{e}\cdot\mathbf{f} = n\kappa \tag{4.49}$$

The factors N and K can be understood as the real and imaginary parts of an effective refractive index. Letting α denote the angle between $\mathbf{e}$ and $\mathbf{f}$, from the first equation in (4.49), we see that

$$N^2 - \frac{n^2\kappa^2}{N^2\cos^2(\alpha)} = n^2 - \kappa^2$$

That is, N satisfies the fourth order equation

$$N^4 - (n^2 - \kappa^2)N^2 - \frac{n^2\kappa^2}{\cos^2(\alpha)} = 0 \tag{4.50}$$

Solving this for N, the imaginary part K can be found from (4.49). Note that in a non-absorbing medium, $\kappa = 0$, this yields two possibilities: $\alpha = \pi/2$ so that $\mathbf{e} \perp \mathbf{f}$, or $N = n$.

Now we move on to Snell's Law. So assume there is a planar interface between two media, the first of which (without loss of generality) will be non-absorbing, so can be characterized by a real refractive index n_1 with (again without loss of generality) we set equal to one. The second medium will be lossy with refractive index $\tilde{n}_2 = n_2 + i\kappa_2$. Consider an incident ray with (unit) direction $\mathbf{e}_i$ (following the notation of [450]) and corresponding wave vector $\mathbf{k}_i = k_0\mathbf{e}_i$. The refracted ray[4] will have component directions $\mathbf{e}_t$ and $\mathbf{f}_t$, respectively, so that the refracted wave vector takes the form $\mathbf{k}_t = k_0(N_t\mathbf{e}_t + iK_t\mathbf{f}_t)$. The corresponding electric field vectors take the form

$$\mathbf{E}_i(\mathbf{x}, t) = \mathbf{E}_i e^{ik_0\mathbf{x}\cdot\mathbf{e}_i - \omega t}, \qquad \mathbf{E}_t(\mathbf{x}, t) = \mathbf{E}_t e^{i(N_t\mathbf{e}_t + iK_t\mathbf{f}_t)\cdot\mathbf{x} - \omega t}$$

Now, at the interface of the two media, the phases of the incident and refracted wave must agree. Letting $\mathbf{x}_s$ denote the position of the surface interface, we must have then that

$$\mathbf{e}_i \cdot \mathbf{x}_s = N_t(\mathbf{e}_t \cdot \mathbf{x}_s) + iK_t(\mathbf{f}_t \cdot \mathbf{x}_s)$$

Since the wave vector for the incident wave is real, we see that

$$\mathbf{e}_i \cdot \mathbf{x}_s = N_t(\mathbf{e}_t \cdot \mathbf{x}_s), \qquad \mathbf{f}_t \cdot \mathbf{x}_s = 0 \tag{4.51}$$

Note that the second equation above implies that $\mathbf{f}_t$ points in the direction normal to the interface. If θ_i denotes the incident angle, i.e., the angle between $\mathbf{e}_i$ and the

[4] The subscripts t which follow stand for "transmitted."

normal $\boldsymbol{\nu}$ to the interface, and θ_t denotes the angle between $\mathbf{e}_t$ and $\boldsymbol{\nu}$, then the first equation in (4.51) actually yields Snell's Law:

$$N_t \sin(\theta_t) = \sin(\theta_i) \tag{4.52}$$

It remains to determine N_t, the real part of the effective refractive index. From the second equation in (4.49), we see that

$$N_t^2 K_t^2 \cos^2(\theta_t) = n_2^2 \kappa_2^2$$

or equivalently, using (4.52),

$$N_t^2 K_t^2 \left(1 - \frac{\sin^2(\theta_i)}{N_t^2}\right) = n_2^2 \kappa_2^2$$

After simplification this yields the following fourth order equation for N_t:

$$N_t^4 - N_t^2 \left(\sin^2(\theta_i) + (n_2^2 - \kappa_2^2)\right) + (n_2^2 - \kappa_2^2)\sin^2(\theta_i) - n_2^2\kappa_2^2 = 0 \tag{4.53}$$

This is quadratic in N_t^2 so (4.53) reduces to

$$N_t^2 = \frac{\sin^2(\theta_i) + (n_2^2 - \kappa_2^2) + \sqrt{(\sin^2(\theta_i) + (n_2^2 - \kappa_2^2))^2 - 4n_2^2\kappa_2^2}}{2} \tag{4.54}$$

Equations (4.53) and (4.54) agree with the literature, see for instance equation (50a) in [450] and equation (10) in [78]. Equation (4.54) yields the real part of the effective refractive index for the Snell's Law (4.52):

$$N_t = \sqrt{\frac{\sin^2(\theta_i) + (n_2^2 - \kappa_2^2) + \sqrt{(\sin^2(\theta_i) + (n_2^2 - \kappa_2^2))^2 - 4n_2^2\kappa_2^2}}{2}} \tag{4.55}$$

Notice that the refractive index N_t not only depends on the material parameters n_2, κ_2, but also the incident angle.

Next we consider an uniaxial anisotropic absorbing medium with (diagonalized) permittivity $\widetilde{\varepsilon}$ given by

$$\widetilde{\varepsilon} = \begin{pmatrix} \widetilde{n}_o^2 & 0 & 0 \\ 0 & \widetilde{n}_o^2 & 0 \\ 0 & 0 & \widetilde{n}_e^2 \end{pmatrix} \tag{4.56}$$

where $\widetilde{n}_o$ and $\widetilde{n}_e$ denote the ordinary and extraordinary refractive indices, respectively. Now, we will use complex vector notation, so in lieu of the wave vector $\mathbf{k}$, to be consistent with the literature we consider a complex unit vector $\widetilde{\mathbf{s}}$. Now, a complex unit vector here means a complex vector whose real and imaginary parts are real vectors and satisfies the unitary condition

$$\widetilde{\mathbf{s}} \cdot \widetilde{\mathbf{s}} = 1$$

Note we are not using the complex conjugate in the dot product above. For an inhomogeneous plane wave, the vector $\widetilde{\mathbf{s}}$ represents the direction of propagation. We represent complex fields via the following ansatz:

$$\widetilde{\mathbf{F}} = \mathbf{F}_0 e^{i\omega(t - \widetilde{n}\mathbf{x}\cdot\widetilde{\mathbf{s}})}$$

Physically, the complex vector $\mathbf{F}_0$ contains information about the polarization of the plane wave [4]. In the above representation, $\widetilde{n}$ denotes the complex refractive index as usual. Taking the fields $\mathbf{E}, \mathbf{H}, \mathbf{D}, \mathbf{B}$ having the above complex form, Maxwell's equations reduce to [108, 118]

$$\begin{aligned} \widetilde{\mathbf{H}} \times \widetilde{\mathbf{s}} &= \frac{1}{\widetilde{n}} \widetilde{\mathbf{D}}, \\ \widetilde{\mathbf{s}} \times \widetilde{\mathbf{E}} &= \frac{1}{\widetilde{n}} \widetilde{\mathbf{H}}, \\ \widetilde{\mathbf{D}} \cdot \widetilde{\mathbf{s}} &= 0, \\ \widetilde{\mathbf{H}} \cdot \widetilde{\mathbf{s}} &= 0 \end{aligned} \tag{4.57}$$

Again for simplicity we have set the speed of light $c = 1$. We are heading toward Snell's Law at the interface of a uniaxial absorbing medium, which as we will see again entails some effective refractive index. To this end, from the first two equations in (4.57), we find that

$$\widetilde{\mathbf{D}} = \widetilde{n}\widetilde{\mathbf{H}} \times \widetilde{\mathbf{s}} = \widetilde{n}\left(\widetilde{n}\widetilde{\mathbf{s}} \times \widetilde{\mathbf{E}}\right) = \widetilde{n}^2\left(\widetilde{\mathbf{E}} - (\widetilde{\mathbf{E}} \cdot \widetilde{\mathbf{s}})\widetilde{\mathbf{s}}\right)$$

Now, from the constitutive relation $\widetilde{\mathbf{D}} = \widetilde{\varepsilon}\widetilde{\mathbf{E}}$, we find that $\widetilde{\mathbf{E}} = \widetilde{\varepsilon}^{-1}\widetilde{\mathbf{D}}$, and so from the above expression we find that, for the jth component of $\widetilde{\mathbf{D}}$, where $j = x, y, z$,[5]

$$\widetilde{D}_j = \widetilde{n}^2\left(\widetilde{n}_j^{-2}\widetilde{D}_j - \widetilde{s}_j(\widetilde{\mathbf{E}} \cdot \widetilde{\mathbf{s}})\right)$$

which implies that $\widetilde{D}_j\left(\widetilde{n}^2/\widetilde{n}_j^2 - 1\right) = \widetilde{n}^2\widetilde{s}_j(\widetilde{\mathbf{E}} \cdot \widetilde{\mathbf{s}})$, or equivalently,

[5] We are assuming the x, y, z, coordinate system here is the one where $\widetilde{\varepsilon}$ is diagonal.

$$\widetilde{D}_j = \frac{\widetilde{s}_j(\widetilde{\mathbf{E}} \cdot \widetilde{\mathbf{s}})}{\frac{1}{\widetilde{n}_j^2} - \frac{1}{\widetilde{n}^2}}, \quad j = x, y, z \tag{4.58}$$

From (4.58) we get the familiar Fresnel equation [428] by using the third equation in (4.57) and using (4.56):

$$\frac{\widetilde{s}_x^2}{\frac{1}{\widetilde{n}_o^2} - \frac{1}{\widetilde{n}^2}} + \frac{\widetilde{s}_y^2}{\frac{1}{\widetilde{n}_o^2} - \frac{1}{\widetilde{n}^2}} + \frac{\widetilde{s}_z^2}{\frac{1}{\widetilde{n}_e^2} - \frac{1}{\widetilde{n}^2}} = 0 \tag{4.59}$$

This allows us to again find the effective refractive index by solving for $\widetilde{n}$, as we have done previously for non-absorbing media in (3.9). Indeed, after some algebra and simplification, (4.59) reduces to

$$\widetilde{n}^2 = \frac{\widetilde{n}_o^2 \widetilde{n}_e^2}{n_o^2(\widetilde{s}_x^2 + \widetilde{s}_y^2) + n_e^2 \widetilde{s}_z^2} \tag{4.60}$$

Indeed, if $\widetilde{s}_z = \mathbf{v} \cdot \mathbf{d}$, where **d** is the uniaxial director and **v** denotes the direction of the extraordinary ray, then (4.60) reduces to (since recall $\widetilde{\mathbf{s}}$ is a unit vector)

$$\widetilde{n}^2 = \frac{\widetilde{n}_o^2 \widetilde{n}_e^2}{n_o^2(\mathbf{v} \cdot \mathbf{d})^2 + n_e^2(\mathbf{v} \cdot \mathbf{d})^2} \tag{4.61}$$

which is the same form as (3.9). Equation (4.61) can be compared for instance with equation (17) in [428], which is the same but written in terms of the angle between **v** and **d**, which is assumed to be constant. The previous Snell's law (4.30) holds in this case, with the effective refractive index here given by (4.61). Thus we have a vector formulation of Snell's Law for uniaxial absorbing media (at least assuming the anchoring condition (4.31) holds for **d**), the scalar form of which can be found for instance in equation (15) in [428]; compare with (4.33).

A similar analysis can be done for biaxial anisotropic media [429]. As we have seen before, the effective refractive index will satisfy a fourth order equation, and there are two waves to consider: the so-called a- and b-waves. Proceeding just as we have done above, now setting

$$\widetilde{\varepsilon} = \begin{pmatrix} \widetilde{n}_x^2 & 0 & 0 \\ 0 & \widetilde{n}_y^2 & 0 \\ 0 & 0 & \widetilde{n}_z^2 \end{pmatrix} \tag{4.62}$$

we find that the effective refractive indices for the a and b-waves in a biaxial absorbing medium are given by (compare with equations (3.9) and (3.9b) in [429])

$$\widetilde{n}_{\mathrm{bi}}^2 = \frac{2}{B \pm \sqrt{B^2 - 4C}} \tag{4.63}$$

where

$$B = \left(\frac{1}{\widetilde{n}_y^2} + \frac{1}{\widetilde{n}_z^2}\right)\widetilde{s}_x^2 + \left(\frac{1}{\widetilde{n}_x^2} + \frac{1}{\widetilde{n}_z^2}\right)\widetilde{s}_y^2 + \left(\frac{1}{\widetilde{n}_x^2} + \frac{1}{\widetilde{n}_y^2}\right)\widetilde{s}_z^2,$$

$$C = \frac{\widetilde{s}_x^2}{\widetilde{n}_y^2\widetilde{n}_z^2} + \frac{\widetilde{s}_y^2}{\widetilde{n}_x^2\widetilde{n}_z^2} + \frac{\widetilde{s}_z^2}{\widetilde{n}_x^2\widetilde{n}_y^2}$$

Finally, we write the previous Snell Laws for absorbing media in vector form in a different way. To do this, we impose that the tangential components of the electric fields are continuous across the interface. For the transmitted field, this means that

$$\boldsymbol{\nu} \times \mathbf{k}_i = \boldsymbol{\nu} \times \mathbf{k}_t \tag{4.64}$$

where $\boldsymbol{\nu}$ as usual denotes the unit normal at the interface. In terms of the unit vectors $\mathbf{e}_i, \mathbf{e}_t, \mathbf{f}_t$, equation (4.64) becomes

$$\boldsymbol{\nu} \times (\mathbf{e}_i - (N_t\mathbf{e}_t + K_t\mathbf{f}_t)) = 0 \tag{4.65}$$

That is, the vector $\mathbf{e}_i - (N_t\mathbf{e}_t + K_t\mathbf{f}_t)$ is parallel to the normal $\boldsymbol{\nu}$. So, there exists some constant $\widetilde{\lambda}$ such that

$$\mathbf{e}_i - (N_t\mathbf{e}_t + K_t\mathbf{f}_t) = \widetilde{\lambda}\boldsymbol{\nu} \tag{4.66}$$

which can be viewed as a vector form of Snell's Law with absorption. It is easy to see how this reduces to the scalar form (4.52), for if the incident wave vector is real, then we must have $\boldsymbol{\nu} \times \mathbf{f}_t = 0$, so that as before $\mathbf{f}_t \parallel \boldsymbol{\nu}$. Thus (4.66) reduces to

$$\boldsymbol{\nu} \times \mathbf{e}_i = N_t \boldsymbol{\nu} \times \mathbf{e}_t,$$

and taking norm of both sides yields (4.52). Let us comment now on how the phase matching condition which we did to obtain (4.51) actually leads to the continuity of tangential components of the wave vectors at the interface. We can decompose $\mathbf{k}$ in terms of its components parallel to $\boldsymbol{\nu}$ and perpendicular to $\boldsymbol{\nu}$:

$$\mathbf{k} = (\boldsymbol{\nu} \cdot \mathbf{k})\boldsymbol{\nu} - \boldsymbol{\nu} \times (\boldsymbol{\nu} \times \mathbf{k})$$

Using the vector identity

$$\mathbf{a} \cdot (\mathbf{b} \times \mathbf{c}) = \mathbf{c} \cdot (\mathbf{a} \times \mathbf{b}) = \mathbf{b} \cdot (\mathbf{c} \times \mathbf{a})$$

with the previous decomposition of **k**, the phase matching condition yields

$$(\boldsymbol{\nu} \times \mathbf{k}_i) \cdot (\boldsymbol{\nu} \times \mathbf{x}_s) = (\boldsymbol{\nu} \times \mathbf{k}_t) \cdot (\boldsymbol{\nu} \times \mathbf{x}_s)$$

where $\mathbf{x}_s$ is as before: a position vector of the interface. But if this equality is to be satisfied for any $\mathbf{x}_s$, we must have the complex (in-plane) components of **k** are conserved [117], i.e.,

$$\boldsymbol{\nu} \times \mathbf{k}_i = \boldsymbol{\nu} \times \mathbf{k}_t$$

as desired.

Remark 4.5 We can also determine the direction of the reflected ray; we have only focused on the transmitted ray. For reflection one has the similar tangential boundary condition:

$$\boldsymbol{\nu} \times \mathbf{k}_i = \boldsymbol{\nu} \times \mathbf{k}_r$$

where $\mathbf{k}_r$ denotes the reflected wave vector: $\mathbf{k}_r = k_0 \mathbf{e}_r$ in the previous notation. Then we also have $\mathbf{e}_r \cdot \mathbf{x}_s = \mathbf{e}_i \cdot \mathbf{x}_s$ in equation (4.51), which eventually leads to the usual scalar Snell Law for the reflected ray:

$$\sin(\theta_i) = \sin(\theta_r)$$

Chapter 5
The Ray Equations

It is well known that light rays in homogeneous, isotropic materials satisfy the classical geodesic equations (see, e.g., [137, Chapter 2])

$$\frac{d^2x_i}{dt^2} + \Gamma^i_{j,k}\frac{dx_j}{dt}\frac{dx_k}{dt} = 0$$

where the $\Gamma^i_{j,k}$ denote the Christoffel symbols. In Euclidean space, these equations can be written in terms of the coefficients of the first fundamental form E, F, G for a given surface $\mathbf{r} = \mathbf{r}(u, v)$:

$$\frac{E_u(u')^2 + 2F_u u'v' + G_u(v')^2}{\sqrt{E(u')^2 + 2Fu'v' + G(v')^2}} - \frac{d}{dt}\frac{2(Eu' + Fv')}{\sqrt{E(u')^2 + 2Fu'v' + G(v')^2}} = 0,$$

$$\frac{E_v(u')^2 + 2F_v u'v' + G_v(v')^2}{\sqrt{E(u')^2 + 2Fu'v' + G(v')^2}} - \frac{d}{dt}\frac{2(Fu' + Gv')}{\sqrt{E(u')^2 + 2Fu'v' + G(v')^2}} = 0$$

where

$$E = \frac{\partial \mathbf{r}}{\partial u}\cdot\frac{\partial \mathbf{r}}{\partial u}, \quad F = \frac{\partial \mathbf{r}}{\partial u}\cdot\frac{\partial \mathbf{r}}{\partial v}, \quad G = \frac{\partial \mathbf{r}}{\partial v}\cdot\frac{\partial \mathbf{r}}{\partial v}$$

But what about in a liquid crystal with director $\mathbf{d}$? In order to determine the trajectory of a light ray in a liquid crystal, we need to understand the governing equations. As usual, in this chapter $\mathbf{x}(t)$ will denote the trajectory of a ray. We will begin by distinguishing between different types of extremals (i.e., weak vs. classical) which depend on the smoothness of $\mathbf{x}(t)$. In particular, given a director $\mathbf{d}$ that is not necessarily differentiable in the classical sense, it is not expected that one can obtain classical (differentiable) extremals of the appropriate functional.

E. Stachura, *Mathematical Methods in Liquid Crystal Optics and Lens Design*,
Springer Tracts in Modern Physics 294,
https://doi.org/10.1007/978-3-031-46614-4_5

Next we discuss the Hamiltonian formulation of the ray equations by use of the Legendre transform. In particular, we will see that the Hamilton–Jacobi equation $H(\mathbf{x}, \mathbf{p}) = 1$ for a certain Hamiltonian recovers the Fresnel surface for the extraordinary ray which was seen in Eq. (2.21) in Chap. 2.

We will further discuss the relationship between the variational approach and an equivalent geometric approach in a so-called Finsler space. In such spaces, the main object is a Finsler function rather than a metric (as in Riemannian geometry). This approach was used in [199] (see also the lensing results in [359] and [360] as well as the discussion in [443] and [335]), but we will see that for the liquid crystal applications we have in mind, this Finsler framework is not needed. Indeed, we will see that the appropriate metric reduces to an essentially Riemannian metric. We will provide a number of examples, including the 2D hedeghog director and more general disclinations.

Finally, we will consider the lensing effects due to point defects as seen in [359] and [360]. By implementing simple deformations of a disclination director, we will be able to modify the amount of bending the light rays satisfy.

5.1 Variational Approach: Review

Throughout this chapter, we consider the usual effective refractive index $n_{\mathrm{eff},r}$ given by (3.11) and drop the index r to simplify the notation. Recall that sufficiently smooth extremals of (4.3) in $\mathbb{R}^3$ satisfy the Euler–Lagrange equations

$$\nabla_{\mathbf{x}} F(\mathbf{x}, \dot{\mathbf{x}}) - \frac{d}{dt} \nabla_{\mathbf{v}} F(\mathbf{x}, \dot{\mathbf{x}}) = 0 \tag{5.1}$$

i.e.,

$$\frac{\partial F}{\partial x_i} - \frac{d}{dt} \frac{\partial F}{\partial v_i} = 0, \qquad i = 1, 2, 3$$

Write $\mathbf{x} = (x_1, x_2, x_3)$ and $\mathbf{v} = (v_1, v_2, v_3)$. Instead of working with F as usual given by the effective index (4.5), we will instead consider

$$Q(\mathbf{x}, \mathbf{v}) = \frac{1}{2} n_{\mathrm{eff}}(\mathbf{x}, \mathbf{v})^2 = \frac{1}{2} \left[n_e^2 |\mathbf{v}|^2 + (n_o^2 - n_e^2)(\mathbf{v} \cdot \mathbf{d}(\mathbf{x}))^2 \right] \tag{5.2}$$

This is justified by the following, which can be found in [139, Proposition 2, page 185]:

Proposition 5.1 *Suppose that* $F(\mathbf{x}, \mathbf{v}) > 0$, *and set* $Q = (1/2) F(\mathbf{x}, \mathbf{v})^2$. *Define*

$$\mathscr{F} = \int_{t_1}^{t_2} F(\mathbf{x}, \dot{\mathbf{x}}) dt, \qquad \mathscr{Q} = \int_{t_1}^{t_2} Q(\mathbf{x}, \dot{\mathbf{x}}) dt$$

Then every $\mathcal{Q}$ extremal $\mathbf{x}(t)$ with $\dot{\mathbf{x}}(t_0) \neq 0$ for some $t_0 \in [t_1, t_2]$ satisfies

$$Q(\mathbf{x}, \dot{\mathbf{x}}) \equiv \frac{1}{2}h^2 \tag{5.3}$$

for some constant $h > 0$ and is an extremal of the parametric integral $\mathcal{F}$. Conversely, if $\mathbf{x}(t)$ is an extremal for $\mathcal{F}$ parameterized in such a way so that (5.3) holds for some $h > 0$, then it is also an extremal of $\mathcal{Q}$.

Let us briefly discuss the different types of extremals that can be obtained; we will take $[t_1, t_2] = [0, 1]$ in what follows below. If $\mathbf{x}(t) \in C^1([0, 1]; \mathbb{R}^3) \cap C^2((0, 1); \mathbb{R}^3)$ and satisfies $\delta\mathcal{Q}(\mathbf{x}, \phi) = 0$ for each $\phi \in C_0^\infty((0, 1); \mathbb{R}^3)$, then $\mathbf{x}$ is called an extremal of $\mathcal{Q}$ and satisfies the Euler–Lagrange equations

$$\frac{d}{dt}\nabla_{\mathbf{v}} n_{\text{eff}}(\mathbf{x}, \mathbf{v})^2 - \nabla_{\mathbf{x}} n_{\text{eff}}(\mathbf{x}, \mathbf{v})^2 = 0 \tag{5.4}$$

Recall that the first variation $\delta\mathcal{Q}(\mathbf{x}, \phi)$ in the direction ϕ means

$$\delta\mathcal{Q}(\mathbf{x}, \phi) = \int_0^1 \left(\nabla_{\mathbf{x}} n_{\text{eff}}(\mathbf{x}, \mathbf{v})^2 \cdot \phi + \nabla_{\mathbf{v}} n_{\text{eff}}(\mathbf{x}, \mathbf{v})^2 \cdot \dot{\phi}\right) dt$$

for each $\phi \in C_0^1\left([0, 1]; \mathbb{R}^3\right)$.

So, we find that for smooth $\mathbf{d}$, there holds

$$\nabla_{\mathbf{v}} n_{\text{eff}}^2 = 2\left(n_e^2\mathbf{v} + (n_o^2 - n_e^2)(\mathbf{v}\cdot\mathbf{d})\mathbf{d}\right)$$

and

$$\nabla_{\mathbf{x}} n_{\text{eff}}^2 = 2(n_o^2 - n_e^2)(\mathbf{v}\cdot\mathbf{d})(J_d^T\mathbf{v})$$

Above, J_d^T denotes the transpose of the Jacobian matrix of $\mathbf{d}$, i.e., in $\mathbb{R}^3$:

$$J_d = \begin{pmatrix} \frac{\partial d_1}{\partial x_1} & \frac{\partial d_1}{\partial x_2} & \frac{\partial d_1}{\partial x_3} \\ \frac{\partial d_2}{\partial x_1} & \frac{\partial d_2}{\partial x_2} & \frac{\partial d_2}{\partial x_3} \\ \frac{\partial d_3}{\partial x_1} & \frac{\partial d_3}{\partial x_2} & \frac{\partial d_3}{\partial x_3} \end{pmatrix}$$

We further see that

$$\frac{d}{dt}\nabla_{\mathbf{v}} n_{\text{eff}}^2 = 2\left[n_e^2\ddot{\mathbf{x}} + (n_o^2 - n_e^2)\left[(\ddot{\mathbf{x}}\cdot\mathbf{d} + \dot{\mathbf{x}}\cdot(J_d\dot{\mathbf{x}}))\,\mathbf{d} + (\dot{\mathbf{x}}\cdot\mathbf{d})(J_d\dot{\mathbf{x}})\right]\right]$$

which yields the following Euler–Lagrange equation:

$$(n_o^2 - n_e^2)(\dot{\mathbf{x}} \cdot \mathbf{d})(J_d^T \dot{\mathbf{x}}) = n_e^2 \ddot{\mathbf{x}} + (n_o^2 - n_e^2)\left[((\ddot{\mathbf{x}} \cdot \mathbf{d}) + \dot{\mathbf{x}} \cdot (J_d \dot{\mathbf{x}}))\,\mathbf{d} + (\dot{\mathbf{x}} \cdot \mathbf{d})(J_d \dot{\mathbf{x}})\right] \tag{5.5}$$

Remark 5.1 If we are in a homogeneous, isotropic medium with $n_e = n_o = n$, then (5.5) reduces to

$$n^2 \ddot{\mathbf{x}} = 0$$

which recovers the well-known result [48] that in such a setting light rays are straight lines.

We also have the notion of weak Lipschitz extremal. We say that $\mathbf{x}(t)$ is a weak Lipschitz extremal of $\mathcal{Q}$ if $\mathbf{x}(t)$ is Lipschitz, $|\dot{\mathbf{x}}| \neq 0$ a.e. on $[0, 1]$, and

$$\delta \mathcal{Q}(\mathbf{x}, \phi) = 0 \qquad \forall\, \phi \in C_0^\infty\left((0, 1); \mathbb{R}^3\right)$$

Remark 5.2 The condition $\dot{\mathbf{x}} \neq 0$ for all $t \in [0, 1]$ is key. This means the curve $\mathbf{x}(t)$ is immersed, so has well-defined inverse. This in particular implies that we may transform $\mathbf{x}(t)$ in terms of arclength s; recall Remark 4.1. From now on we only consider such curves.

A natural question is under what conditions will there exist weak Lipschitz extremals or classical extremals. So, suppose that $n_o > n_e$ and let $\Omega \subset \mathbb{R}^3$ be a container filled with the liquid crystal. For reasons to be seen below, we will assume that Ω is closed and connected. We see that there exist constants $C_1, C_2 > 0$ such that

$$C_1|\mathbf{v}| \leq n_{\text{eff}}(\mathbf{x}, \mathbf{v}) \leq C_2|\mathbf{v}| \quad \forall\, (\mathbf{x}, \mathbf{v}) \in \Omega \times \mathbb{R}^3 \tag{5.6}$$

We will appeal to a method using lower semicontinuity to address existence of minimizers of $\mathcal{F}$. To this end, fix two points $P_1, P_2 \in \Omega$ which can be connected by some Lipschitz curve. Then we define the set

$$C = \{\text{Lipschitz curves } \mathbf{x} \text{ such that } \mathbf{x}([0, 1]) \subset \Omega, \text{ and } \mathbf{x}(0) = P_1,\ \mathbf{x}(1) = P_2\} \tag{5.7}$$

Then, Theorem 1, page 261 in [139] gives the existence of a quasi-normal curve $\mathbf{x} \in C$ which minimizes $\mathcal{F}$ in C, provided n_{eff} is convex with respect to $\mathbf{v}$. This method uses lower semicontinuity of $\mathcal{F}$, which follows easily in this smooth setting by convexity (see below and, e.g., Lemma 5, Section 4.2 in [139]). Note that a Lipschitz curve $\mathbf{x}$ in this context is called quasi-normal if

$$n_{\text{eff}}(\mathbf{x}(t), \dot{\mathbf{x}}(t)) \equiv \text{constant a.e. on } [0, 1] \tag{5.8}$$

for some positive constant.

Remark 5.3 The quasi-normal condition arises in the following way. Defining $Q(\mathbf{x}, \mathbf{v}) = \frac{1}{2} n_{\text{eff}}(\mathbf{x}, \mathbf{v})^2$, we see this is homogeneous of second order in $\mathbf{v}$. Define $\Phi(\mathbf{x}, \mathbf{v}) = \mathbf{v} \cdot \nabla_{\mathbf{v}} Q(\mathbf{x}, \mathbf{v}) - Q(\mathbf{x}, \mathbf{v})$. It follows by homogeneity that $Q(\mathbf{x}, \mathbf{v}) = \mathbf{v} \cdot \nabla_{\mathbf{v}} Q(\mathbf{x}, \mathbf{v}) - Q(\mathbf{x}, \mathbf{v}) = \Phi(\mathbf{x}, \mathbf{v})$, and so if $\Phi(\mathbf{x}(t), \dot{\mathbf{x}}(t)) =$ constant a.e. on $[0, 1]$, then $Q(\mathbf{x}(t), \dot{\mathbf{x}}(t)) =$ constant a.e. on $[0, 1]$, and the condition that $\Phi =$ constant is a "conservation law" which is satisfied if

$$\frac{\partial}{\partial \epsilon} \mathcal{Q}(\mathbf{x}(\xi(t, \epsilon)), \dot{\mathbf{x}}(\xi(t, \epsilon)))|_{\epsilon=0} = 0$$

for any admissible variation $\mathbf{x}(\xi(t, \epsilon))$ of $\mathbf{x}$, see Proposition 1.15 in [60]. Finally, note that $Q(\mathbf{x}(t), \dot{\mathbf{x}}(t)) =$ constant a.e. along a minimizer $\mathbf{x}$ of $\mathcal{Q}$ is equivalent to $n_{\text{eff}}(\mathbf{x}(t), \dot{\mathbf{x}}(t)) =$ constant a.e. for some other constant.

Since $n_{\text{eff}}(\mathbf{x}, \cdot)$ is C^1 (away from $\mathbf{v} = 0$), then convexity with respect to $\mathbf{v}$ (at least away from 0) is equivalent to the fact that the excess function

$$\mathcal{E}(\mathbf{x}, \mathbf{v}, \mathbf{w}) := n_{\text{eff}}(\mathbf{x}, \mathbf{w}) - n_{\text{eff}}(\mathbf{x}, \mathbf{v}) - (\mathbf{w} - \mathbf{v}) \cdot \nabla_{\mathbf{v}} n_{\text{eff}}(\mathbf{x}, \mathbf{v})$$

satisfies

$$\mathcal{E}(\mathbf{x}, \mathbf{v}, \mathbf{w}) \geq 0, \qquad \forall\, \mathbf{v}, \mathbf{w} \in \mathbb{R}^3 \setminus \{0\} \tag{5.9}$$

We will discuss the excess function further in Sect. 5.6.1. We only mention the following. We say an element $(\mathbf{x}, \mathbf{v})$ is elliptic if $Q = \frac{1}{2} F^2$ satisfies

$$\frac{\partial^2 Q}{\partial v_i \partial v_j}(\mathbf{x}, \mathbf{v}) \xi_i \xi_j > 0, \qquad \forall\, \xi \in \mathbb{R}^3 \setminus \{0\}$$

Then, (5.9) follows from ellipticity for all directions $\mathbf{v}$ with fixed supporting point $\mathbf{x}$ (see, e.g., Remark 1, page 259 in [139]). Thus we only need to show ellipticity for all directions $\mathbf{v}$. We find that

$$\frac{\partial^2 Q}{\partial v_i \partial v_j} = (n_o^2 - n_e^2) d_i d_j + n_e^2 \delta_{ij}$$

Since this matrix is symmetric, it will be positive definite if and only if all its eigenvalues are positive. We find the eigenvalues in 2D are given by $\lambda_1 = n_e^2$, $\lambda_2 = n_o^2$ since $\mathbf{d}$ is a unit vector, so indeed we have ellipticity. We can see this similarly in 3D as well (the eigenvalues of the resulting symmetric matrix are n_e^2, n_e^2, n_o^2).

Thus we obtain the basic existence theorem:

Theorem 5.1 *Consider n_{eff} given by (3.11). Suppose that $n_o > n_e$. Let P_1, P_2 be two points in Ω such that the set C defined in (5.7) is not empty. Then there exists a quasi-normal curve $\mathbf{x}(t)$ which minimizes both $\mathcal{F}$ and $\mathcal{Q}$ in the class C.*

Remark 5.4 The assumption that $n_o > n_e$ in Theorem 5.1 above is not needed. The result holds even if $n_e > n_o$. Indeed, if $n_e > n_o$, as is frequently the case in various applications [220], then we still actually have (5.6) as well as convexity as long as C_2 is chosen large enough (e.g., one can take $C_1 = n_o^2$ and $C_2 > n_e^2 > n_o^2$). We have seen above that (5.9) also holds, so we have the basic existence theorem in this case as well (in 2D and 3D). Moreover, it is straightforward to see that, regardless of the sign of $n_o - n_e$, there holds

$$n_{\text{eff}}^2 \geq \min(n_e^2, n_o^2)|\dot{\mathbf{x}}|^2$$

Since no smoothness on Ω has been imposed, in general we cannot expect that a minimizer of $\mathcal{F}$ in C is a classical extremal . But, if $\mathbf{x}(t)$ stays away from the boundary of Ω, we can say the following (see, e.g., Proposition 1, page 262, [139]).

Proposition 5.2 *If $\mathbf{x}(t)$ is a quasi-normal minimizer of $\mathcal{F}$ in C such that $\mathbf{x}([0, 1]) \in$ int(Ω), then $\mathbf{x}(t)$ is also a weak Lipschitz extremal of $\mathcal{F}$.*

Let us briefly comment on the requirement that $\mathbf{x}(t)$ stay away from the boundary of Ω. Clearly, this is trivially satisfied if $\Omega = \mathbb{R}^3$ but can also be satisfied if $\Omega \subset \mathbb{R}^3$ is a compact, connected submanifold without boundary (such as the sphere S^2).

5.2 Variational Approach: Minimizers in the Class of Rectifiable Curves

In this section we will work in $\mathbb{R}^2$ and assume that $\mathbf{d}$ is smooth (at least C^1). We will look at rectifiable curves $\mathbf{x}$. Recall that a curve $\mathbf{x} : [0, T] \to \mathbb{R}^n$ is rectifiable if and only if the coordinate functions of $\mathbf{x}$ are continuous and of bounded variation. If $\mathbf{x}$ is not continuous, then one can define the length (total variation) of $\mathbf{x}$ in a similar way, i.e.,

$$\ell(\mathbf{x}) = \sup \sum_{i=1}^{N} |\mathbf{x}(t_i) - \mathbf{x}(t_{i-1})| \tag{5.10}$$

where the sup is taken over all subdivisions $0 = t_0 < t_1 < \ldots < t_N = T$ of $[0, T]$. For this we require at most countably many points of discontinuity.

Thus we will consider curves $\mathbf{x}$ which are rectifiable. More precisely, fix a compact set $A_1 \subset \mathbb{R}^2$, and let

$$[\mathbf{x}] = \left\{ \mathbf{x} \in \mathbb{R}^2 : \mathbf{x} = \mathbf{x}(\tau),\ 0 \leq \tau \leq T \right\}$$

denote the set covered by $\mathbf{x}$ (which is a set of measure zero in $\mathbb{R}^2$). We will consider the minimization problem in the set

$$C_{\text{rect}} = \{\mathbf{x} = \mathbf{x}(t) \text{ rectifiable } : [\mathbf{x}] \subset A_1,\ \mathbf{x}(0) = P_1,\ \mathbf{x}(T) = P_2;\ P_1, P_2 \in A_1\} \tag{5.11}$$

We list the following useful properties of rectifiable curves:

Theorem 5.2 ([72], Theorem 14.1.i)

1. *$\mathbf{x} = \mathbf{x}(t)$ is rectifiable if and only if the component functions $x_i(t)$ are of bounded variation, i.e., $x_i(t) \in BV(0, T)$.*
2. *If $\mathbf{x} = \mathbf{x}(t)$ is rectifiable, then there is always an absolutely continuous representative. In particular, the arclength parameter s for $0 \le s \le \ell(\mathbf{x})$ yields a unique absolutely continuous representative $\mathbf{x} = X(s)$ for $0 \le s \le \ell(\mathbf{x})$ with $|X'(s)| = 1$ a.e. in $[0, \ell(\mathbf{x})]$.*

In particular, the arclength parameterization X of a rectifiable curve $\mathbf{x}$ is actually Lipschitz continuous. The problem we consider is the following:

$$\min\left\{\int_0^T \sqrt{n_e^2|\dot{\mathbf{x}}|^2 + (n_o^2 - n_e^2)(\dot{\mathbf{x}} \cdot \mathbf{d}(\mathbf{x}))^2}dt : \mathbf{x} \in C_{\text{rect}},\ \mathbf{d} \in C^1(\Omega; S^1)\right\} \tag{5.12}$$

where the class C_{rect} is given by (5.11).

We easily see that the integrand $f(\mathbf{x}, \mathbf{v}) := n_{\text{eff}}(\mathbf{x}, \mathbf{v})$ satisfies the following, which are conditions (a)–(g) in Chapter 14, Section 1.C of [72]:

1. $f(\mathbf{x}, \mathbf{v}) > 0$ for all $x \in A_1, \mathbf{v} \in S^1$.
2. There exists a vector $\mathbf{b} = (b_1, b_2)$ such that $f(\mathbf{x}, \mathbf{v}) + \mathbf{b} \cdot \mathbf{v} > 0$ for all $(\mathbf{x}, \mathbf{v}) \in A_1 \times S^1$.
3. There exists a vector function $\mathbf{G}(\mathbf{x}) = (G_1, G_2)$, $\mathbf{x} \in A_1$, such that $f(\mathbf{x}, \mathbf{v}) + \mathbf{G} \cdot \mathbf{v} > 0$ for all $(\mathbf{x}, \mathbf{v}) \in A_1 \times S^1$.
4. $f(\mathbf{x}, \mathbf{v}) \ge 0$ in $A_1 \times \mathbb{R}^2$, and there exists[1] a fixed unit vector $\mathbf{w} = (w_1, w_2) \in S^1$ and a number $\delta > 0$ such that $f(\mathbf{x}, \mathbf{v}) > 0$ for $\mathbf{w} \cdot \mathbf{v} \ge \delta$ for $(\mathbf{x}, \mathbf{v}) \in A_1 \times S^1$.
5. f has a finite number of zeros in A_1, say $\{x_1, \ldots, x_N\}$. Moreover, given $x_i \in \{x_1, \ldots, x_N\}$, there is a vector $w_i \in S^1$ and a number $\delta_i > 0$ such that $f(x_i, \mathbf{v}) > 0$ for $\mathbf{w}_i \cdot \mathbf{v} \ge \delta_i$ for $\mathbf{v} \in S^1$.
6. f has a set Z of zeros which are contained in the finite union of simple continuous curves in A_1 but form no closed curve in $\mathbb{R}^2$. Moreover, for each $z \in Z$, there are a vector $\mathbf{w} \in S^1$ and a number $\delta > 0$ such that $f(z, \mathbf{w}) > 0$ for $\mathbf{w} \cdot \mathbf{v} > \delta$ with $\mathbf{v} \in S^1$.

[1] That is, f can vanish at any point say x, but only in those directions $\mathbf{v}$ which form an angle $\theta \in [0, \frac{\pi}{2} - \epsilon]$ with the fixed direction $\mathbf{w}$ and $\delta = \sin(\epsilon)$.

Items (5) and (6) above can be replaced by the following actually:

7. f has a finite number of zeros in A_1, all within the interior of A_1. For each zero x_i, in a neighborhood $N \ni x_i$, there exist constants $0 < c < C$ and $\gamma > 0$ such that $c|\mathbf{x} - x_i|^\gamma \leq f(\mathbf{x}, \mathbf{v}) \leq C|\mathbf{x} - x_i|^\gamma$ for all $\mathbf{x} \in N, \mathbf{v} \in S^1$.

All of these conditions imply the existence of a real-valued, monotone, nondecreasing function $\phi(s)$, $-\infty < s < \infty$, such that

$$\ell(\mathbf{x}) \leq \phi\left(\mathscr{F}[\mathbf{x}, \dot{\mathbf{x}}]\right) \tag{5.13}$$

for all $\mathbf{x} \in A_1$. Above, $\ell(\mathbf{x})$ is the total variation of $\mathbf{x}$, see (5.10). The fact that the previous conditions are equivalent to the existence of such ϕ was shown in a series of papers by McShane [284–287].

So we set $m = \inf_{C_{\text{rect}}} \mathscr{F}[\mathbf{x}]$.

We may assume there exists $M_1 > 0$ such that $\mathscr{F}[\mathbf{x}] \leq M_1$ for all $\mathbf{x} \in C_{\text{rect}}$. Indeed, since we are interested in the minimizer of $\mathscr{F}$, it is enough to look at

$$\widetilde{C}_{\text{rect}} = \{\mathbf{x} \in C_{\text{rect}} : \mathscr{F}[\mathbf{x}] \leq M_1\}$$

Then, (5.13) implies that

$$\ell(\mathbf{x}) \leq \phi(\mathscr{F}[\mathbf{x}]) \leq \phi(M_1) := M_2 \tag{5.14}$$

We next see that on the set $A_1 \times B(0, 1)$, we have

$$\begin{aligned}
|\mathscr{F}[\mathbf{x}]| &\leq \int_0^T \sqrt{n_e^2|\dot{\mathbf{x}}|^2}dt + \sqrt{n_o^2 - n_e^2} \int_0^T \sqrt{|\dot{\mathbf{x}}|^2|\mathbf{d}(\mathbf{x})|^2}dt \\
&\leq C\ell(\mathbf{x}) + C||\mathbf{d}(\mathbf{x})||_{L^1(0,\ell(\mathbf{x});\mathbb{R}^2)} \\
&\leq C\ell(\mathbf{x}) + C||\mathbf{d}(\mathbf{x})||_{L^2(0,\ell(\mathbf{x});\mathbb{R}^2)}\ell(\mathbf{x}) \\
&\leq C\ell(\mathbf{x}) \leq CM_2
\end{aligned}$$

since $\dot{\mathbf{x}} \in B(0, 1)$ and using (5.14) and the fact that $\mathbf{d}$ is bounded. Thus $m < \infty$.

We need a metric to measure the distance between trajectories. The ρ metric describes closeness of curves which may be defined on distinct time intervals:

$$\rho(\mathbf{x}, \mathbf{y}) = |a - c| + |b - d| + \sup_t |\mathbf{x}(t) - \mathbf{y}(t)|$$

where $\mathbf{x}(t)$ is defined on $a \leq t \leq b$, $\mathbf{y}(t)$ is defined on $c \leq t \leq d$, and we think of $\mathbf{x}, \mathbf{y}$ as being extended to the entire real line and being constant outside their original interval of definition. Thus we say a sequence $\mathbf{x}_k(t)$ for $t_{1k} \leq t \leq t_{2k}$ is said to converge to some element $\mathbf{x}(t)$ for $t_1 \leq t \leq t_2$ provided $\rho(\mathbf{x}_k, \mathbf{x}) \to 0$ as $k \to \infty$. Equivalently, $\mathbf{x}_k \to \mathbf{x}$ in the ρ metric if $t_{1k} \to t_1$, $t_{2k} \to t_2$ as $k \to \infty$ and $\mathbf{x}_k \to \mathbf{x}$

uniformly in some interval $[a_0, b_0]$ which contains all the $[t_{1k}, t_{2k}]$. Convergence in the ρ metric is essentially uniform convergence of trajectories.

Remark 5.5 If $\mathbf{x}(t)$ and $\mathbf{y}(t)$ are defined in the same time interval, we can take the metric $d(\mathbf{x}, \mathbf{y}) = \sup |\mathbf{x}(t) - \mathbf{y}(t)|$.

Take a minimizing sequence (whose absolutely continuous representative is parameterized by arclength) $\mathbf{X}_k(s)$ so that $\mathbf{X}_k(s) \to m$. Then $\mathbf{X}_k(s) \in A_1$ which is compact and $|\mathbf{X}_k'(s)| \leq 1$. Here $0 \leq s \leq L_k = \ell(\mathbf{x}_k) \leq M_2$ by (5.14). Since $\{\mathbf{X}_k\}$ are uniformly bounded and equicontinuous, by the Arzela–Ascoli Theorem,[2] we obtain a subsequence, still denoted $\mathbf{X}_k$, such that $\mathbf{X}_k \to \mathbf{x}_0$ in the ρ metric.

The limit $\mathbf{x}_0$ we have found is absolutely continuous,[3] so in particular $\mathbf{x}_0'(s) \in B(0, 1)$ for a.e. $s \in [0, L_0]$ with $L_0 \leq M_2$, where $L_0 = \ell(\mathbf{x}_0)$. Next we need to show lower semicontinuity of $\mathscr{F}$:

$$\mathscr{F}[\mathbf{x}_0, \dot{\mathbf{x}}_0] \leq \liminf_{k\to\infty} \mathscr{F}[\mathbf{x}_k, \dot{\mathbf{x}}_k] \tag{5.15}$$

Since the length ℓ (total variation) from (5.10) is lower semicontinuous with respect to the previous convergence (see, e.g., Theorem 14.1.ii in [72]; this also follows from the Arzela–Ascoli Theorem)

$$\ell(\mathbf{x}_0) \leq \liminf_{k\to\infty} \ell(\mathbf{X}_k)$$

we have $\mathbf{x}_0$ belongs to C. Thus the only thing remaining is to prove (5.15).

First recall that a family $\{x(t) : a \leq t \leq b\}$ of absolutely continuous functions is called equi-absolutely continuous if given $\epsilon > 0$ there exists $\delta > 0$ such that

$$\sum_{j=1}^{N} |x(\beta_j) - x(\alpha_j)| \leq \epsilon$$

for all functions x in the family and for all nonoverlapping intervals $[\alpha_j, \beta_j]$ in $[a, b]$ such that $\sum_{j=1}^{N}(\beta_j - \alpha_j) \leq \delta$. Further, a family $\{y(t) : a \leq t \leq b\}$ of integrable functions is called equi-absolutely integrable if, given $\epsilon > 0$, there exists $\delta > 0$ such that $\int_J |y(t)| dt \leq \epsilon$ for every element $y(t)$ of the family and all measurable subsets $J \subset [a, b]$ with $|J| \leq \delta$.

Now, consider the family $\{\mathbf{X}_k'(s)\}$. We claim that this family is equi-absolutely integrable since $\{\mathbf{X}_k\}$ is equi-absolutely continuous.[4] To see this, let $\epsilon > 0$ and take $\delta = \epsilon/6$. Let $J \subset [0, M_2]$ be any measurable subset with $|J| \leq \delta/2 = \epsilon/12$.

[2] If a sequence $\{\mathbf{X}_k\}$ of Lipschitz continuous functions is uniformly bounded and equicontinuous, or equivalently, the Lipschitz constant for each $\mathbf{X}_k$ is the same for each s, then there exists a subsequence which converges uniformly.

[3] Since the absolutely continuous parameterization $\mathbf{X}_k$ is actually Lipschitz and thus absolutely continuous, we have the uniform limit is also Lipschitz (and thus also absolutely continuous).

[4] Actually these are equivalent for absolutely continuous functions; see, e.g., Theorem 10.2.i. in [72].

Consider the sets

$$J_+ = \{s \in J : \mathbf{X}_k'(s) \geq 0\}, \qquad J_- = \{s \in J : \mathbf{X}_k'(s) \leq 0\}$$

Then certainly we have $|J_+|, |J_-| \leq \delta/2$. Since $\mathbf{X}_k'(s)$ is absolutely integrable, we can find an $\eta > 0$, with $\eta < \delta/2$, such that $\int_K |\mathbf{X}_k'(s)|ds \leq \epsilon/6$ for $|K| \leq \eta \leq \delta/2$. Now, let Ω be an open set which covers J_+ such that $|\Omega| < |J_+|+\eta$, and let (α_j, β_j) be disjoint intervals (the components of Ω). Then we find

$$\sum_{j=1}^{\infty}(\beta_j - \alpha_j) = |\Omega| \leq \delta$$

Let $\Omega_N = \cup_{j=1}^{n}(\alpha_j, \beta_j)$; for N sufficiently large we see that $|\Omega \setminus \Omega_N| \leq \eta$, and so

$$\begin{aligned}\int_{J_+} |\mathbf{X}_k'(s)|ds &= \int_{J_+\cap\Omega} \mathbf{X}_k'(s)ds + \int_{J_+\cap(\Omega\setminus\Omega_N)} \mathbf{X}_k'(s)ds \\ &\leq \int_{\Omega_N} \mathbf{X}_k'(s)ds + \int_{\Omega_N\setminus J_+} |\mathbf{X}_k'(s)|ds + \int_{\Omega\setminus\Omega_N} |\mathbf{X}_k'(s)|ds \\ &\leq \sum_{j=1}^{N}(\mathbf{X}_k(\beta_j) - \mathbf{X}_k(\alpha_j)) + \frac{\epsilon}{6} + \frac{\epsilon}{6} \leq \frac{\epsilon}{2}\end{aligned}$$

since $\{\mathbf{X}_k(s)\}$ is equi-absolutely continuous. Similarly, we find that $\int_{J_-} |\mathbf{X}_k'(s)|ds \leq \epsilon/2$. Since J was arbitrary, we conclude that indeed $\{\mathbf{X}_k'(s)\}$ is equi-absolutely integrable. This (sub) sequence can be chosen such that $\mathbf{X}_k'(s)$ converges weakly in L^1 to some integrable function $\xi(s)$ as equi-absolutely integrable implies sequential weak compactness in L^1 by the Dunford–Pettis Theorem.[5] Then (5.15) follows from Theorem 10.8.ii in [72] since $\mathbf{d}$ is smooth. We recall (one version of) this result here:

Theorem 5.3 (Theorem 10.8.ii in [72]) *Suppose that in the functional*

$$I = \int_0^T F_0(t, \mathbf{x}(t), \mathbf{v}(t))dt, \qquad \mathbf{v}(t) \in S^1$$

the function F_0 is bounded below by an integrable function and is a Carathéodory function. That is, F_0 is a measurable function in t for every $(\mathbf{x}, \mathbf{v})$ and is a continuous function in $(\mathbf{x}, \mathbf{v})$ for almost every t. Suppose further that F_0 is convex in $\mathbf{v}$ for almost all $t \in [0, T]$ and all $\mathbf{x} \in A_1$. Then F_0 is lower semicontinuous with respect to the topology of weak convergence of the derivatives of $\mathbf{x}$. That is, if $\mathbf{x}_k \to \mathbf{x}$ in measure and $\xi_k \rightharpoonup \xi$ weakly in L^1, then

[5] See, e.g., Theorem 10.3.i in [72]; in fact this is an equivalence.

$$\int_0^T F_0(t, \mathbf{x}(t), \xi(t))dt \leq \liminf_{k\to\infty} \int_0^T F_0(t, \mathbf{x}_k, \xi_k)dt$$

Note that convexity of n_{eff} holds provided that $\mathbf{v} \neq \mathbf{0}$, continuity holds since $\mathbf{d}$ is smooth, and as mentioned before we have $n_{\text{eff}}(\mathbf{x}, \dot{\mathbf{x}}) \geq C(n_o, n_e)|\dot{\mathbf{x}}|$. Moreover, since the $\mathbf{X}_k$ are absolutely continuous, we also have that $\xi(s) = \mathbf{x}_0'(s)$ for a.e. $s \in [0, L_0]$.

Thus we have:

Theorem 5.4 *Let A_1 be a fixed compact set such that $P_1, P_2 \in A_1$ and let $\mathbf{d}$ be at least C^1 smooth. Then, there exists an absolute minimum of $\mathcal{F}[\mathbf{x}, \dot{\mathbf{x}}]$ in the class C_{rect} of all rectifiable curves in A_1.*

Before discussing the light ray approach in a Finsler space, we discuss how the previous Lagrangian formulation is equivalent to an appropriate Hamiltonian formulation.

5.3 Hamiltonian Formulation

In this section we will again assume the director $\mathbf{d}$ is sufficiently smooth, i.e., at least C^1. Now, the Hamiltonian formulation is obtained through the Legendre transform, which we recall below:

Definition 5.1 Given a Banach space X with dual[6] Y, and a continuous bilinear form $\langle\cdot, \cdot\rangle$ on $X \times Y$, consider a convex function $f : X \to (-\infty, \infty]$. The Legendre transform of f, denoted f^*, is the function defined on Y given by

$$f^*(y) = \sup_{x\in X} (\langle x, y\rangle - f(x))$$

In our setting, we take $f(x) = Q(\mathbf{x}, \mathbf{v})$ as defined in (5.2). Playing the role of the dual variable y in the previous definition, we introduce the so-called conjugate momentum $\mathbf{p}$ by

$$p_i = \frac{\partial Q}{\partial v_i}, \qquad i = 1, 2, 3 \tag{5.16}$$

Since we see that

$$\det \frac{\partial^2 Q}{\partial v_i \partial v_j} \neq 0$$

[6] The dual of a Banach space X is the set of continuous linear functionals on X.

by the Implicit Function Theorem, we may solve for the v_i in terms of the p_i. This allows us to form the Legendre transform

$$\Phi(\mathbf{x}, \mathbf{p}) = \sum_{i=1}^{3} p_i v_i - Q(\mathbf{x}, \mathbf{v}) \tag{5.17}$$

which is consistent with Definition 5.1. Indeed, for sufficiently regular f, e.g., if f satisfies

$$\lim_{x \to \infty} \frac{f(x)}{|x|} = +\infty$$

then the sup in Definition 5.1 is really a maximum, so there exists some $x^* \in X$ such that

$$f^*(y) = \langle x^*, y \rangle - f(x^*)$$

Thus, if f is actually differentiable, then $y = f'(x^*)$, so the equation $y = f'(x)$ is solvable for x in terms of y.

Now, returning to Q, from (5.16), we compute

$$p_i = n_e^2 v_i + (n_o^2 - n_e^2)(\mathbf{v} \cdot \mathbf{d}(\mathbf{x}))d_i$$

from which we obtain

$$\mathbf{v} \cdot \mathbf{d} = \frac{\mathbf{p} \cdot \mathbf{d}}{n_o^2}$$

which in turn allows us to solve for $\mathbf{v}$ in terms of $\mathbf{p}$:

$$\mathbf{v} = \frac{n_o^2 \mathbf{p} - (n_o^2 - n_e^2)(\mathbf{p} \cdot \mathbf{d})\mathbf{d}}{n_o^2 n_e^2}$$

Thus we plug this into (5.17) and simplify to obtain

$$\Phi(\mathbf{x}, \mathbf{p}) = \frac{n_o^2 |\mathbf{p}|^2 + (n_e^2 - n_o^2)(\mathbf{p} \cdot \mathbf{d})^2}{2 n_o^2 n_e^2} \tag{5.18}$$

It is straightforward to see that Φ is indeed positive homogeneous of degree 2 with respect to p and satisfies

$$\Phi(\mathbf{x}, \mathbf{p}) = Q(\mathbf{x}, \mathbf{v}), \qquad \mathbf{p} = \nabla_{\mathbf{v}} Q(\mathbf{x}, \mathbf{v})$$

This allows us to define the Hamiltonian $H(\mathbf{x}, \mathbf{p})$ of F through the formula

$$H(\mathbf{x}, \mathbf{p}) = F(\mathbf{x}, \mathbf{v}), \qquad \mathbf{v} = \nabla_{\mathbf{p}} \Phi(\mathbf{x}, \mathbf{p}) \tag{5.19}$$

This directly implies that H and Φ are related by

$$\Phi(\mathbf{x}, \mathbf{p}) = \frac{1}{2} H(\mathbf{x}, \mathbf{p})^2$$

and since $F > 0$, we obtain

$$H(\mathbf{x}, \mathbf{p}) = \sqrt{2\Phi(\mathbf{x}, \mathbf{p})}$$

i.e.,

$$H(\mathbf{x}, \mathbf{p}) = \sqrt{\frac{n_o^2|\mathbf{p}|^2 + (n_e^2 - n_o^2)(\mathbf{p} \cdot \mathbf{d}(\mathbf{x}))^2}{n_o^2 n_e^2}} \tag{5.20}$$

Notice in particular that the Hamilton–Jacobi equation $H(\mathbf{x}, \mathbf{p}) = 1$ corresponds to the Fresnel surface for the extraordinary ray, recall (2.21) in Chap. 2.

Now we consider the quadratic functional $\mathcal{Q}$ given by

$$\mathcal{Q} = \int_{t_1}^{t_2} Q(\mathbf{x}, \dot{\mathbf{x}}) dt$$

which we now may write as

$$\mathcal{Q} = \int_{t_1}^{t_2} (\mathbf{p} \cdot \mathbf{v} - \Phi(\mathbf{x}, \mathbf{p}))\, dt$$

in light of (5.17). Then, Hamilton's equations arise as the Euler–Lagrange equations for the functional above:

$$\dot{\mathbf{p}} = -\nabla_{\mathbf{x}} \Phi, \qquad \dot{\mathbf{x}} = \nabla_{\mathbf{p}} \Phi \tag{5.21}$$

i.e., in terms of **d**:

$$\begin{aligned} \dot{\mathbf{p}} &= \frac{-n_o^2 \mathbf{p} \cdot \nabla_{\mathbf{x}} \mathbf{p} + (n_o^2 - n_e^2)(\mathbf{p} \cdot \mathbf{d})\,(\nabla_{\mathbf{x}} \mathbf{d} \cdot \mathbf{p} + \nabla_{\mathbf{x}} \mathbf{p} \cdot \mathbf{d})}{n_o^2 n_e^2}, \\ \dot{\mathbf{x}} &= \frac{n_o^2 \mathbf{p} + (n_e^2 - n_o^2)(\mathbf{p} \cdot \mathbf{d})\mathbf{d}}{n_o^2 n_e^2} \end{aligned} \tag{5.22}$$

Notice that

$$\nabla_{\mathbf{x}} \mathbf{p} = \frac{\partial^2 Q}{\partial \mathbf{x} \partial \mathbf{v}} = (n_o^2 - n_e^2)\,[(\mathbf{v} \cdot \mathbf{d})\nabla_{\mathbf{x}} \mathbf{d} + (\mathbf{v} \cdot \nabla_{\mathbf{x}} \mathbf{d})\mathbf{d}]$$

$$= (n_o^2 - n_e^2)\left[\frac{(\mathbf{p}\cdot\mathbf{d})}{n_o^2}\nabla_{\mathbf{x}}\mathbf{d} + \left(\frac{n_o^2\mathbf{p}\cdot\nabla_{\mathbf{x}}\mathbf{d} + (n_o^2 - n_e^2)(\mathbf{p}\cdot\mathbf{d})(\mathbf{d}\cdot\nabla_{\mathbf{x}}\mathbf{d})}{n_o^2 n_e^2}\right)\mathbf{d}\right]$$

which implies that

$$\frac{-n_o^2\mathbf{p}\cdot\nabla_{\mathbf{x}}\mathbf{p}}{n_o^2 n_e^2} = \frac{(n_e^2 - n_o^2)}{n_o^2 n_e^4}\left[(\mathbf{p}\cdot\mathbf{d})(\mathbf{p}\cdot\nabla_{\mathbf{x}}\mathbf{d})(n_e^2 + n_o^2) + (\mathbf{p}\cdot\mathbf{d})^2(\mathbf{d}\cdot\nabla_{\mathbf{x}}\mathbf{d})(n_o^2 - n_e^2)\right]$$

and

$$\frac{(n_o^2 - n_e^2)(\mathbf{p}\cdot\mathbf{d})\,(\nabla_{\mathbf{x}}\mathbf{d}\cdot\mathbf{p} + \nabla_{\mathbf{x}}\mathbf{p}\cdot\mathbf{d})}{n_o^2 n_e^2} =$$

$$\frac{(n_o^2 - n_e^2)}{n_o^2 n_e^2}(\mathbf{p}\cdot\mathbf{d})\times$$

$$\left\{(\mathbf{p}\cdot\nabla_{\mathbf{x}}\mathbf{d}) + \left[(n_o^2 - n_e^2)\left(\frac{(\mathbf{p}\cdot\mathbf{d})}{n_o^2}(\mathbf{d}\cdot\nabla_{\mathbf{x}}\mathbf{d}) + \frac{(\mathbf{p}\cdot\nabla_{\mathbf{x}}\mathbf{d})}{n_e^2}\right.\right.\right.$$

$$\left.\left.\left. + (n_o^2 - n_e^2)(\mathbf{p}\cdot\mathbf{d})(\mathbf{d}\cdot\nabla_{\mathbf{x}}\mathbf{d})\right)\right]\right\}$$

Thus Hamilton's equations (5.22) become

$$\dot{\mathbf{x}} = \frac{n_o^2\mathbf{p} + (n_o^2 - n_e^2)(\mathbf{p}\cdot\mathbf{d})\mathbf{d}}{n_o^2 n_e^2},$$

$$\begin{aligned}
\dot{\mathbf{p}} = &\frac{(n_e^2 - n_o^2)}{n_o^2 n_e^4}\left[(\mathbf{p}\cdot\mathbf{d})(\mathbf{p}\cdot\nabla_{\mathbf{x}}\mathbf{d})(n_e^2 + n_o^2) + (\mathbf{p}\cdot\mathbf{d})^2(\mathbf{d}\cdot\nabla_{\mathbf{x}}\mathbf{d})(n_o^2 - n_e^2)\right]\\
&+ \frac{(n_o^2 - n_e^2)}{n_o^2 n_e^2}(\mathbf{p}\cdot\mathbf{d}) \times \left\{(\mathbf{p}\cdot\nabla_{\mathbf{x}}\mathbf{d}) + \right.\\
&+ (n_o^2 - n_e^2)\left(\frac{(\mathbf{p}\cdot\mathbf{d})}{n_o^2}(\mathbf{d}\cdot\nabla_{\mathbf{x}}\mathbf{d}) + \frac{(\mathbf{p}\cdot\nabla_{\mathbf{x}}\mathbf{d})}{n_e^2}\right.\\
&\left.\left. + (n_o^2 - n_e^2)(\mathbf{p}\cdot\mathbf{d})(\mathbf{d}\cdot\nabla_{\mathbf{x}}\mathbf{d})\right)\right\}
\end{aligned} \tag{5.23}$$

Remark 5.6 If we assume that the $\mathbf{x}$ dependence of $\mathbf{p}$ is negligible so that $\nabla_{\mathbf{x}}\mathbf{p} \approx 0$, then Eq. (5.23) simplifies to

$$\dot{\mathbf{p}} = \frac{(n_o^2 - n_e^2)(\mathbf{p}\cdot\mathbf{d})\nabla_{\mathbf{x}}\mathbf{d}\cdot\mathbf{p}}{n_o^2 n_e^2}, \qquad \dot{\mathbf{x}} = \frac{n_o^2\mathbf{p} + (n_e^2 - n_o^2)(\mathbf{p}\cdot\mathbf{d})\mathbf{d}}{n_o^2 n_e^2} \tag{5.24}$$

which is consistent with, e.g., equations (8)–(9) in [343].

Given a director **d**, the system (5.23) can be solved via, e.g., Runge–Kutta methods, see, for instance, Section 4.2.4 in [384] or [386].

Now we turn to the approach in a Finsler space and see how for liquid crystals the Riemannian framework is actually sufficient.

5.4 Geometric Approach

Before discussing the geometric approach to the Euler–Lagrange equations, we review some basic Riemannian geometry notions and introduce the analogous ideas for Finsler geometry.

5.4.1 Reminder of Riemannian Geometry

Let us recall some basic notions from Riemannian geometry; this material is largely from [244], and more details can be found in Appendix A.1.1. Below we employ the summation convention, so that repeated indices imply a sum over those indices. The starting point in Euclidean space is the square of the infinitesimal distance between points x^i and $x^i + dx^i$:

$$ds^2 = dx^2 + dy^2 + dz^2 = \delta_{ij} dx^i dx^j \tag{5.25}$$

where δ_{ij} denotes the Kronecker delta. The quantity ds^2 is called the line element. To express the line element in arbitrary coordinates, we introduce the transformation matrices

$$\Lambda^i_{i'} = \frac{\partial x^i}{\partial x^{i'}}, \qquad \Lambda^{i'}_i = \frac{\partial x^{i'}}{\partial x^i}$$

which go between the coordinate systems $\{x^i\}$ and $\{x^{i'}\}$. Then, we can transform coordinates as

$$dx^i = \Lambda^i_{i'} dx^{i'}, \qquad dx^{i'} = \Lambda^{i'}_i dx^i$$

With this in hand, can write the line element ds^2 in an arbitrary coordinate system $x^{i'}$ as

$$ds^2 = \delta_{ij} \Lambda^i_{i'} \Lambda^j_{j'} dx^{i'} dx^{j'} \tag{5.26}$$

where we are taking the x^i to be Cartesian coordinates (so that (5.25) holds). The object

$$g_{i'j'} = \delta_{ij}\Lambda^i_{i'}\Lambda^j_{j'}$$

is called the metric tensor and allows us to compute distances in space. Clearly in Cartesian coordinates we have

$$g_{ij} = \delta_{ij}$$

which recovers the usual Euclidean metric. Now, we can compute the length of a curve in space in arbitrary coordinates using the metric tensor. Indeed, since we can write

$$ds^2 = g_{i'j'}dx^{i'}dx^{j'}$$

the length of the curve between A with coordinates $x^i(\xi_1)$ and B with coordinates $x^i(\xi_2)$ is obtained by integrating ds along the curve, so that

$$s(A,B) = \int_{\xi_1}^{\xi_2} ds = \int_{\xi_1}^{\xi_2} \sqrt{g_{ij}dx^i(\xi)dx^j(\xi)} = \int_{\xi_1}^{\xi_2} \sqrt{g_{ij}\frac{dx^i(\xi)}{d\xi}\frac{dx^j(\xi)}{d\xi}}d\xi \tag{5.27}$$

Finally let us consider Fermat's principle . Let A, B be two points in some Riemannian manifold with metric g_{ij}. Then Fermat's principle [48] says that the path actually followed by light rays is the one that minimizes the optical path length

$$\int_A^B n d\ell$$

where n denotes the refractive index of the material and $d\ell$ is an element of arclength along the path. In Riemannian geometry, we have seen that the line element ds depends on the position coordinates x^i of the manifold:

$$ds^2 = g_{ij}dx^i dx^j \tag{5.28}$$

The geodesic joining the points A and B in such a manifold is obtained by minimizing $\int_B^A ds$; see, e.g., [138, Chapter 2, page 128]. Thus, light rays can be viewed as geodesics in an effective geometry [199, 359]. In particular, the geodesic equations arise as the Euler–Lagrange equation for the length functional s:

$$\frac{d^2x_i}{dt^2} + \Gamma^i_{jk}\frac{dx_j}{dt}\frac{dx_k}{dt} = 0 \tag{5.29}$$

where

$$\Gamma^i_{jk} = \frac{g^{il}}{2}\left(\frac{\partial g_{lj}}{\partial x_k} + \frac{\partial g_{lk}}{\partial x_j} - \frac{\partial g_{jk}}{\partial x_l}\right) \tag{5.30}$$

are the classical Christoffel symbols. Indeed, the Euler–Lagrange equation takes the form

$$\frac{d}{d\xi}\frac{\partial \mathcal{L}}{\partial \dot{x}^i} - \frac{\partial \mathcal{L}}{\partial x^i} = 0$$

where

$$\mathcal{L} = \sqrt{g_{ij}dx^i(\xi)dx^j(\xi)}$$

Since $\mathcal{L} = \dfrac{\partial s}{\partial \xi}$ where s is the arclength along the curve, the Euler–Lagrange equation above can be written

$$\frac{d}{ds}\left(g_{ij}\frac{dx^j}{ds}\right) - \frac{1}{2}\frac{\partial g_{jk}}{\partial x^i}\frac{dx^j}{ds}\frac{dx^k}{ds} = 0$$

Using the fact that we can write

$$\frac{\partial g_{ij}}{\partial x^k}\frac{dx^k}{ds}\frac{dx^j}{ds} = \frac{1}{2}\left(\frac{\partial g_{ij}}{\partial x^k} + \frac{\partial g_{ik}}{\partial x^j}\right)\frac{dx^j}{ds}\frac{dx^k}{ds}$$

then we find that

$$g_{ij}\frac{d^2x^j}{ds^2} + \frac{1}{2}\left(\frac{\partial g_{ij}}{\partial x^k} + \frac{\partial g_{ik}}{\partial x^j} - \frac{\partial g_{jk}}{\partial x^i}\right)\frac{dx^j}{ds}\frac{dx^k}{ds} = 0$$

Multiply both sides by $g^{\ell i}$ now and contract on i to see that (after relabeling) we get exactly the Christoffel symbols (5.30) in the second term.

Remark 5.7 Later on, we will not distinguish between vectors and one-forms (covariant vectors) since we will assume Cartesian coordinates (where these objects are indeed indistinguishable). Indeed, given a contravariant vector x^i and covariant vector x_i, we can relate the two through the inverse metric tensor

$$x^k = g^{ki}x_i$$

Since later we will study geodesics in Euclidean space, it is worth briefly recalling the simpler idea behind geodesics on surfaces on $\mathbb{R}^3$ (see, e.g., [109] for more details). So, suppose $\sigma : U \to \mathbb{R}^3$ is a parameterized surface and let

$\gamma : I \to \mathbb{R}^3$ be a parameterized curve in σ. Here, $U \subset \mathbb{R}^2$ is an open set and $I \subset \mathbb{R}$ is an open interval. For $\sigma = (\sigma_1, \sigma_2, \sigma_3)$, put

$$\sigma'_u = \begin{pmatrix} \frac{\partial \sigma_1}{\partial u} \\ \frac{\partial \sigma_2}{\partial u} \\ \frac{\partial \sigma_3}{\partial u} \end{pmatrix}$$

and similar for σ'_v. Then denote the unit normal of σ by

$$\mathbf{N} = \frac{\sigma'_u \times \sigma'_v}{||\sigma'_u \times \sigma'_v||}$$

and let $\mathbf{m}(t) = \mathbf{N}(\mu(t))$, where $\mu : I \to U$ is a regular plane curve, be the unit normal vector of σ along γ. Then, we can define the geodesic curvature of γ at t with respect to σ via

$$\kappa_g(t) = \frac{\det\left(\gamma'(t)\gamma''(t)\mathbf{m}(t)\right)}{||\gamma'(t)||^3} \tag{5.31}$$

Then, we define a geodesic on σ as a regular parameterized curve γ whose geodesic curvature (5.31) is identically zero. Let us next obtain the equations satisfied by the geodesics, analogous to (5.29).

To this end, let $\sigma'_1 = \sigma'_u$ and $\sigma'_2 = \sigma'_v$, and similar for higher derivatives. Define

$$\mathfrak{g}_{ij} = \sigma'_i \cdot \sigma'_j, \qquad i = 1, 2 \tag{5.32}$$

Then, it is straightforward to see that

$$\sigma''_{ij} \cdot \sigma'_k = \frac{1}{2}\left(\frac{\partial g_{ik}}{\partial u_j} + \frac{\partial g_{jk}}{\partial u_i} - \frac{\partial g_{ij}}{\partial u_k}\right), \qquad i, j, k = 1, 2$$

by using the symmetry $\sigma''_{ij} = \sigma''_{ji}$. In this way we can define the Christoffel symbols $\Gamma^i_{j,k} : U \to \mathbb{R}$ associated with σ by

$$\begin{pmatrix} \Gamma^1_{j,k} \\ \Gamma^2_{j,k} \end{pmatrix} = \mathfrak{g}^{-1} \begin{pmatrix} \sigma''_{jk} \cdot \sigma'_1 \\ \sigma''_{jk} \cdot \sigma'_2 \end{pmatrix} \tag{5.33}$$

from which we obtain the formula

$$\Gamma^i_{j,k} = \frac{1}{2}\mathfrak{g}^{il}\left(\frac{\partial \mathfrak{g}_{jl}}{\partial u_k} + \frac{\partial \mathfrak{g}_{kl}}{\partial u_i} - \frac{\partial \mathfrak{g}_{jk}}{\partial u_l}\right) \tag{5.34}$$

with $\mathfrak{g}$ given by (5.32); compare (5.33) with (5.30). Finally we are arriving at the geodesic equations. To this end define the function

$$F_i(t) = (u_i)''(t) + \Gamma^i_{j,k}(\mu(t))(u_j)'(t)(u_k)'(t), \qquad i = 1, 2$$

with $\Gamma^i_{j,k}$ given by (5.34), for μ such that $\gamma = \sigma \circ \mu$ a regular parameterized curve on σ. Then a calculation shows that the geodesic curvature given in (5.31) satisfies

$$\kappa_g = \frac{\sqrt{\det \mathfrak{g}}\left((u_1)'F_1 - (u_2)'F_2\right)}{||\gamma'||^3}$$

with $\mathfrak{g}$ given as usual by (5.32). Now, we use the following characterization of geodesics on a surface, see, e.g., Theorem 4.10 in [364]:

Theorem 5.5 *Let $\gamma = \sigma \circ \mu$ be a regular parameterized curve on σ. Then γ is a geodesic and has constant speed if and only if $\gamma''(t) \cdot \gamma'(t) = 0$ for all t.*

We compute

$$\gamma''(t) = u_i''\sigma_i' + u'_j u'_k \sigma''_{jk}$$

Now, a quick calculation shows that

$$\sigma''_{jk} = \Gamma^1_{jk}\sigma_1' + \Gamma^2_{jk}\sigma_2' + (\sigma''_{jk} \cdot \mathbf{N})\mathbf{N}$$

Thus in terms of the F_1, F_2, we find that

$$\gamma''(t) = F_1\sigma_1' + F_2\sigma_2' + c\mathbf{m}, \qquad c \text{ constant}$$

If $\gamma''(t) \cdot \gamma'(t) = 0$, i.e., γ'' is normal to the tangent space of $\mu(t)$, the above calculation means we must have $F_1 = F_2 = 0$. Thus $\kappa_g = 0$, i.e., γ is a geodesic, and so u_1, u_2 satisfy the geodesic equations $F_1 = F_2 = 0$, i.e.,

$$(u_i)''(t) + \Gamma^i_{j,k}(\mu(t))(\mu(t))(u_j)'(t)(u_k)'(t), \qquad i = 1, 2 \tag{5.35}$$

Compare (5.35) with (5.29).

5.4.2 Remarks on Finsler Geometry and Geodesics

Starting from the work of Joets and Ribotta in [199], the path of a light ray which minimizes a certain distance leads to the interpretation of the rays as geodesics in a so-called Finsler space. The idea is to model a liquid crystal as an inhomogeneous optical medium, where via Fermat's principle one is interested in minimizing the usual functional

$$\mathcal{F}[\mathbf{x}, \dot{\mathbf{x}}] = \int_0^1 n_{\text{eff}}(\mathbf{x}, \dot{\mathbf{x}})dt$$

This leads interpreting light paths as geodesics in some effective geometry. In Finsler geometry, a general line element depends on both position and direction. The arclength can then be written as

$$ds = F(\mathbf{x}, \dot{\mathbf{x}})dt$$

where F is a so-called Finsler function. To be precise, let us define such a function rigorously here (see, e.g., [374]).

Definition 5.2 A Minkowski norm on a vector space V is a nonnegative function $F : V \to [0, \infty)$ such that:

(M1) F is C^∞ on $V \setminus \{0\}$.
(M2) $F(\lambda y) = \lambda F(y)$ for all $\lambda > 0$, $y \in V$.
(M3) For any $y \in V \setminus \{0\}$, the symmetric bilinear form g_y on V is positive definite, where

$$g_y(u, v) := \frac{1}{2}\frac{\partial^2}{\partial s \partial t}\left(F^2(y + su + tv)\right)\Big|_{s=t=0}$$

The pair (V, F) is called a Minkowski space.

From Lemma 1.2.5 in [374], any Minkowski norm F satisfies

$$F(u + v) \leq F(u) + F(v), \qquad u, v \in V \tag{5.36}$$

Thus, F induces a metric d on V by

$$d(u, v) := F(v - u), \qquad u, v \in V \tag{5.37}$$

The metric d is called a Minkowski metric if it is induced by a Minkowski norm F. From here we can define a Finsler metric on a manifold M (below, TM denotes the tangent space of M):

Definition 5.3 A function $F : TM \to [0, \infty)$ is called a Finsler metric if it satisfies:

(F1) F is C^∞ on $TM \setminus \{0\}$.
(F2) For each $x \in M$, $F_x := F|_{T_xM}$ is a Minkowski norm on T_xM.

The pair (M, F) is called a Finsler space.

Given a Finsler metric F on M, for a Lipschitz continuous curve $\gamma : [a, b] \to M$, we can define the length of γ by

$$L_F(\gamma) = \int_a^b F(\dot{\gamma}(t))dt$$

which induces a function

$$d_F(p, q) := \inf_{\gamma} L_F(\gamma)$$

where the infimum is taken over all Lipschitz curves γ such that $\gamma(0) = p, \gamma(1) = q$. It is straightforward to see that d_F is a metric on M and the manifold topology coincides with the metric topology.

Let $g = \{g_x\}_{x \in M}$ be a Riemannian metric, such that

$$g_{ij}(x) = g_x \left(\left.\frac{\partial}{\partial x^i}\right|_x, \left.\frac{\partial}{\partial x^j}\right|_x \right)$$

are C^∞. Define

$$F_x(y) = \sqrt{g_x(y, y)}, \qquad y \in T_xM$$

Then F_x is an Euclidean norm, and the family $\{F_x\}_{x \in M}$ is a Finsler metric on M.

Since the light rays are interpreted as geodesics in a Finsler space, we should define precisely the notion of geodesic.

Define the quantities

$$g_{ij}(y) = \frac{1}{2} \frac{\partial^2 F^2}{\partial y^i \partial y^j}$$

and

$$G^i(y) := \frac{1}{4} g^{il}(y) \left\{ \frac{\partial^2 F^2}{\partial x_k \partial y^l}(y) y^k - \frac{\partial F^2}{\partial x^l}(y) \right\} \tag{5.38}$$

where $(g^{ij}(x)) = (g_{ij}(x))^{-1}$. Finally define

$$\kappa(t) := \frac{1}{F(\dot{\gamma}(t))^2}\left[\ddot{\gamma}^i + 2G^i(\dot{\gamma})\right]\frac{\partial}{\partial x^i}\bigg|_{\gamma(t)} \tag{5.39}$$

where γ is a constant speed piecewise C^∞ curve $F(\dot{\gamma}) = \lambda =$ constant. Similar to the Riemannian case, geodesics are length minimizing curves, i.e., they are critical points of the functional $L_F(\gamma)$.

The function $\kappa(t)$ in (5.39) is called the geodesic curvature of γ at $\gamma(t)$. We see that in local coordinates, $\kappa = 0$ is equivalent to

$$\ddot{\gamma}^i + 2G^i(\dot{\gamma}) = 0 \tag{5.40}$$

This leads us to:

Definition 5.4 A C^∞ curve in a Finsler space (M, F) is called a geodesic if it has constant speed and its geodesic curvature $\kappa = 0$.

Compare this definition with the one in the previous section for geodesics on surfaces. For a full derivation of (5.40) assuming all functions are smooth, see [374, Chapter 5].

Now, we can write the coefficients G^i in (5.38) as

$$G^i(y) = \frac{1}{4}g^{il}(y)\left[2\frac{\partial g_{jl}}{\partial x^k}(y) - \frac{\partial g_{jk}}{\partial x^l}(y)\right]y^j y^k \tag{5.41}$$

and so they are sometimes called the geodesic coefficients. Notice that, unlike the classical Christoffel symbols (5.30), the coefficients involving the metric in $G^i(y)$ are functions on the tangent bundle rather than the manifold M. Indeed, we define the general Christoffel symbols in a Finsler space by

$$\Gamma^i_{j,k}(x, y) := \frac{\partial^2 G^i}{\partial y_j \partial y_k} - g^{il}L_{jkl}, \tag{5.42}$$

where

$$L_{ijk} := \frac{\partial C_{ijk}}{\partial x_l}y_l - 2C_{ijkl}G^l - C_{ljk}\frac{\partial G^l}{\partial y_i} - C_{ilk}\frac{\partial G^l}{\partial y_j} - C_{ijl}\frac{\partial G^l}{\partial y_k} \tag{5.43}$$

and

$$C_{ijk} = \frac{1}{4}\frac{\partial^3 F^2}{\partial y_i \partial y_j \partial y_k} \tag{5.44}$$

and

$$C_{ijkl} = \frac{\partial C_{ijk}}{\partial y_l} = \frac{1}{4}\frac{\partial^4 F^2}{\partial y_i \partial y_j \partial y_k \partial y_l} \tag{5.45}$$

Explicitly, we find that

$$\begin{aligned}\Gamma^{i}_{j,k}(x,y) = \frac{1}{2}g^{il}\left(\frac{\partial g_{jl}}{\partial x_k} + \frac{\partial g_{kl}}{\partial x_j} - \frac{\partial g_{jk}}{\partial x_l}\right) + \frac{\partial C_{jlm}}{\partial x_k} y_k \\ - 2C_{jklm}G^k - 2C_{jkl}\frac{\partial G^k}{\partial y_m} - 2C_{klm}\frac{\partial G^k}{\partial y_j} - L_{jlm}\end{aligned} \tag{5.46}$$

which clearly depends on y in general.

Now we will see that, for the main effective refractive index we have been working with, this framework is actually unnecessary since the metric will be independent of y.

5.4.3 Application to Liquid Crystal Optics

Associate to Ω, a domain in $\mathbb{R}^3$ (i.e., a container filled with a liquid crystal), a Finsler function $F(x, v)$ which defines [199] the Finslerian arclength ds by $ds = F(x, dx)$. The metric tensor is induced from the Finsler function F by

$$g_{ij} = \frac{1}{2}\frac{\partial^2 F^2}{\partial v_i \partial v_j} \tag{5.47}$$

In our case, the Finsler function is given by

$$F(x, v) = F(x_i, v_i) = n_{\text{eff}}(x_i, v_i) = \sqrt{n_e^2|\mathbf{v}|^2 + (n_o^2 - n_e^2)(\mathbf{v}\cdot\mathbf{d}(\mathbf{x}))^2} \tag{5.48}$$

so that

$$g_{ij} = (n_o^2 - n_e^2)d_i d_j + n_e^2\delta_{ij} \tag{5.49}$$

where δ_{ij} denotes the Kronecker delta. Note in particular that with this metric, the coefficients L_{ijk} in (5.43) vanish, so the Christoffel symbols in (5.46) reduce to

$$\Gamma^{i}_{j,k}(x,y) = \frac{1}{2}g^{il}\left(\frac{\partial g_{jl}}{\partial x_k} + \frac{\partial g_{lk}}{\partial x_j} - \frac{\partial g_{jk}}{\partial x_l}\right) \tag{5.50}$$

Compare (5.50) with the Riemannian case (5.30). In particular, with g_{ij} given by (5.49), the symbols $\Gamma^{i}_{j,k}$ just depend on x.

Remark 5.8 In fact, since the g_{ij} from (5.49) are independent of v, the resulting Finsler function F is called Riemannian [29]. In particular, since $C_{ijk} = 0$ for all i, j, k, we have that F is Euclidean, i.e., arises from an inner product [104]. This is straightforward to see from the definition of C_{ijk}. Indeed, C_{ijk} is proportional to the vertical derivative of g_{ij}, which vanishes if and only if g_{ij} has no y independence. That is, F arises from an inner product.

For the metric (5.49), we find its inverse explicitly:

$$g^{-1} = \begin{pmatrix} \frac{(n_o^2-n_e^2)(d_2^2+d_3^2)+n_e^2}{n_e^2 n_o^2} & \frac{(n_e-n_o)(n_e+n_o)d_1 d_2}{n_e^2 n_o^2} & \frac{(n_e-n_o)(n_e+n_o)d_1 d_3}{n_e^2 n_o^2} \\ \frac{(n_e-n_o)(n_e+n_o)d_1 d_2}{n_e^2 n_o^2} & \left(\frac{1}{n_o^2}-\frac{1}{n_e^2}\right)d_2^2+\frac{1}{n_e^2} & \frac{(n_e-n_o)(n_e+n_o)d_2 d_3}{n_e^2 n_o^2} \\ \frac{(n_e-n_o)(n_e+n_o)d_1 d_3}{n_e^2 n_o^2} & \frac{(n_e-n_o)(n_e+n_o)d_2 d_3}{n_e^2 n_o^2} & \left(\frac{1}{n_o^2}-\frac{1}{n_e^2}\right)d_3^2+\frac{1}{n_e^2} \end{pmatrix} \tag{5.51}$$

As expected, in the isotropic limit $n_o \to n_e := n$, we obtain

$$g^{-1} = \begin{pmatrix} \frac{1}{n^2} & 0 & 0 \\ 0 & \frac{1}{n^2} & 0 \\ 0 & 0 & \frac{1}{n^2} \end{pmatrix}$$

We can also explicitly compute the Christoffel symbols (recall that $\Gamma^i_{j,k} = \Gamma^i_{k,j}$): See Appendix A.2. Note that, as we have seen previously, in this specific case the symbols only depend on $\mathbf{x}$, not $\dot{\mathbf{x}}$.

The ray equations again just take the form

$$\frac{d^2 x_i}{ds^2} + \Gamma^i_{jk}\frac{dx_j}{ds}\frac{dx_k}{ds} = 0 \tag{5.52}$$

With the Christoffel coefficients given in Appendix A.2, we can compute the three geodesic equations coming from (5.52):

$$\begin{aligned} \frac{d^2 x_1}{ds^2} &= -\frac{F_1(\mathbf{d}, x_1, x_2, x_3)}{n_e^2 n_o^2} \\ \frac{d^2 x_2}{ds^2} &= -\frac{F_2(\mathbf{d}, x_1, x_2, x_3)}{n_e^2 n_o^2} \\ \frac{d^2 x_3}{ds^2} &= -\frac{F_3(\mathbf{d}, x_1, x_2, x_3)}{n_e^2 n_o^2} \end{aligned} \tag{5.53}$$

where F_1, F_2, F_3 are given in Appendix A.2, see Eqs. (A.7)–(A.9).

Note that in the case of a constant director (and the isotropic limit $n_e \to n_o$), we have

$$F_1|_{\text{constant director}} = F_2|_{\text{constant director}} = F_3|_{\text{constant director}} = 0$$

and the geodesic equations (5.53) yield the expected result that geodesics are straight lines, i.e., we obtain

$$\frac{d^2x_1}{ds^2} = 0$$
$$\frac{d^2x_2}{ds^2} = 0$$
$$\frac{d^2x_3}{ds^2} = 0$$

Remark 5.9 Consider instead the effective refractive index given by (3.9): the effective refractive index associated to the phase velocity (rather than the energy velocity). This will yield a metric g_{ij} which genuinely depends on $\mathbf{v}$, so does not merely reduce to the Riemannian case. Indeed, we find that

$$\frac{\partial n^2_{\text{eff},p}}{\partial v_i} = \frac{-2d_i n_e^2 n_o^2 (n_e^2 - n_o^2)(\mathbf{v} \cdot \mathbf{d})}{\left(n_o^2 + (n_e^2 - n_o^2)(\mathbf{v} \cdot \mathbf{d})^2\right)^2}$$

and so

$$g_{ij}(\mathbf{x}, \mathbf{v}) = \frac{1}{2}\frac{\partial n^2_{\text{eff},p}}{\partial v_i \partial v_j} = \frac{d_i d_j n_e^2 n_o^2 (n_e^2 - n_o^2)\left(-n_o^3 + 3(n_e^2 - n_o^2)(\mathbf{v} \cdot \mathbf{d})^2\right)}{\left(n_o^2 + (n_e^2 - n_o^2)(\mathbf{v} \cdot \mathbf{d})^2\right)^3} \tag{5.54}$$

However, we cannot really proceed since $n_{\text{eff},p}$ is not homogeneous of degree one in $\mathbf{v}$, so one cannot place this into the Finsler geometry framework.

5.5 Examples

5.5.1 *Hedgehog Director*

This first example was considered in part in [359]. Consider spherical coordinates (r, θ, ϕ) and a director given by the Hedgehog:

$$\mathbf{d}(r, \theta, \phi) = \widehat{r} \tag{5.55}$$

where recall

$$\widehat{r} = \sin(\theta)\cos(\phi)\mathbf{e}_x + \sin(\theta)\sin(\phi)\mathbf{e}_y + \cos(\theta)\mathbf{e}_z$$

where the $\mathbf{e}_j$ denote the unit vectors in the x, y, z Cartesian directions. Writing $\mathbf{x}(s) = r\widehat{r}(s)$, we find

$$\mathbf{v} = \dot{\mathbf{x}}(s) = \dot{r}\widehat{r} + r\left(\dot{\theta}\widehat{\theta} + \dot{\phi}\sin(\theta)\widehat{\phi}\right) = \dot{r}\widehat{r} + r\frac{\partial\widehat{r}}{\partial s}$$

Now, since $\widehat{r} \perp \dfrac{\partial\widehat{r}}{\partial s}$, we see that

$$\mathbf{v}\cdot\mathbf{d} = \dot{r}$$

and thus

$$n_{\text{eff},r}(r,\theta,\phi,\dot{r},\dot{\theta},\dot{\phi}) = \sqrt{n_o^2(\dot{r})^2 + n_e^2 r^2\left(\dot{\theta}^2 + \sin^2(\theta)\dot{\phi}^2\right)} \tag{5.56}$$

Note that if the angle between $\mathbf{v}$ and $\mathbf{d}$ is denoted by β, then (5.56) is precisely the effective refractive index in equation (1) from [359]:

$$n_{\text{eff},r}^2 = n_o^2\cos^2(\beta) + n_e^2\sin^2(\beta)$$

We know that

$$g_{ij} = \frac{1}{2}\frac{\partial^2 n_{\text{eff},r}^2}{\partial v_i \partial v_j}, \qquad v_i \in \{\dot{r},\dot{\theta},\dot{\phi}\}$$

We can rescale the radial coordinate making $\rho = n_o r$, which leads to

$$g_{ij} = \begin{pmatrix} 1 & 0 & 0 \\ 0 & \frac{n_e^2}{n_o^2}\rho^2 & 0 \\ 0 & 0 & \frac{n_e^2}{n_o^2}\rho^2\sin^2(\theta) \end{pmatrix} \tag{5.57}$$

with inverse

$$\begin{pmatrix} 1 & 0 & 0 \\ 0 & \frac{n_o^2}{n_e^2\rho^2} & 0 \\ 0 & 0 & \frac{n_o^2\csc^2(\theta)}{n_e^2\rho^2} \end{pmatrix}$$

From here we can find the nonzero $\Gamma^i{}_{jk}$:

$$\Gamma^1_{2,2} = -\frac{n_e^2 r}{n_o^2}\rho, \quad \Gamma^1_{3,3} = -\frac{n_e^2 \rho \sin^2(\theta)}{n_o^2}, \quad \Gamma^2_{2,1} = \frac{1}{\rho}, \quad \Gamma^2_{3,3} = -\cos(\theta)\sin(\theta),$$

$$\Gamma^3_{3,1} = \frac{1}{\rho}, \quad \Gamma^3_{3,2} = \cot(\theta)$$

Noticing that $\Gamma^i_{j,k} = \Gamma^i_{k,j}$ we find the ray equations (using that the Euclidean line element in spherical coordinates is $dr^2 + r^2(d\theta^2 + \sin^2(\theta)d\phi^2)$):

$$\frac{d^2\rho}{ds^2} = \frac{n_e^2}{n_o^2}\rho\left(\left(\frac{d\theta}{ds}\right)^2 + \sin(\theta)^2\left(\frac{d\phi}{ds}\right)^2\right) \tag{5.58}$$

$$\frac{d^2\theta}{ds^2} = -\frac{2}{\rho}\frac{d\rho}{ds}\frac{d\theta}{ds} + \cos(\theta)\sin(\theta)\left(\frac{d\phi}{ds}\right)^2 \tag{5.59}$$

and

$$\frac{d^2\phi}{ds^2} = -\frac{2}{\rho}\left(\frac{d\rho}{ds} + \rho\cot(\theta)\frac{d\theta}{ds}\right)\frac{d\phi}{ds} \tag{5.60}$$

This system can be solved numerically, e.g., with a Runge–Kutta method, see, e.g., [359, Section 3]. For initially parallel light beams show lensing behavior similar to cosmic defects: The effective geometry for the hedgehog director is the same as that of the global monopole [420].

5.5.2 Disclinations

In addition to the hedgehog point defect, we can look at the disclination configurations in the $z =$ constant plane, which have director of the form

$$\mathbf{d} = (\cos(\theta(x, y)), \sin(\theta(x, y)), 0) \tag{5.61}$$

where $\theta(x, y)$ is the polar angle in $\mathbb{R}^2$, i.e.,

$$\theta(x, y) = \arctan(y/x)$$

Recall that disclinations are line defects along which a symmetry of rotation is broken; see [220, Chapter 11].

More generally, (planar) wedge disclinations are given by

$$\mathbf{d} = (\cos(k\theta(x, y) + c), \sin(k\theta(x, y) + c), 0) \tag{5.62}$$

for various k, c. The parameter k measures how much the director rotates around the defect. Here, wedge means that the rotation axis is along the disclination line. This planar model was considered by Frank [131], and in this case, under the simplified model when the Frank constants $K_1 = K_2 = K_3$ and $K_4 = 0$, with general director $\mathbf{d} = (\cos(\phi), \sin(\phi), 0)$, the Euler–Lagrange equation (3.24) becomes the Laplace equation for ϕ

$$\frac{\partial^2 \phi}{\partial x^2} + \frac{\partial^2 \phi}{\partial y^2} = 0 \tag{5.63}$$

By seeking solutions to (5.63) of the form $\phi = a\theta + \text{harmonic}$, $a \in \mathbb{R}$, i.e., a sum of $\theta(x, y)$ plus a harmonic function, one is led to $\phi = k\theta + c$, yielding the director (5.62).

Let us consider (5.61), i.e., the 2D hedgehog. In this case we let $\mathbf{x}(s) = r(s)\cos(\theta(s))\mathbf{e}_x + r(s)\sin(\theta(s))\mathbf{e}_y$, combining polar coordinates (r, θ) with the Cartesian basis in $\mathbb{R}^2$: $\{\mathbf{e}_x, \mathbf{e}_y\}$. Then we compute

$$\mathbf{v}(s) = \dot{\mathbf{x}}(s) = \left(\dot{r}\cos(\theta) - r\dot{\theta}\sin(\theta)\right)\mathbf{e}_x + \left(\dot{r}\sin(\theta) + r\dot{\theta}\cos(\theta)\right)\mathbf{e}_y$$

and find that

$$\mathbf{v} \cdot \mathbf{d} = \dot{r}$$

Then the effective index is given by

$$n_{\text{eff},r} = \sqrt{n_e^2\left(\dot{r}^2 + r^2\dot{\theta}^2\right) + (n_o^2 - n_e^2)\dot{r}^2} = \sqrt{n_o^2\dot{r}^2 + n_e^2 r^2\dot{\theta}^2}$$

Again we know that

$$g_{ij} = \frac{1}{2}\frac{\partial^2 n_{\text{eff},r}^2}{\partial v_i \partial v_j}, \qquad v_i \in \{\dot{r}, \dot{\theta}\}$$

which leads to

$$g_{ij} = \begin{pmatrix} 2n_o^2 & 0 \\ 0 & 2n_e^2 r^2 \end{pmatrix}$$

This leads to the following nonzero Christoffel symbols:

$$\Gamma^1_{2,2} = -\frac{n_e^2 r}{n_o^2}, \qquad \Gamma^2_{2,1} = \frac{1}{r}$$

which yield the following geodesic equations for the rays in 2D:

$$\begin{aligned}\frac{d^2r}{ds^2} &= \frac{n_e^2 r}{n_o^2}\left(\frac{d\theta}{ds}\right)^2 \\ \frac{d^2\theta}{ds^2} &= -\frac{2}{r}\frac{dr}{ds}\frac{d\theta}{ds}\end{aligned} \tag{5.64}$$

which is precisely the system (10)–(11) obtained in [360]. The solution of the system (5.64) can be found to be:[7]

$$\begin{aligned}r(s) &= \frac{\sqrt{c_1^2 n_o^2 + 4c_2^2 n_e^2 (s + n_e c_3)^2}}{n_e\sqrt{2c_2}} \\ \theta(s) &= \frac{n_o}{n_e}\arctan\left(\frac{2c_2 n_e (s + n_e c_3)}{c_1 n_o}\right) + c_4\end{aligned} \tag{5.65}$$

for particular integration constants c_1, c_2, c_3, c_4. Now, to show (5.65), from the second equation in (5.64) we see that

$$\left(\frac{n_e}{n_o}\right)^2\left(2r\frac{dr}{ds}\frac{d\theta}{ds} + r^2\frac{d^2\theta}{ds^2}\right) = \left(\frac{n_e}{n_o}\right)^2\frac{d}{ds}\left(r^2\frac{d\theta}{ds}\right) = 0$$

which after integration yields

$$\left(\frac{n_e}{n_o}\right)^2 r^2\frac{d\theta}{ds} = c_1 \tag{5.66}$$

Now, consider

$$\frac{d}{ds}\left(\frac{1}{2}\left(\frac{dr}{ds}\right)^2 + \frac{c_1^2 n_o^2}{2n_e^2}r^{-2}\right)$$

We will show this term is equal to zero, and thus there exists a constant c_2 such that

$$\frac{1}{2}\left(\frac{dr}{ds}\right)^2 + \frac{c_1^2 n_o^2}{2n_e^2}r^{-2} = c_2 \tag{5.67}$$

[7] This solution was also obtained in [100], where similar equations arose from the study of disclinations in elasticity. This is perhaps not unexpected since we understand disclinations as dislocations of rotation, and dislocations in elastic solids are well studied [217, 220, 230, 306]; they are responsible for plastic deformation of crystalline materials and are examples of topological singularities.

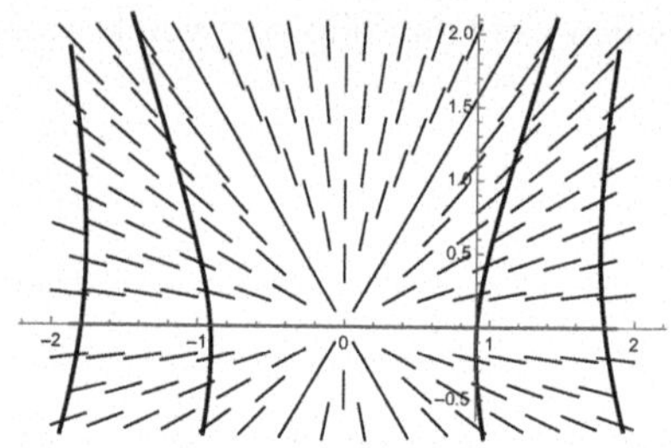

Fig. 5.1 Rays are diverging away from the origin. Here we have taken $n_e = 1.9$ and $n_o = 1.6$, which is similar to the parameters for the nematic Merck Phase V

Indeed, we have

$$\frac{d}{ds}\left(\frac{1}{2}\left(\frac{dr}{ds}\right)^2 + \frac{c_1^2 n_o^2}{2n_e^2} r^{-2}\right) = \frac{dr}{ds}\frac{d^2r}{ds^2} - \frac{c_1^2 n_o^2}{n_e^2 r^3}\frac{dr}{ds}$$

$$= \frac{dr}{ds}\left[\frac{d^2r}{ds^2} - \frac{c_1^2 n_o^2}{n_e^2 r^3}\right]$$

$$= \frac{dr}{ds}\left[\frac{n_e^2}{n_o^2} r\left(\frac{d\theta}{ds}\right)^2 - \frac{n_o^2}{n_e^2 r^3}\left(\frac{d\theta}{ds}\right)^2 r^4 \frac{n_e^4}{n_o^4}\right] = 0$$

where in the second to last equality we have used the first equation in (5.64) as well as (5.66). Thus it remains to solve (5.66) and (5.67). The solution of (5.67) is given by

$$r(s) = \frac{\sqrt{c_1^2 n_o^2 + 4c_2^2 n_e^2 (s + n_e c_3)^2}}{n_e \sqrt{2c_2}} \tag{5.68}$$

where c_3 is an integration constant. We plug $r(s)$ given by (5.68) into (5.66) to obtain

$$\frac{d\theta}{ds} = \frac{2c_1 c_2 n_o^2}{c_1^2 n_o^2 + 4c_2^2 n_e^2 (s + n_e c_3)^2}$$

Integrating yields

$$\theta(s) = \frac{n_o}{n_e} \arctan\left(\frac{2c_2 n_e (s + n_e c_3)}{c_1 n_o}\right) + c_4 \tag{5.69}$$

for another integration constant c_4; thus (5.65) indeed holds and agrees with equations (12)–(13) in [360].

Bending of rays can be seen through the numerical solution of (5.64). Indeed, in Fig. 5.1, one can see the defect at the origin acting as a diverging lens for the rays, which has been observed in [359], see in particular their Figure 5.

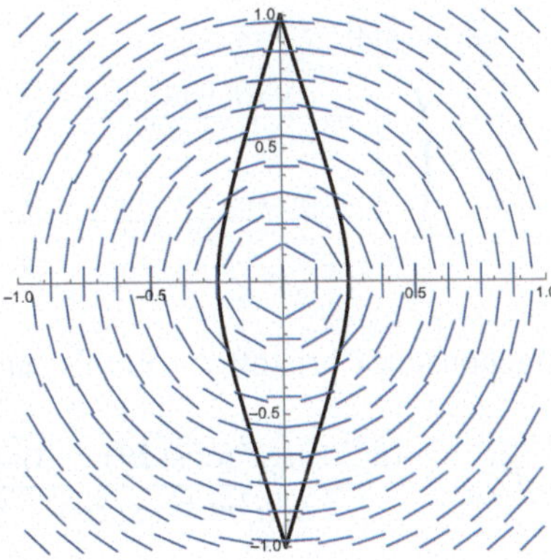

Fig. 5.2 Rays starting from the point $(0, -1)$ are now converging. Again we have taken $n_e = 1.9$ and $n_o = 1.6$

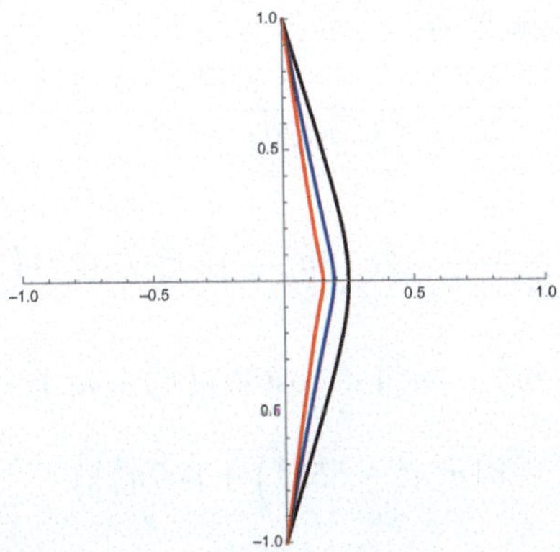

Fig. 5.3 Rays starting from the point $(0, -1)$ are bent in different amounts depending on how much the director is stretched. In this example, the director is stretched in the y direction. The black curve is the original director (5.70), the blue curve is a result of the director (5.70) being stretched by 2 in the y direction, and the red curve is a result of the director (5.70) being stretched by 4 in the y direction

Note that lensing behavior is also observed for disclinations [359], and when $k = 1$ the effective geometry is seen to be that of the cosmic string [147]. Indeed, in Fig. 5.2, we see a converging lens behavior for the $k = 1, c = \pi/2$ disclination (as seen in [359]).

It is interesting to note that a simple stretching of the $k = 1, c = \pi/2$ director, which recall has the form

$$\mathbf{d} = \left(\cos\left(\theta + \frac{\pi}{2}\right), \sin\left(\theta + \frac{\pi}{2}\right), 0\right) \tag{5.70}$$

results in different amounts of bending of the rays, see in particular Fig. 5.3.

Consider for instance the blue curve in Fig. 5.3, which results from the director (5.70) being stretched by two in the y direction. The new metric is found to be

$$g_{\text{stretched}}(r,\theta) = \begin{pmatrix} n_e^2 + \frac{2\sin^2(\theta)\cos^2(\theta)(n_e^2-n_o^2)}{3\cos(2\theta)-5} & \frac{r\sin(2\theta)(\cos(2\theta)-3)(n_e^2-n_o^2)}{6\cos(2\theta)-10} \\ \frac{r\sin(2\theta)(\cos(2\theta)-3)(n_e^2-n_o^2)}{6\cos(2\theta)-10} & \frac{r^2(-n_e^2+\cos(4\theta)(n_e^2-n_o^2)+12n_o^2\cos(2\theta)-19n_o^2)}{4(3\cos(2\theta)-5)} \end{pmatrix} \tag{5.71}$$

The nonzero Christoffel symbols can then be computed:

$$\Gamma^1_{1,1} = \frac{(n_e^2-n_o^2)^2(\cos(2\theta)-3)^2\sin(2\theta)^2}{n_e^2 n_o^2 r(3\cos(2\theta)-5)^3},$$

$$\Gamma^1_{2,1} = \frac{(\sin(4\theta)-6\sin(2\theta))(n_e^2-n_o^2)\,(-n_e^2+\cos(4\theta)(n_e^2-n_o^2)+12n_o^2\cos(2\theta)-19n_o^2)}{4n_e^2 n_o^2(3\cos(2\theta)-5)^3},$$

$$\begin{aligned}\Gamma^1_{2,2} = \frac{1}{8n_e^2 n_o^2(3\cos(2\theta)-5)^3}\Big(r\Big(3n_e^4 - 18n_o^2\cos(2\theta)\left(57n_e^2+52n_o^2\right) \\ -6n_o^2\cos(6\theta)\left(5n_e^2+4n_o^2\right)+\cos(8\theta)\left(n_e^2-n_o^2\right)^2+670n_e^2n_o^2 \\ -4\cos(4\theta)\left(n_e^4-81n_e^2n_o^2-55n_o^4\right)+867n_o^4\Big)\Big),\end{aligned}$$

$$\Gamma^2_{1,1} = \frac{(\sin(4\theta)-6\sin(2\theta))(n_e^2-n_o^2)\,(-12n_e^2\cos(2\theta)+19n_e^2+\cos(4\theta)(n_e^2-n_o^2)+n_o^2)}{4n_e^2 n_o^2 r^2(3\cos(2\theta)-5)^3},$$

$$\begin{aligned}\Gamma^2_{2,1} = \frac{1}{8n_e^2 n_o^2 r(3\cos(2\theta)-5)^3}\Big(-37n_e^4+\cos(8\theta)\left(n_e^2-n_o^2\right)^2-1466n_e^2n_o^2 \\ +6\cos(2\theta)\left(2n_e^4+323n_e^2n_o^2+2n_o^4\right)+36\cos(4\theta)\left(n_e^4-17n_e^2n_o^2+n_o^4\right) \\ -6\cos(6\theta)\left(2n_e^4-13n_e^2n_o^2+2n_o^4\right)-37n_o^4\Big)\end{aligned}$$

$$\Gamma^2_{2,2} = -\frac{(\sin(4\theta)-6\sin(2\theta))(n_e^2-n_o^2)\,(-n_e^2+\cos(4\theta)(n_e^2-n_o^2)+12n_o^2\cos(2\theta)-19n_o^2)}{4n_e^2 n_o^2(3\cos(2\theta)-5)^3} \tag{5.72}$$

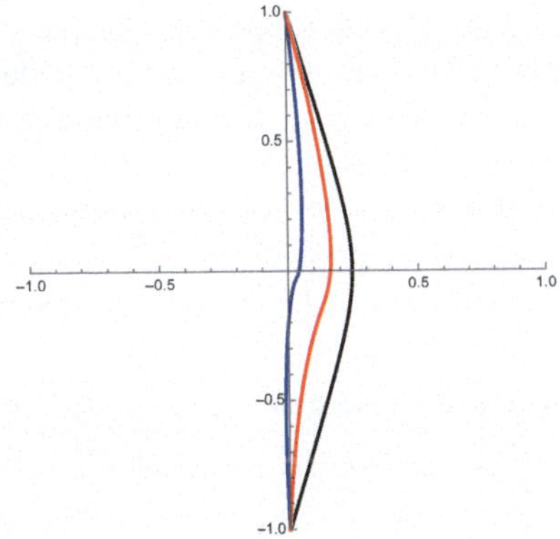

Fig. 5.4 The affect on the light ray trajectories from a shear transformation of (5.70). The black line comes from the original director, while the blue ($k = 1$) and red ($k < 1$) trajectories arise from two different horizontal shear transformations

From here, the geodesic ray equations for $r(s)$ and $\theta(s)$ can be determined explicitly, and the blue curve in Fig. 5.3 is a result of the numerical solution of the resulting system (in the figure we again took $n_e = 1.9, n_o = 1.6$ as we have been).

Since such a simple transformation of the $k = 0, c = \pi/2$ director can produce different bending of the light rays, one might wonder if other transformations will affect the converging lens phenomenon that we see. For example, one could consider a (horizontal) shear transformation; this is the 2D transformation which maps $(x, y) \mapsto (x + ky, y)$ for some scalar k. In Fig. 5.4 one can see the affect of taking different shear transformations of (5.70).

A natural continuation to this exploration is to include more standard director deformations: bend, splay, and twist.

5.6 Broken Extremals

It is sometimes useful to construct solutions to the Euler–Lagrange equation which may not be fully C^1, but rather, in some sense, are piecewise C^1. In particular, we can allow for corners; this leads us to the notion of broken extremals which frequently have concrete physical interpretations. For instance, in phase transition theory, broken extremals correspond to multiphase equilibria (see, e.g., [148] and the references therein). In the context of geometric optics, broken extremals arise naturally when there is a surface discontinuity through which rays pass, e.g., from air to glass. Thus, it is further expected that the notion of broken extremal is related to the Weierstrass–Erdmann condition seen before in (4.8). Finally, if we do not allow solutions to freely vary in every direction, discontinuous solutions may be expected, e.g., if we attempt to find the shortest distance in the plane between two points in a domain Ω which is non-convex. We will begin with some notation and discuss the relevant form of the Weierstrass–Erdmann corner condition for the so-called weak D^1 extremals.

Let D^1 denote the class of piecewise smooth functions $\phi : [a, b] \to \mathbb{R}^3$. In particular, $\phi \in D^1$ if ϕ is continuous and there exists a decomposition $a = t_0 < t_1 \cdots < t_k < t_{k+1} = b$ into finitely many intervals $I_j = [t_{j-1}, t_j]$ such that $u|_{I_j}$ is of class C^1.

A point $t = \tau_j$ is a point of discontinuity of $\dot{\mathbf{x}}$ if and only if

$$\dot{\mathbf{x}}(\tau_j^-) \neq \dot{\mathbf{x}}(\tau_j^+) \tag{5.73}$$

where $\dot{\mathbf{x}}(\tau_j^{\pm})$ denote the one-sided limits of $\dot{\mathbf{x}}$; see [139, Section 1.3].

We say that $\mathbf{x}(t)$ is a weak D^1 extremal of the functional

$$\mathcal{F}(\mathbf{x}) = \int_{t_1}^{t_2} F(\mathbf{x}(t), \dot{\mathbf{x}}(t))dt$$

if $\mathbf{x}(t)$ is of class D^1 satisfying

$$\delta\mathcal{F}(\mathbf{x}, \phi) = 0 \qquad \forall\, \phi \in C_0^\infty(I; \mathbb{R}^3)$$

which is equivalent to the existence of a constant vector λ such that

$$\nabla_{\mathbf{v}} F(\mathbf{x}, \dot{\mathbf{x}}) = \lambda + \int_{t_1}^{t_2} \nabla_{\mathbf{x}} F(\mathbf{x}, \dot{\mathbf{x}})d\tau, \qquad \forall\, t \in [t_1, t_2] \tag{5.74}$$

which is essentially an integrated form of the Euler–Lagrange equations. In fact, if $\mathbf{x}$ is a weak C^1 extremal (rather than just D^1), then $\mathbf{x}$ satisfies the Euler–Lagrange system

$$\frac{d}{dt}\nabla_{\mathbf{v}} F(\mathbf{x}, \dot{\mathbf{x}}) = \nabla_{\mathbf{x}} F(\mathbf{x}, \dot{\mathbf{x}}) \tag{5.75}$$

Indeed, if $\mathbf{x}$ is C^1, then the right side of (5.75) is of class C^1, and so $\nabla_{\mathbf{v}} F$ is also a C^1 function of time. Thus we may differentiate (5.74), which yields (5.75).

Now, the corner condition takes the following form and can be found in [139] (compare with Eq. (4.8)):

Theorem 5.6 *Let $\mathbf{x}(t)$ be a weak D^1 extremal of $\mathcal{F}$. Then, if τ is a point of discontinuity of $\dot{\mathbf{x}}$, there holds*

$$\nabla_{\mathbf{v}} F(\mathbf{x}(\tau), \dot{\mathbf{x}}(\tau^-)) = \nabla_{\mathbf{v}} F(\mathbf{x}(\tau), \dot{\mathbf{x}}(\tau^+)) \tag{5.76}$$

5.6.1 Indicatrix and Excess Function

For a parametric Lagrangian $F(\mathbf{x}, \mathbf{v})$, we introduce the indicatrix surface $I_{\mathbf{x}}$, which can help to visualize various properties of F. This idea goes back to Carathéodory [69].

Given $\mathbf{x}$, the indicatrix $I_{\mathbf{x}}$ of $\mathcal{F}$ is the set of all tangent vectors $\mathbf{v}$ satisfying $F(\mathbf{x}, \mathbf{v}) = 1$, i.e.,

$$I_{\mathbf{x}} = \left\{\mathbf{v} \in \mathbb{R}^3 : F(\mathbf{x}, \mathbf{v}) = 1\right\} \tag{5.77}$$

This definition easily extends to $\mathbb{R}^n$ as well.

Now we discuss another relevant function: the excess function $\mathcal{E}$. For line elements $\ell = (\mathbf{x}, \mathbf{v})$ and $\ell' = (\mathbf{x}, \mathbf{v}')$ with the same supporting point $\mathbf{x}$, define the excess function

$$\mathcal{E}(\mathbf{x}, \mathbf{v}, \mathbf{v}') = \mathbf{v}' \cdot \left[\nabla_{\mathbf{v}} F(\mathbf{x}, \mathbf{v}') - \nabla_{\mathbf{v}} F(\mathbf{x}, \mathbf{v})\right] \tag{5.78}$$

We note that the following homogeneity condition of $\mathcal{E}$ holds:

$$\mathcal{E}(\mathbf{x}, \lambda\mathbf{v}, \mu\mathbf{v}') = \mu\mathcal{E}(\mathbf{x}, \mathbf{v}, \mathbf{v}'), \qquad \forall\, \lambda, \mu > 0 \tag{5.79}$$

Indeed, this follows from the homogeneity condition satisfied by F, i.e., that

$$F(\mathbf{x}, \lambda\mathbf{v}) = \lambda F(\mathbf{x}, \mathbf{v}), \qquad \forall\, \lambda > 0$$

In addition to homogeneity, we assume that F is positive definite, i.e.,

$$F(\mathbf{x}, \mathbf{v}) > 0, \qquad \text{if } \mathbf{v} \neq 0$$

Convexity of the indicatrix turns out to be an important property and can be related to the excess function. Indeed, $I_{\mathbf{x}}$ is strictly convex if and only if

$$\mathcal{E}(\mathbf{x}, \mathbf{v}, \mathbf{v}') > 0, \quad \forall\, \mathbf{v}, \mathbf{v}' \in I_{\mathbf{x}} \text{ with } \mathbf{v} \neq \mathbf{v}' \tag{5.80}$$

This can be seen by writing the excess function as

$$\mathcal{E}(\mathbf{x}, \mathbf{v}) = F(\mathbf{x}, \mathbf{v}') - F(\mathbf{x}, \mathbf{v}) - (\mathbf{v}' - \mathbf{v}) \cdot \nabla_{\mathbf{v}} F(\mathbf{x}, \mathbf{v}) \tag{5.81}$$

To see this, we note that homogeneity of F implies that

$$F(\mathbf{x}, \mathbf{v}) = v_i \frac{\partial F}{\partial v_i}(\mathbf{x}, \mathbf{v}), \qquad \mathbf{v} \neq 0 \tag{5.82}$$

Indeed, let $g(\lambda) = F(\mathbf{x}, \lambda\mathbf{v}) - \lambda F(\mathbf{x}, \mathbf{v})$. Then homogeneity of F implies that $g \equiv 0$, so $g'(\lambda) = 0$. But we can compute

$$g'(\lambda) = \frac{\partial}{\partial v_i} F(\mathbf{x}, \lambda\mathbf{v}) v_i - F(\mathbf{x}, \mathbf{v})$$

Setting $\lambda = 1$ then gives (5.82). Thus we write

$$\begin{aligned} F(\mathbf{x}, \mathbf{v}') - F(\mathbf{x}, \mathbf{v}) - (\mathbf{v}' - \mathbf{v}) \cdot \nabla_{\mathbf{v}} F(\mathbf{x}, \mathbf{v}) \\ &= F(\mathbf{x}, \mathbf{v}') - \mathbf{v} \cdot \nabla_{\mathbf{v}} F(\mathbf{x}, \mathbf{v}) - (\mathbf{v}' - \mathbf{v}) \cdot \nabla_{\mathbf{v}} F(\mathbf{x}, \mathbf{v}) \\ &= F(\mathbf{x}, \mathbf{v}') - \mathbf{v}' \cdot \nabla_{\mathbf{v}} F(\mathbf{x}, \mathbf{v}) \\ &= \mathbf{v}' \cdot (\nabla_{\mathbf{v}} F(\mathbf{x}, \mathbf{v}') - \nabla_{\mathbf{v}} F(\mathbf{x}, \mathbf{v})) \end{aligned}$$

Finally, from (5.81) then it is clear to see that the indicatrix is convex provided $\mathcal{E}(\mathbf{x}, \mathbf{v}, \mathbf{v}') > 0$, for this implies that

$$F(\mathbf{x}, \mathbf{v}) > F(\mathbf{x}, \mathbf{v}) + (\mathbf{v}' - \mathbf{v}) \cdot \nabla_{\mathbf{v}} F(\mathbf{x}, \mathbf{v})$$

We collect a number of useful results relating extremals to the indicatrix:

Theorem 5.7 *Let $F(\mathbf{x}, \mathbf{v})$ be homogeneous of degree 1 in $\mathbf{v}$. Then:*

1. *If all indicatrices of F are strictly convex, then every weak D^1 extremal of $\mathcal{F}$ must also be of class C^1.*
2. *In Riemannian Geometry with line element $ds^2 = g_{ij}dx^i dx^j$, there are no broken extremals.*
3. *There are no broken extremals in isotropic media, i.e., for $F(\mathbf{x}, \mathbf{v}) = \omega(\mathbf{x})|\mathbf{v}|$. That is, every weak extremal is classical.*

Proof First, suppose that all indicatrices of F are strictly convex, and suppose $\mathbf{x}(t)$ is a weak D^1 extremal of $\mathcal{F}$ for $t_1 \leq t \leq t_2$. Let us normalize F such that $F(\mathbf{x}(t), \dot{\mathbf{x}}(t)) \equiv 1$. For any $\tau \in (t_1, t_2)$, set

$$\mathbf{x} = \mathbf{x}(\tau),\ \mathbf{v}^- = \dot{\mathbf{x}}(\tau^-),\ \mathbf{v}^+ = \dot{\mathbf{x}}(\tau^+),\ \mathbf{y}^- = \nabla_{\mathbf{v}} F(\mathbf{x}, \mathbf{v}^-),\ \mathbf{y}^+ = \nabla_{\mathbf{v}} F(\mathbf{x}, \mathbf{v}^+)$$

Then we have that $\mathbf{v}^-, \mathbf{v}^+ \in I_{\mathbf{x}}$, and the corner condition (5.76) implies that $\mathbf{y}^- = \mathbf{y}^+$. Now, define the so-called figuatrix of F at $\mathbf{x}$ by

$$f_{\mathbf{x}} = \left\{ y \in T^*_{\mathbf{x}}\mathbb{R}^n : y = \nabla_{\mathbf{v}} F(\mathbf{x}, \mathbf{v}),\ \mathbf{v} \in I_{\mathbf{x}} \right\} \tag{5.83}$$

Here, the space $T^*_{\mathbf{x}}\mathbb{R}^n$ denotes the cotangent space to $\mathbb{R}^n$ at $\mathbf{x}$. An element of this cotangent space is a linear functional on the tangent space $T_{\mathbf{x}}\mathbb{R}^n$. Now, we see that $\mathbf{v}^- = \mathbf{v}^+$ if $\nabla_{\mathbf{v}} F(\mathbf{x}, \cdot) : I_{\mathbf{x}} \to f_{\mathbf{x}}$ is one-to-one. But this is true since $\mathcal{E} > 0$. Indeed, suppose $\mathbf{v}^- \neq \mathbf{v}^+$. Then since $\mathcal{E} > 0$, we have that

$$\nabla_{\mathbf{v}} F(\mathbf{x}, \mathbf{v}^-) \cdot (\mathbf{v}^+ - \mathbf{v}^-) < F(\mathbf{x}, \mathbf{v}^+) - F(\mathbf{x}, \mathbf{v}^-) \tag{5.84}$$

and similarly

$$\nabla_{\mathbf{v}} F(\mathbf{x}, \mathbf{v}^+) \cdot (\mathbf{v}^- - \mathbf{v}^+) < F(\mathbf{x}, \mathbf{v}^-) - F(\mathbf{x}, \mathbf{v}^+)$$

Multiplying this last equation by -1 and replacing $\nabla_{\mathbf{v}} F(\mathbf{x}, \mathbf{v}^+)$ by $\nabla_{\mathbf{v}} F(\mathbf{x}, \mathbf{v}^-)$ yield

$$\nabla_{\mathbf{v}} F(\mathbf{x}, \mathbf{v}^-) \cdot (\mathbf{v}^+ - \mathbf{v}^-) > F(\mathbf{x}, \mathbf{v}^+) - F(\mathbf{x}, \mathbf{v}^-)$$

which contradicts (5.84); thus $\mathbf{v}^- = \mathbf{v}^+$, and so every weak D^1 extremal is also classical, i.e., C^1.

Next we prove there are no broken extremals in Riemannian geometry. Consider the line element $ds^2 = g_{ij} dx^i dx^j$ and the associated Lagrangian

$$F(\mathbf{x}, \mathbf{v}) = \sqrt{g_{ij} v^i v^j}$$

Then it is straightforward to show that such F satisfies (5.80), and so from the previous item we have no broken extremals. Finally, the third item follows immediately from the first since the Lagrangian

$$F(\mathbf{x}, \mathbf{v}) = \omega(x)|\mathbf{v}|, \qquad \omega(x) > 0$$

certainly satisfies (5.80).

Remark 5.10 If $\mathbf{x}(t)$ is a weak D^1 extremal with $F(\mathbf{x}, \dot{\mathbf{x}}) = 1$ such that

$$\mathcal{E}(\mathbf{x}, \dot{\mathbf{x}}, \mathbf{w}) \geq 0 \qquad \forall\, \mathbf{w} \in I_{\mathbf{x}} \tag{5.85}$$

then $\mathbf{x}(t)$ can be a discontinuous extremal. Indeed, let $\mathbf{y}^- = \nabla_{\mathbf{v}} F(\mathbf{x}, \mathbf{v}^-)$ and $\mathbf{y}^+ = \nabla_{\mathbf{v}} F(\mathbf{x}, \mathbf{v}^+)$. If $\mathbf{v}^- \neq \mathbf{v}^+$, the corner condition that $\mathbf{y}^- = \mathbf{y}^+$. Now, we can write the excess function as

$$\mathcal{E}(\mathbf{x}, \mathbf{v}^-, \mathbf{v}^+) = 1 - \mathbf{y}^- \cdot \mathbf{v}^+ = (\mathbf{y}^+ - \mathbf{y}^-) \cdot \mathbf{v}^+$$

But since $\mathbf{y}^- = \mathbf{y}^+$, we see that $\mathcal{E}(\mathbf{x}, \mathbf{v}^-, \mathbf{v}^+) = 0$. Thus we conclude that the indicatrix $I_{\mathbf{x}}$ has a double tangent plane touching $I_{\mathbf{x}}$ at $\mathbf{v}^-$ and $\mathbf{v}^+$.

Finally, if for fixed $\mathbf{x}$, $F(\mathbf{x}, \mathbf{v})$ is elliptic, then $I_{\mathbf{x}}$ is strictly convex, so we cannot expect discontinuous extremals. Recall from Chap. 5.1 that a line element $(\mathbf{x}, \mathbf{v})$ is elliptic if

$$Q(\mathbf{x}, \mathbf{v}) := \frac{1}{2} F^2(\mathbf{x}, \mathbf{v})$$

satisfies

$$\frac{\partial^2 Q}{\partial v_i \partial v_k}\xi_i\xi_k > 0 \qquad \forall\, \xi \in \mathbb{R}^n,\ \xi \neq 0$$

In fact, we have the following, which can be found in [139, Chapter 8, Section 2.3]:

Theorem 5.8 *Suppose F is positive definite and satisfies the usual homogeneity condition. For a line element $\ell = (\mathbf{x}, \mathbf{v})$, the following are equivalent:*

1. *$\nabla^2_{\mathbf{v}} Q$ is positive definite, i.e., ℓ is elliptic.*
2. *$\nabla^2_{\mathbf{v}} F$ is positive definite, i.e., ℓ satisfies the parametric Legendre condition for F:*

$$\frac{\partial^2 F}{\partial v_i \partial v_k}\xi_i\xi_k > 0 \qquad \forall\, \xi \in \mathbb{R}^n \text{ with } \xi \neq 0 \text{ and } \xi \cdot \mathbf{v} = 0 \tag{5.86}$$

Finally, if F is elliptic for all directions $\mathbf{v}$, then we can write the excess function as

$$\mathcal{E}(\mathbf{x}, \mathbf{v}, \mathbf{w}) = F(\mathbf{x}, \mathbf{w}) \cdot [1 - \cos(\alpha(\mathbf{v}, \mathbf{w}))], \qquad \mathbf{v} \in I_{\mathbf{x}} \tag{5.87}$$

where α is the angle between $\mathbf{v}$ and $\mathbf{w}$ at $\mathbf{x}$:

$$\cos(\alpha) = \frac{F(\mathbf{x}, \mathbf{v})\nabla_{\mathbf{v}} F(\mathbf{x}, \mathbf{v}) \cdot \mathbf{w}}{F(\mathbf{x}, \mathbf{w})}, \qquad \mathbf{v} \in I_{\mathbf{x}} \tag{5.88}$$

Consider as usual $F(\mathbf{x}, \mathbf{v}) = n_{\text{eff}}(\mathbf{x}, \mathbf{v}) = \sqrt{n_e^2|\mathbf{v}|^2 + (n_o^2 - n_e^2)(\mathbf{v} \cdot \mathbf{d}(\mathbf{x}))^2}$. Then

$$Q(\mathbf{x}, \mathbf{v}) = \frac{1}{2}\left[n_e^2|\mathbf{v}|^2 + (n_o^2 - n_e^2)(\mathbf{v} \cdot \mathbf{d}(\mathbf{x}))^2\right]$$

We compute

$$\frac{\partial Q}{\partial v_k} = (n_o^2 - n_e^2)(\mathbf{v} \cdot \mathbf{d}(\mathbf{x}))d_k + n_e^2 v_k$$

and

$$\frac{\partial^2 Q}{\partial v_i \partial v_k} = (n_o^2 - n_e^2)d_i d_k + n_e^2 \delta_{ik}$$

Then we have

$$\frac{\partial^2 Q}{\partial v_i \partial v_k}\xi_i\xi_k = (\mathbf{d} \cdot \xi)^2(n_o^2 - n_e^2) + n_e^2|\xi|^2 > 0, \quad \forall\, \xi \in \mathbb{R}^n,\ n = 2, 3;\ \xi \neq 0$$

Thus we can define the angle between $\mathbf{v}$ and $\mathbf{w}$, for $\mathbf{v} \in I_{\mathbf{x}}$, by

$$\cos(\alpha(\mathbf{v}, \mathbf{w})) = \frac{n_e^2(\mathbf{v} \cdot \mathbf{w}) + (n_o^2 - n_e^2)(\mathbf{v} \cdot \mathbf{d}(\mathbf{x}))(\mathbf{w} \cdot \mathbf{d}(\mathbf{x}))}{\sqrt{n_e^2|\mathbf{w}|^2 + (n_o^2 - n_e^2)(\mathbf{w} \cdot \mathbf{d}(\mathbf{x}))^2}}$$

Thus in light of (5.87) we may write the excess function as

$$\begin{aligned}\mathcal{E}(\mathbf{x}, \mathbf{v}, \mathbf{w}) = &\sqrt{n_o^2|\mathbf{w}|^2 + (n_o^2 - n_e^2)(\mathbf{w} \cdot \mathbf{d}(\mathbf{x}))^2} \\ &\cdot \left[1 - \frac{n_e^2(\mathbf{v} \cdot \mathbf{w}) + (n_o^2 - n_e^2)(\mathbf{v} \cdot \mathbf{d}(\mathbf{x}))(\mathbf{w} \cdot \mathbf{d}(\mathbf{x}))}{\sqrt{n_e^2|\mathbf{w}|^2 + (n_o^2 - n_e^2)(\mathbf{w} \cdot \mathbf{d}(\mathbf{x}))^2}}\right]\end{aligned}$$

if $\mathbf{v} \in I_{\mathbf{x}}$. Thus we have $\mathcal{E}(\mathbf{x}, \mathbf{v}, \mathbf{w}) \geq 0$ if $\mathbf{v} \in I_{\mathbf{x}}$. In the next section we will explore the possibility of obtaining discontinuous extremals in 2D.

5.6.2 *Liquid Crystal Application*

Let us work in 2D, say $\mathbf{x} = (x, y)$ and $\mathbf{v} = (\dot{x}, \dot{y})$. Then we consider the indicatrix $F(x_0, y_0, \dot{x}, \dot{y}) = 1$ in polar coordinates:

$$\dot{x} = r\cos(\theta), \qquad \dot{y} = r\sin(\theta)$$

so that

$$F(x_0, y_0, r\cos(\theta), r\sin(\theta)) = 1$$

which by homogeneity gives the curve in polar form

$$r = \frac{1}{F(x_0, y_0, \cos(\theta), \sin(\theta))} \tag{5.89}$$

In this setting, a corner corresponds to a double tangency in θ. Indeed, assuming merely (5.85), let $x(\tau)$ be a corner with one-sided tangent vectors $v^{\pm} = \dot{\mathbf{x}}(\tau^{\pm})$. The corner condition (5.76) implies

$$\mathcal{E}(\mathbf{x}, \mathbf{v}^-, \mathbf{v}^+) = 0$$

which means the indicatrix has a double tangent touching $I_{\mathbf{x}}$ at $\mathbf{v}^-$ and $\mathbf{v}^+$.

In polar coordinates, assume we have a corner at (x_0, y_0) and the angle of the incoming extremal is θ_1 and the outgoing extremal is θ_2. Write $\mathbf{v}^- = (\dot{x_0}, \dot{y_0})$ and similar for $\mathbf{v}^+$. Then, the corner condition (5.76) implies that, using that $\mathbf{v}^-$ is on the indicatrix so that $F(\mathbf{x}, \mathbf{v}^+) = 1$, together with the fact that

$$\mathbf{v}^- \cdot \nabla_{\mathbf{v}} F(\mathbf{x}, \mathbf{v}^-) = F(\mathbf{x}, \mathbf{v}^-)$$

we obtain

$$\begin{aligned} \frac{\partial F}{\partial \dot{x}}(x_0, y_0, \cos(\theta_1), \sin(\theta_1)) &= \frac{\partial F}{\partial \dot{x}}(x_0, y_0, \cos(\theta_2), \sin(\theta_2)), \\ \frac{\partial F}{\partial \dot{y}}(x_0, y_0, \cos(\theta_1), \sin(\theta_1)) &= \frac{\partial F}{\partial \dot{y}}(x_0, y_0, \cos(\theta_2), \sin(\theta_2)) \end{aligned} \tag{5.90}$$

Since we can write the equation of the tangent line to the indicatrix as

$$\dot{x}\frac{\partial F}{\partial \dot{x}}(x_0, y_0, \cos(\theta_0), \sin(\theta_0)) + \dot{y}\frac{\partial F}{\partial \dot{y}}(x_0, y_0, \cos(\theta_0), \sin(\theta_0)) = 1 \tag{5.91}$$

we see that points on the indicatrix corresponding to angles θ_1, θ_2 have the same tangent line, and so a corner appears as a line of double tangency.

Now we seek the polar form of the indicatrix. Assuming the effective refractive index given by (3.11), we again write $F(\mathbf{x}, \mathbf{v}) = n_{\text{eff}}(\mathbf{x}, \mathbf{v}) = \sqrt{n_e^2|\mathbf{v}|^2 + (n_o^2 - n_e^2)(\mathbf{v} \cdot \mathbf{d})^2}$. Then in polar coordinates we have

$$\begin{aligned} F &= (x_0, y_0, \cos(\theta), \sin(\theta)) \\ &= \sqrt{n_e^2 + (n_o^2 - n_e^2)\left(d_1^2 \cos^2(\theta) + 2d_1 d_2 \cos(\theta)\sin(\theta) + d_2^2 \sin^2(\theta)\right)} \end{aligned} \tag{5.92}$$

where each $d_j = d_j(x_0, y_0)$. Thus the indicatrix takes the form

$$r = \frac{1}{\sqrt{n_e^2 + (n_o^2 - n_e^2)\left(d_1^2 \cos^2(\theta) + 2d_1 d_2 \cos(\theta)\sin(\theta) + d_2^2 \sin^2(\theta)\right)}} \tag{5.93}$$

To find corners, we find points of double tangency in θ of (5.93). Thus we solve

$$r = \frac{a}{\cos(\theta)} = \frac{1}{F(x_0, y_0, \cos(\theta), \sin(\theta))} \tag{5.94}$$

for θ and determine which value of a provides a double root. Now, writing out (5.94) after squaring yields

$$a^2\left[n_e^2 + (n_o^2 - n_e^2)\left(d_1^2 \cos^2(\theta) + 2d_1 d_2 \cos(\theta)\sin(\theta) + d_2^2 \sin^2(\theta)\right)\right] = \cos^2(\theta)$$

Rearranging terms, we see that this equation becomes

$$\cos^2(\theta)\left[a^2(n_o^2-n_e^2)(d_1^2-d_2^2)-1\right]+a^2(n_o^2-n_e^2)$$
$$\times\left(2d_1d_2\cos(\theta)\sin(\theta)+d_2^2\right)+a^2n_e^2=0$$

which can be written in the form

$$A\cos^2(\theta)+B\cos(\theta)\sin(\theta)+C=0 \tag{5.95}$$

where

$$A=a^2(n_o^2-n_e^2)(d_1^2-d_2^2)-1,\ B=2a^2(n_o^2-n_e^2)d_1d_2,\ C=a^2n_e^2+a^2(n_o^2-n_e^2)d_2^2$$

The solution of (5.95) is given by

$$\tan(\theta)=\frac{-B\pm\sqrt{B^2-4(A+C)C}}{2C} \tag{5.96}$$

which has a double root precisely when

$$B^2-4(A+C)C=0$$

Solving this yields the following (nonzero) solutions for a (recalling that $\mathbf{d}$ is a unit vector)

$$a=\pm\frac{\sqrt{d_1^2n_e^2+d_2^2n_o^2}}{n_en_o} \tag{5.97}$$

This yields

$$\tan(\theta)=\frac{d_1(x_0,y_0)d_2(x_0,y_0)(n_e^2-n_o^2)}{d_1(x_0,y_0)^2n_e^2+d_2(x_0,y_0)^2n_o^2} \tag{5.98}$$

Letting

$$d_1(x,y)=\cos(\alpha(x,y)),\qquad d_2(x,y)=\sin(\alpha(x,y))$$

then (5.98) implies

$$\tan(\theta)=\frac{(n_e^2-n_o^2)\sin(\alpha(x_0,y_0))\cos(\alpha(x_0,y_0))}{n_e^2\cos^2(\alpha(x_0,y_0))+n_o^2\sin^2(\alpha(x_0,y_0))} \tag{5.99}$$

Denote by θ_1,θ_2 the solutions of (5.98). Note that since there will only be one solution within an interval of length π, generally we have $\theta_2=\theta_1\pm\pi$.

Then, a corner is possible at each point (x_0, y_0) in the (x, y) plane if the extremal goes into the corner at an angle θ_1 relative to the x-axis and leaves the corner at an angle θ_2 (or vice versa). This means that at such a corner, the extremal is either "perfectly transmitted" through the corner or "perfectly reflected" back.

Remark 5.11 Note that if $n_o = n_e = n$, then the indicatrix (5.93) takes the form

$$r = \frac{1}{n}$$

and the right hand side is independent of θ, so clearly no double tangency in θ can occur, i.e., no corners exist.

Note also that we lost the possible solution $\theta = \pi/2$ in solving (5.95). Supposing $\theta = \pi/2$, this forces $C = 0$. Thus, $\theta = \pi/2$ could be a solution if

$$d_2^2 = \frac{n_e^2}{n_e^2 - n_o^2}$$

But this does not correspond to a point of double tangency in θ.

As an example, consider the hedgehog director in 2D:

$$\mathbf{d}(x, y) = \frac{1}{\sqrt{x^2 + y^2}}(x, y) \tag{5.100}$$

Then with this director (5.98) becomes

$$\tan(\theta) = \frac{(n_e^2 - n_o^2)x_0 y_0}{n_e^2 x_0^2 + n_o^2 y_0^2}$$

Suppose we wish the rays to reflect back with angle $\theta = 0$. This means that $\tan(\theta) = 0$, and so (5.99) implies that

$$\sin(\alpha(x_0, y_0)) \cos(\alpha(x_0, y_0)) = 0 \tag{5.101}$$

This gives a condition on the director, and the allowable points for which corners could exist. Namely, the director should satisfy (5.101) at those points (x_0, y_0) where one looks for a corner. In particular, for this case we see that in the interval $[0, \pi]$, we need one of

$$\alpha(x_0, y_0) = 0, \frac{\pi}{2}, \pi$$

to hold. Suppose we wish to construct a parabolic "mirror," say $x = y^2$. Thus, along the mirror the director should satisfy $\alpha(y_0^2, y_0) = 0$. For instance, one can take $\alpha(x, y) = x - y^2$ and this will be satisfied. Thus the director in this case takes the form

$$d(x, y) = \begin{pmatrix} \cos(x - y^2) \\ \sin(x - y^2) \end{pmatrix} \tag{5.102}$$

Chapter 6
Aberrations and a Lens Design Problem

The concept of perfect imaging is fundamental in geometrical optics due to the inherent limitations of various optical instruments. A typical goal is to minimize the optical aberrations of the device (the precise definition of aberrations will be given later in this chapter). Informally, devices which can eliminate aberrations altogether are called absolute instruments [415].

To give a more rigorous definition of absolute instrument, we need the notion of stigmatic imaging, for which we appeal to the classical definitions in [48], Section 4.2. Given two points $A, B \in \mathbb{R}^3$, we say that B is a stigmatic image of A if an infinity of rays emerging from A passes through B. We may imagine a point source is located at A. We may then define:

Definition 6.1 An absolute instrument is a device which stigmatically images a region $\Omega \subset \mathbb{R}^3$.

As we will see, finding absolute instruments is rather challenging. We begin by reviewing optical aberrations.

6.1 Aberrations

Let us start from the classical Snell's Law in isotropic media:

$$n_1 \sin(\theta_1) = n_2 \sin(\theta_2)$$

and the corresponding Fermat principle:

$$\delta \int_A^B n(s)ds = 0$$

E. Stachura, *Mathematical Methods in Liquid Crystal Optics and Lens Design*, Springer Tracts in Modern Physics 294,
https://doi.org/10.1007/978-3-031-46614-4_6

In an anisotropic, inhomogeneous system, it is not practical to use this Snell's Law (at least, alone) to trace rays through an optical system. Indeed, we have seen that rays typically will not trace out linear trajectories! A common technique to find a focal location (i.e., the location where rays focus) is to use the *paraxial approximation*, sometimes known as the small-angle approximation. In such an approximation, it is assumed that angles are small with respect to the optical axis in a system with certain symmetry:

$$\sin(\theta) \approx \theta, \quad \cos(\theta) \approx 1, \quad \tan(\theta) \approx \theta \tag{6.1}$$

This is known as **first order geometric optics**. We can also talk about higher order geometric optics, and mathematically we can think about this via the Taylor expansion of $\sin(\theta)$:

$$\sin(\theta) = \theta - \frac{\theta^3}{3!} + \frac{\theta^5}{5!} - \dots$$

Thus first order optics corresponds to the first term in the Taylor expansion, third order optics corresponds to the second term, etc. In first order optics, imaging in optical systems is ideal: All lenses image perfectly. When third order optics is considered, imaging is no longer perfect; *aberrations* exist. Aberrations can be seen as deviations from idealized imaging. It turns out there are various types of aberrations, which we will discuss below.

Now, to quantify the effects of aberrations, we will discuss an aberration function. The idea behind this function is the following: In ideal systems (first order optics), spherical wavefronts should be transformed into another spherical wavefront after passing through the optical system. But, once aberrations are considered, the transformed wavefront is no longer spherical. Thus, to measure the effect of aberrations (in some sense), typically one considers a reference sphere and defines the wavefront aberration function as the optical path difference between the aberrated wavefront and the spherical reference wave [48]. When the aberration function is expanded into a power series, various terms correspond to different aberrations.

There are actually two different types of aberrations: wave aberrations and ray aberrations, and they are related to each other. We have described previously the wave aberrations. Ray aberrations, on the other hand, describe the deviation between aberrated rays and the ideal rays as a transverse distance measured in the image plane [349]. Typically these two aberrations are related by the formulas [442]

$$\frac{\partial W}{\partial x} \approx -\frac{\epsilon_x}{r}, \qquad \frac{\partial W}{\partial y} \approx -\frac{\epsilon_y}{r} \tag{6.2}$$

where $W = W(x, y)$ is the wave aberration function, $\epsilon_{x,y}$ represent the ray aberrations in the x and y directions, and r is the radius of the ideal wavefront/reference sphere. The approximation (6.2) is valid for small numerical apertures[1] and small aberrations. Exact relations between the two aberrations have been obtained [290, 347] as well. Indeed, in [347] it was shown that the following exact relations hold:

$$\frac{\partial W}{\partial x} = -\frac{\epsilon_x}{r - W}, \qquad \frac{\partial W}{\partial y} = -\frac{\epsilon_y}{r - W} \tag{6.3}$$

In the case of a system where $W \ll r$, neglecting W in the denominator above recovers the approximation (6.2).

By interpreting wavefronts as an envelope of spheres à la Huygens (see Chap. 9.2.5 below and e.g. Section 3.3. in [48]), a more rigorous definition of W has been given in [349] for the optical path difference W as a phase difference along the aberrated ray. This interpretation, sometimes known as Huygens' principle, asserts that each element of a wavefront can be regarded as the center of a secondary disturbance which gives rise to spherical wavelets and the position of the wavefront at any time is the envelope of all such wavelets.

Let us make this a bit more clear for the context of liquid crystals we are interested in. Recall that the effective index we consider is related to the direction of the time averaged Poynting vector **S**. Thus, our starting point is the definition of light rays; we said that light rays are integral curves of $\langle \mathbf{S} \rangle$. We saw in Chap. 2 that $\langle \mathbf{S} \rangle$ is parallel to $\nabla_{\mathbf{p}}\mathcal{H}(\mathbf{r}, \mathbf{p})$, where $\mathcal{H}$ defines the optical indicatrix. Thus we can understand light rays having trajectories of a vector which is orthogonal to $\mathcal{H} = 0$, i.e., in the direction of $\nabla_{\mathbf{p}}\mathcal{H}(\mathbf{r}, \mathbf{p})$. Recall the optical indicatrix:

$$\mathcal{H}(\mathbf{r}, \mathbf{p}) = \left(|\mathbf{p}|^2 - n_o^2\right)\left(n_o^2|\mathbf{p}|^2 + (n_e^2 - n_o^2)(\mathbf{p} \cdot \mathbf{d}(\mathbf{r}))^2 - n_e^2 n_o^2\right)$$

To define the wave aberration function W, we let A^* be a point on the surface $\mathcal{H} = 0$ and B^* a point on a reference sphere such that rays pass through A^* and B^*. See Fig. 6.1.

Then we define W to be the line integral along the trajectory from A^* to B^*:

$$W = \int_{A^*}^{B^*} n_{\text{eff}} ds = \int_0^1 n_{\text{eff}}(\mathbf{x}(s))ds, \qquad \mathbf{x}(0) = A^*,\ \mathbf{x}(1) = B^* \tag{6.4}$$

That is, W is the distance (as measured in optical length, along the light ray) between the actual "wavefront" and the reference sphere.

Notice that for ordinary rays, the wavefront $\mathcal{H}_o = 0$ is a sphere; thus, ordinary rays behave as if there is an ideal lens . However, the surface $\mathcal{H}_e = 0$ is elliptical,

[1] The numerical aperture of an optical system is a number which characterizes the angles through which the system can emit light. In microscopy and many other areas of classical optics, the numerical aperture is defined by $n \sin(\theta)$, where n is the refractive index of the medium and θ is half the angle of the maximum cone of light which can exit or enter a lens.

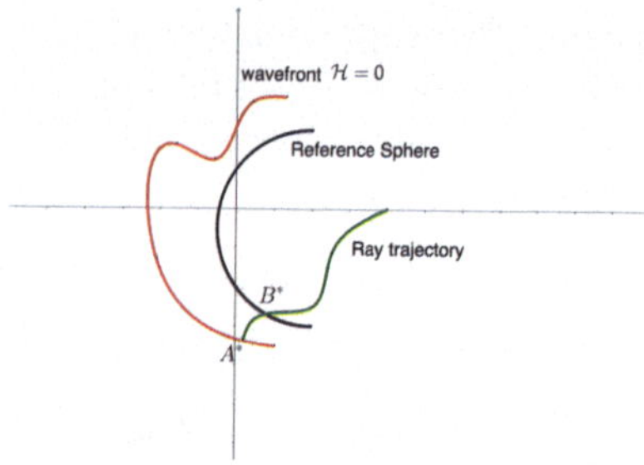

Fig. 6.1 Illustration of the definition of W as the optical path difference between a reference sphere and the actual wavefront

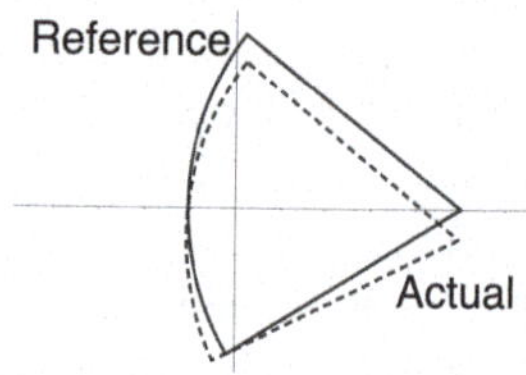

Fig. 6.2 An example of tilt

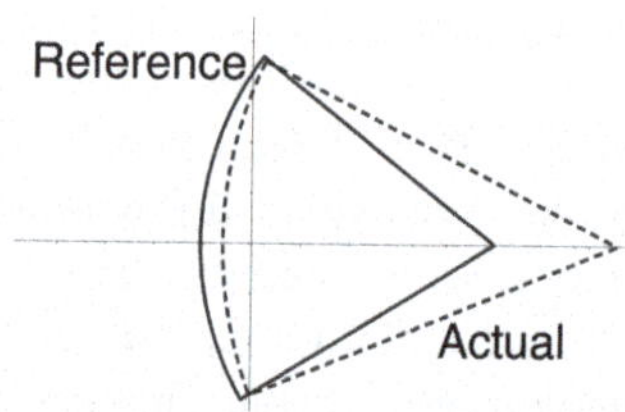

Fig. 6.3 An example of defocus

and the shape depends on whether $n_o > n_e$ or vice versa (i.e., it depends on the birefringence).

Next let us discuss the different kinds of aberrations which can exist, in both first and third order optics. Note that the first order aberrations correspond to making the paraxial approximation, and third order aberrations are sometimes called primary aberrations [390] or Seidel aberrations. Moreover, since we will discuss such aberrations with respect to the wave aberration function W, we are considering *wave* aberrations.

6.1.1 First Order Aberrations

One first order aberration is known as *tilt* and is caused by the actual wavefront being tilted relative to the reference wavefront. *Defocus* is caused by an incorrect reference image and results in an image in the wrong image plane. See Figs. 6.2 and 6.3; the actual (aberrated) wavefront is denoted with the dashed curve, whereas the reference wavefront is the solid line. Finally, *piston* represents a time delay/advance in the arrival of the wavefront as it propagates from the object.

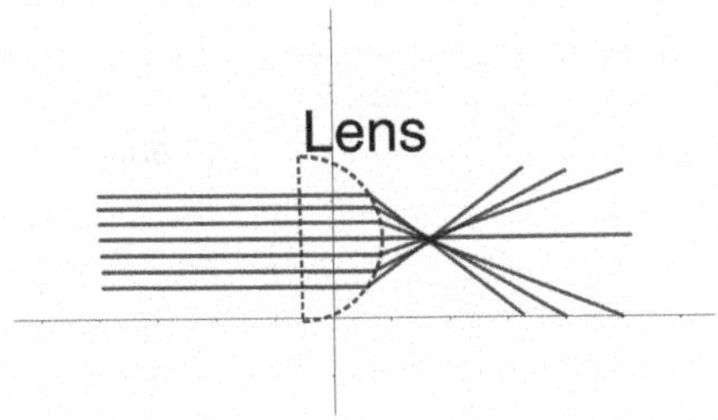

Fig. 6.4 Spherical aberration: The rays do not focus at the same point after exiting the lens

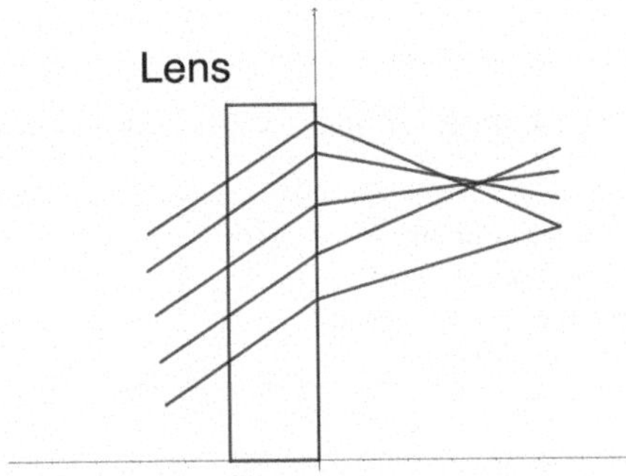

Fig. 6.5 Coma: The rays image at different heights depending on where they pass through the lens, especially when entering the lens through an angle

When considering monochromatic light, typically these terms are ignored since they have limited effect. However, for polychromatic light these effects play a role and give rise to the so-called chromatic aberrations [442].

6.1.2 Third Order Aberrations

The first of the third order aberrations is *spherical aberration* and occurs with all spherical objects. The image will appear blurred, and the light rays from the edge of the lens focus at a different point from rays from the center of the lens; see Fig. 6.4.

Next is *coma* which occurs when magnification changes with respect to location on the image. In particular, rays incident on the edge of a lens with coma might be imaged at a different height than those passing through the center (Fig. 6.5).

Astigmatism causes two focal points (one horizontal and the other vertical); the image of a point turns into two separate rays and a blur in between. A nice illustration of astigmatism, which we have not attempted to reproduce, can be found in Figure 9.8 in [358]. *Field curvature* results in a perfect image, but it lies on a curved image plane rather than a flat plane; see Fig. 6.6.

If there is also astigmatism present, then this surface is a paraboloid. If no astigmatism is present, then the surface is known as the Petzval surface. *Distortion* results in a quadratic magnification error; points on the image are too close or too far from the center. Magnification is a lens property which describes how an large (or small) an object can be reproduced on the image plane. Simply put, if an object

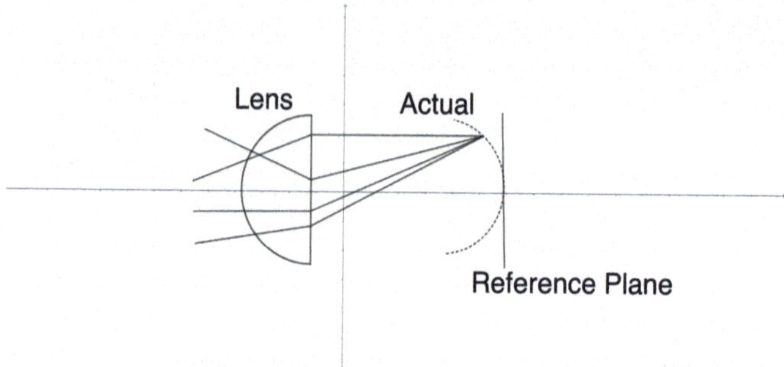

Fig. 6.6 Field curvature results in a perfect image on a curved image plane rather than a flat plane

Fig. 6.7 Chromatic aberration: Different wavelengths of light are focused differently

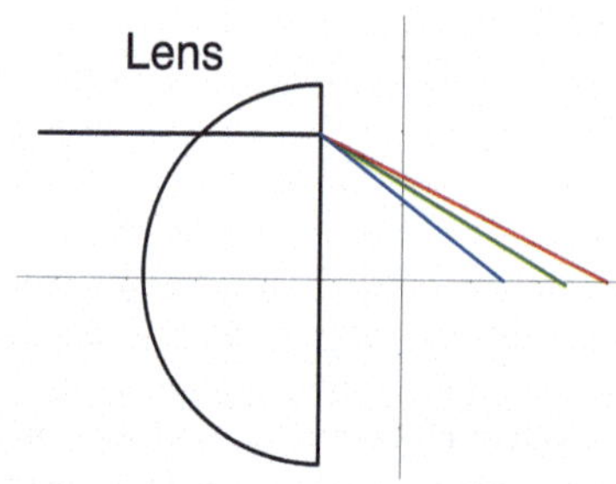

of length x forms an image of length y in the image, then the magnification is just y/x. For instance, *elliptic* distortion of images can be represented through a scaling transform of the image:

$$\begin{pmatrix} x' \\ y' \end{pmatrix} = S_{\text{avg.}} \mathcal{R} \begin{pmatrix} \frac{1}{1+\Delta/2} & 0 \\ 0 & 1+\Delta/2 \end{pmatrix} \mathcal{R}^T \begin{pmatrix} x \\ y \end{pmatrix}, \qquad \mathcal{R} = \begin{pmatrix} \cos(\theta) & -\sin(\theta) \\ \sin(\theta) & \cos(\theta) \end{pmatrix}$$

where $\mathcal{R}$ is a rotation matrix, $S_{\text{avg.}}$ is the average magnification, θ is the angle of the minor axis, and Δ is the amount of distortion. In the case of this elliptic distortion, the parameter $S_{\text{avg.}}$ is provided by the user based on the calibrated lens instrument. In general, the amount of distortion increases as image size increases. Distortion represents a rigid tilt of the wavefront; see for instance Figure 8.16 in [358] for a sequence of images illustrating the effects of distortion on the image of a square grid.

Finally, *chromatic aberration* occurs due to the wavelength frequency of the refractive index. Shorter wavelengths are refracted stronger due to the higher index of refraction; see Fig. 6.7.

Recent work in [167] has provided a mathematical approach to study chromatic aberration in the so-called metalenses. The key there is the development of a generalized Snell's Law, similar to that from [164]. Indeed, if a light wave with frequency ω emitted from a point A in some medium strikes an interface surface, separating two media A and B, at a point P, it is then refracted into a point B in a second medium. In the context of metasurfaces, one has a phase discontinuity ϕ (mathematically, this is just a C^1 function defined on a neighborhood of the surface). If **a** denotes the incident ray direction and **m** the refracted ray direction, then from Fermat's principle and using Lagrange multipliers, one obtains the generalized Snell's Law

$$n_1\mathbf{a} - n_2\mathbf{m} = \mu\nu + \nabla\left(\frac{\lambda_0\phi}{2\pi}\right), \qquad \mu \in \mathbb{R},$$

where ν is a unit normal vector; see equation (2.4) in [167]. Above, n_1 denotes the refractive index of the first medium, n_2 the refractive index of the second medium, and $\lambda_0 = 2\pi c/\omega$, where c denotes the speed of light in vacuum. The authors then go on to analyze chromatic dispersion in such metalenses. This is one of the only mathematically rigorous results available for chromatic aberration, so this area is quite rich in open problems.

6.1.3 *Pupil Coordinates*

Now, as a particular example, we work in an appropriate coordinate system to study the wave aberration function W. For this we follow [442]. All that follows below comes from taking a particular parameterization of the unit sphere.

As mentioned above, an ideal lens would transform a spherical wavefront into another spherical wave. To develop the appropriate coordinate system, we need to understand the notion of center of curvature. To this end, given a curve $y = f(x)$, suppose we draw a circle which touches the curve at one point P and has the same curvature. The center of this circle is called the center of curvature (at P).

This becomes particularly simple if the curve is part of a spherical wavefront (which it will be). Now, for an ideal lens , suppose the transformed spherical wavefront has radius of curvature R (this is the reciprocal of its curvature [109]), and its center of curvature lies at the origin of the coordinate system (x_0, y_0, z_0). Let (x, y, z) denote the coordinate system whose origin is precisely the center of where the spherical wave exits the lens (more precisely, this is the so-called exit pupil [442]); see Figs. 6.8 and 6.9.

With this setup, we see that $x = x_0$, $y = y_0$ and $z = z_0 - R$; thus, the equation of the transformed wavefront in the (x, y, z) coordinate system is

$$x^2 + y^2 + (z - R)^2 = R^2 \tag{6.5}$$

If $x, y \ll R$ (i.e., the size of the so-called exit pupil is small compared to the radius of curvature of the spherical wave), and z is small enough so that $z^2 \approx 0$, then the equation of the spherical wavefront can be approximated by a parabola

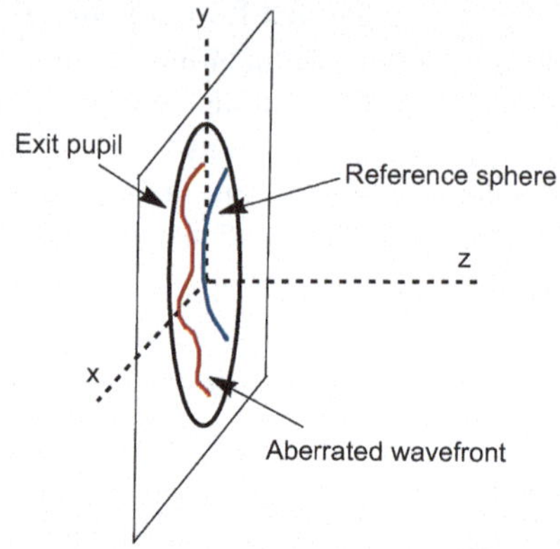

Fig. 6.8 Exit pupil coordinate system

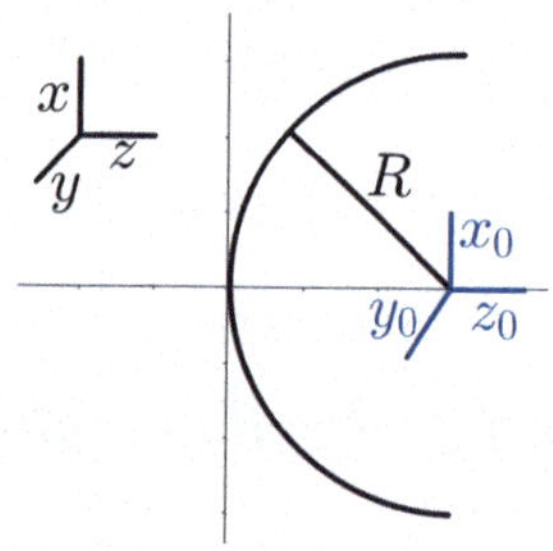

Fig. 6.9 Coordinate system

$$z = \frac{x^2 + y^2}{2R} \tag{6.6}$$

Thus an optical path difference of

$$W_0(x, y) = \frac{x^2 + y^2}{R^2} \tag{6.7}$$

represents a spherical wavefront which converges to a point at distance R from the exit of the lens . Of course, for the systems we are interested in, aberrations will be present, and the exiting wavefront is not necessarily spherical.

Let us now consider the series expansion of W in terms of the pupil coordinates (x, y) and the coordinates of the image point (x_0, y_0); this is justified in [434]. Assuming rotational symmetry of the lens system, we may expand

$$\begin{aligned} W(x, y, x_0) &= W(x^2 + y^2, xx_0, x_0^2) \\ &= a_1(x^2 + y^2) + a_2xx_0 + a_3x_0^2 + b_1(x^2 + y^2)^2 \\ &\quad + b_2xx_0(x^2 + y^2) + b_3x^2x_0^2 \\ &\quad + b_4x_o^2(x^2 + y^2) + b_5xx_0^3 + b_6x_0^4 + \text{H. O. T.} \end{aligned} \tag{6.8}$$

Table 6.1 List of wavefront coefficients and corresponding aberrations

Wavefront coefficient	Name
W_{200}	Piston
W_{111}	Tilt
W_{020}	Focus
W_{040}	Spherical
W_{131}	Coma
W_{222}	Astigmatism
W_{220}	Field curvature
W_{311}	Distortion

where H. O. T. denotes higher order terms. The reason for expansion in the term xx_0 is that the function W should not change under a rigid rotation of the x_0, y_0, z_0 axes about the z-axis. As such W will be a function of combinations of xx_0 and yy_0 which are rotationally invariant. By choosing the coordinate system such that image point lies in the $x - z$ plane, y can be taken to be zero. For this reason only xx_0 terms appear in the expansion of W.

The first five terms with $b_j^{\prime}s$ in the expansion of W are the important ones as they correspond to the Seidel aberrations. Using polar coordinates in the pupil variables:

$$x = \rho\cos(\theta), \qquad y = \rho\sin(\theta)$$

then we can expand

$$\begin{aligned} W(x_0, \rho, \theta) &= \sum_{j,m,n} W_{klm} x_0^k \rho^l \cos^m(\theta), \qquad k = 2j + m,\ l = 2n + m \\ &= W_{200}x_0^2 + W_{111}x_0\rho\cos(\theta) + W_{020}\rho^2 + W_{040}\rho^4 \\ &\quad + W_{131}x_0\rho^3\cos(\theta) + W_{222}x_0^2\rho^2\cos^2(\theta) + W_{220}x_0^2\rho^2 \\ &\quad + W_{311}x_0^3\rho\cos(\theta) \end{aligned} \tag{6.9}$$

where the coefficients above correspond to the aberrations (see also [358]) in Table 6.1.

Now, if $W = W(\rho)$ is circularly symmetric, then (6.9) becomes

$$W(x_0, \rho) = W_{200}x_0^2 + W_{020}\rho^2 + W_{040}\rho^4 + W_{220}x_0^2\rho^2 \tag{6.10}$$

Now that we know what aberrations are, we move to a simple lens design problem which aims to remove aberrations generated from a given surface.

6.2 Lens Design Problem

The problem we consider in this section is the following: Given the first surface of a lens in 2D, find the second surface, such that the lens containing a liquid crystal layer between the surfaces refracts the extraordinary ray into a given fixed point. In particular, the second surface will be constructed in such a way as to correct aberration generated by the first surface. The method we use below is essentially that from [169], but we provide all the relevant details here.

In the problem below we assume that the director field **d** of the liquid crystal makes a constant angle with the angle of the refracted ray coming from the exterior of the lens. This is a restrictive assumption but serves as a starting point for the liquid crystal lens design problem. See Fig. 6.10 for a setup of the problem. We also assume that within the liquid crystal layer the light ray trajectories are linear.

Consider a given first surface of the LC slab lens in 2D, say parametrically by $G(u, v) = g(u)x(u) = g(u)(\cos(u), \sin(u)) = (a, a\tan(u))$, where $x = a$ is the location of the first surface of the lens and $g(u)$ is a given smooth function (the polar radius), in this case given by

$$g(u) = \frac{a}{\cos(u)}$$

We have assumed without much loss of generality that the first surface is a line (in 3D we would assume a planar surface). Suppose that rays originate from the origin. Let us first also assume homeotropic anchoring of the director at the given surface, whose normal is $\nu_1 = (1, 0)$. Thus

$$\mathbf{d}(a, y) = (1, 0)$$

Consider now a general director that makes a constant angle θ with the refracted ray, so that

$$n_{\text{eff}} = \sqrt{n_o^2 \cos^2(\theta) + n_e^2 \sin^2(\theta)} \tag{6.11}$$

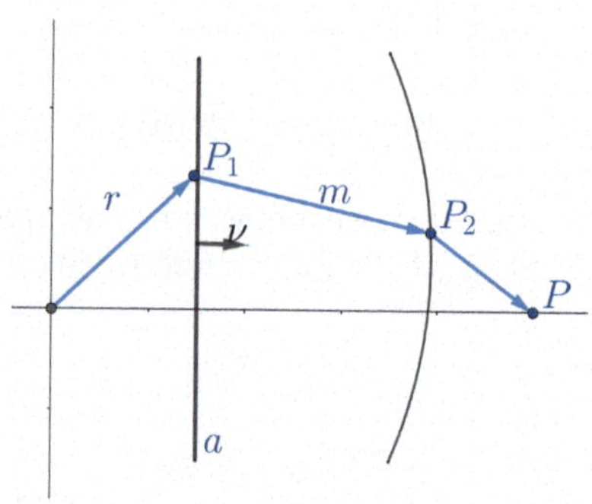

Fig. 6.10 Setup of the lens design problem

This follows from our original definition of the effective refractive index (3.11) after parameterizing the ray in such a way such that $|\mathbf{v}| = 1$.

Applying Snell's Law (4.10) at the first surface yields

$$\boldsymbol{r} - n_{\text{eff}}\mathbf{m} = \lambda_1(1, 0), \qquad \lambda_1 \in \mathbb{R} \tag{6.12}$$

where $\boldsymbol{r}$ is the unit vector of the incident ray and $\mathbf{m}$ is the unit direction of the refracted ray. See Fig. 6.10.

From (6.12) we can find λ_1 explicitly:

$$\lambda_1 = \boldsymbol{r} \cdot (1, 0) - \sqrt{(\boldsymbol{r} \cdot (1, 0))^2 - (1 - n_{\text{neff}}^2)} \tag{6.13}$$

Indeed, this comes from dotting (6.12) with $\boldsymbol{r}$, then with $\mathbf{m}$, and solving the resulting quadratic equation for λ_1. Note that for the radicand will be positive when $n_{\text{eff}} > 1$, which holds provided that

$$n_o > 1 \qquad \text{and} \qquad n_e > 1 \tag{6.14}$$

which is a standard assumption; for instance, $n_o = 1.65$ and $n_e = 1.95$ for the nematic Merck Phase V (see, e.g., [359]). Then, in terms of the angle u, λ_1 from (6.13) reduces to

$$\lambda_1 = \cos(u) - \sqrt{n_o^2 \cos^2(\theta) + n_e^2 \sin^2(\theta) - \sin^2(u)}$$

and the radicand above is positive for small enough angles u, e.g., for $u \in [-\pi/2, \pi/2]$.

Remark 6.1 The maximum angle of refraction, denoted θ_c, is achieved not when $\theta_1 = \pi/2$, but rather when $n_{\text{eff}}(\theta)\sin(\theta)$ is maximum since on the surface, the angle between $\mathbf{m}$ and $\mathbf{d}$ is the same as the angle between $\mathbf{m}$ and $\boldsymbol{v}$ due to the anchoring condition. Now, from Snell's Law , we can solve for θ in terms of θ_1, i.e., $\sin(\theta_1) = n_{\text{eff}}(\theta)\sin(\theta)$ implies that

$$\theta = \arccos\left(-\sqrt{\frac{2n_e^2 - n_o^2 - \sqrt{2n_e^2 - 2n_o^2 + n_o^4 - 2(n_e^2 - n_o^2)\cos(2\theta_1)}}{2(n_e^2 - n_o^2)}}\right)$$

Note that the integrand is positive, e.g., if we take the material parameters n_o, n_e for Merck Phase V as above. This function $\theta = \theta(\theta_1)$ is the function we wish to maximize; it appears that the maximum occurs when $\theta_1 = 0$. In this case, the maximum angle of refraction is then

$$\theta_c = \arccos(-1) = \pi$$

which corresponds to reflection for $\pi/2 < \theta_c < \pi$, meaning the maximum angle for refraction should be $\pi/2$. This is consistent with the critical angle for refraction in isotropic media [449].

At the second surface, since we have assumed linear ray trajectories, we may again apply Snell's Law (4.10) at this surface:

$$n_{\text{eff}}\mathbf{m} - \frac{P - P_2(\boldsymbol{r}(u))}{|P - P_2(\boldsymbol{r}(u))|} = \lambda_2 \nu_2 \tag{6.15}$$

where ν_2 is the unit normal at the unknown second surface, P_2 is the intersection point of the ray with the second surface, and P is the final destination of the ray (say, along the x-axis, so we write $P = (b, 0)$, where $b > a$). Similarly we find

$$\lambda_2 = n_{\text{eff}}(\mathbf{m} \cdot \boldsymbol{\nu}_2) - \sqrt{n_{\text{eff}}^2((\mathbf{m} \cdot \boldsymbol{\nu}_2)^2 - 1) + 1} \tag{6.16}$$

Since rays are linear in the liquid crystal layer, we can define the distance between the points P_1 and P_2, where P_1 is the intersection of the incident ray with the given slab:

$$D(\boldsymbol{r}(u)) = |P_1(\boldsymbol{r}(u)) - P_2(\boldsymbol{r}(u))|$$

We can see that P_2 is described by the vector

$$f(u) = (a, a\tan(u)) + D(\boldsymbol{r}(u))\mathbf{m} \tag{6.17}$$

Now, it is clear from (6.12) and (6.15) that the vector $\boldsymbol{w}$ given by

$$\boldsymbol{w} := \boldsymbol{r} - \lambda_1(1, 0) - \frac{P - f(u)}{|P - f(u)|}$$

satisfies $\boldsymbol{w} \parallel \nu_2$. Since $\boldsymbol{w} \parallel \nu_2$, we have that

$$\boldsymbol{w} \cdot \frac{\partial f}{\partial u} = 0$$

We next see that

$$\frac{d}{du}|P - f(u)| = -\frac{P - f(u)}{|P - f(u)|} \cdot f'(u)$$

and

$$(\boldsymbol{r} - \lambda_1(1, 0)) \cdot f'(u) = g'(u) + n_{\text{eff}} D'(u)$$

Using the definition of f from (6.17), we then obtain (6.20). Now, the discriminant of (6.20) is equal to

$$\begin{aligned}\Delta &= \beta^2 - 4\alpha\delta \\ &= 4\Big[(n_{\text{eff}}\mathbf{m}\cdot(P-G(u)) + g - C)^2 + \left(n_{\text{eff}}^2 - 1\right) \\ &\times \left(|P-G(u)|^2 - (\mathbf{m}\cdot(P-G(u)))^2\right)\Big]\end{aligned}$$

In order to ensure existence of solutions D, we must have the discriminant Δ be positive. This will hold if n_o, n_e are sufficiently large, i.e., (6.14) is satisfied.

Now, notice that we have

$$P - G = (b - a, -a\tan(u))$$

so

$$\begin{aligned}\mathbf{m}\cdot(P-G) &= (b-a)m_1 - a\tan(u)m_2 \\ &= \frac{(b-a)\sqrt{n_o^2\cos^2(\theta) + n_e^2\sin^2(\theta) - \sin^2(u)} - a\sin(u)\tan(u)}{\sqrt{n_o^2\cos^2(\theta) + n_e^2\sin^2(\theta)}}\end{aligned}$$

Next we find that

$$|P-G|^2 = (b-a)^2 + a^2\tan^2(u) = b^2 - 2ab + \frac{a^2}{\cos^2(u)}$$

so

$$\begin{aligned}\beta &= 2\left[\mathbf{m}\cdot(P-G) + n_{\text{eff}}(g-C)\right] \\ &= 2\left(n_o^2\cos^2(\theta) + n_e^2\sin^2(\theta)\right)^{-1/2}\Big\{(a\sec(u) - C)(n_o^2\cos^2(\theta) + n_e^2\sin^2(\theta)) \\ &\quad + (b-a)\sqrt{n_o^2\cos^2(\theta) + n_e^2\sin^2(\theta) - \sin^2(u)} - a\sin(u)\tan(u)\Big\}\end{aligned} \tag{6.21}$$

and

$$\alpha = n_o^2\cos^2(\theta) + n_e^2\sin^2(\theta) - 1$$

and finally

Thus

$$g'(u) + n_{\text{eff}} D'(u) + \frac{d}{du}|P - f(u)| = 0$$

Integrating yields

$$g(u) + n_{\text{eff}} D(u) + |P - f(u)| = C$$

i.e.,

$$\frac{a}{\cos(u)} + n_{\text{eff}} D(u) + |P - f(u)| = C \tag{6.18}$$

for some constant C.

Then, from (6.12) we can solve for the refracted direction, which after normalization becomes

$$m_1 = \frac{\sqrt{n_o^2 \cos^2(\theta) + n_e^2 \sin^2(\theta) - \sin^2(u)}}{\sqrt{n_o^2 \cos^2(\theta) + n_e^2 \sin^2(\theta)}}$$
$$m_2 = \frac{\sin(u)}{\sqrt{n_o^2 \cos^2(\theta) + n_e^2 \sin^2(\theta)}} \tag{6.19}$$

Now, we start from (6.18) and obtain the following quadratic equation for D:

$$\alpha D^2 + \beta D + \delta = 0 \tag{6.20}$$

where

$$\alpha = n_{\text{eff}}^2 - 1,$$
$$\beta = 2\left[\mathbf{m} \cdot (P - G(u)) + n_{\text{eff}}(g - C)\right],$$
$$\delta = (C - g)^2 - |P - G(u)|^2$$

where recall $g(u) = a\sec(u)$. Indeed, from (6.18), we write

$$|P - f(u)| = C - g(u) - n_{\text{eff}} D(u)$$

Squaring this equation yields

$$|P - f(u)|^2 = (C - g(u) - n_{\text{eff}} D(u))(C - g(u) - n_{\text{eff}} D(u))$$

$$\delta = \left(C - \frac{a}{\cos(u)}\right)^2 - \left[b^2 - 2ab + \frac{a^2}{\cos^2(u)}\right] = 2ab - b^2 + C^2 - \frac{2aC}{\cos(u)}$$

Then, the formula for D is given by the quadratic formula:

$$D(\boldsymbol{x}) = \frac{-\beta \pm \sqrt{\Delta}}{2(n_{\text{eff}}^2 - 1)} \tag{6.22}$$

We seek a solution for D in a neighborhood of $u = 0$. Notice that

$$\alpha_0 = \alpha|_{u=0} = n_o^2 \cos^2(\theta) + n_e^2 \sin^2(\theta) - 1$$

and

$$\beta_0 = \beta|_{u=0} = 2\left(b - a + (a - C)\sqrt{n_o^2 \cos^2(\theta) + n_e^2 \sin^2(\theta)}\right)$$

and

$$\delta_0 = \delta|_{u=0} = (2a - b - C)(b - C) = (C - a)^2 - (b - a)^2$$

To obtain a solution we must have the discriminant be positive:

$$\Delta_0 := \beta_0^2 - 4\alpha_0\delta_0 > 0$$

We find that

$$\Delta_0 = 4(a - C)^2 + 4(b - a)\times$$
$$\left[(b - a)\left(n_o^2 \cos^2(\theta) + n_e^2 \sin^2(\theta)\right) + 2(a - C)\sqrt{n_o^2 \cos^2(\theta) + n_e^2 \sin^2(\theta)}\right]$$

We cannot have $C < a$ since from below we see that $b < C$, and this would imply that $b < C < a$, which is impossible since $b > a$. Thus we must have

$$C > a$$

But this means we must have conditions in order to have $\Delta_0 > 0$; namely, that

$$\cos(\theta) < \frac{a - C}{(b - a)n_o} \tag{6.23}$$

Now we check which sign ($\pm$) to take for the square root in D in (6.22). First we plug $u = 0$ into (6.18) to obtain

$$a + n_{\text{eff}} D((1,0)) + (b - f(0)) = C$$

Since $f(0) = a + D((1,0))$, we get

$$(n_{\text{eff}} - 1)\, D((1,0)) + b = C$$

Since $0 < D((1,0)) < b - a$, this implies that

$$b < C < a + n_{\text{eff}}(b - a)$$

Now, from above, we see that we must have

$$D((1,0)) = \frac{C - b}{n_{\text{eff}} - 1} = \frac{-\beta_0 \pm \sqrt{\Delta_0}}{2(n_{\text{eff}}^2 - 1)}$$

which is only possible if we take the negative square root.

Thus, for u in a neighborhood of 0, we obtain the following formula for D:

$$D(\boldsymbol{r}) = \frac{-\beta - \sqrt{\Delta}}{2(n_{\text{eff}}^2 - 1)} \tag{6.24}$$

where β is given by (6.21), n_{eff} is given by (6.11), and

$$\Delta = \beta^2 - 4\left(n_o^2 \cos^2(\theta) + n_e^2 \sin^2(\theta) - 1\right)\left(2ab - b^2 + C^2 - \frac{2aC}{\cos(u)}\right) \tag{6.25}$$

This implies that, for u in a neighborhood of 0,

$$f(u) = G(u) + D(u)\mathbf{m} = (a, a\tan(u)) + D(u)\mathbf{m} \tag{6.26}$$

with the components of $\mathbf{m}$ given by (6.19), and D given by (6.24). A sample resulting lens can be found below in Fig. 6.11.

Remark 6.2 Next let us discuss the strong anchoring assumption. This means we have the following Snell's Laws at the interfaces:

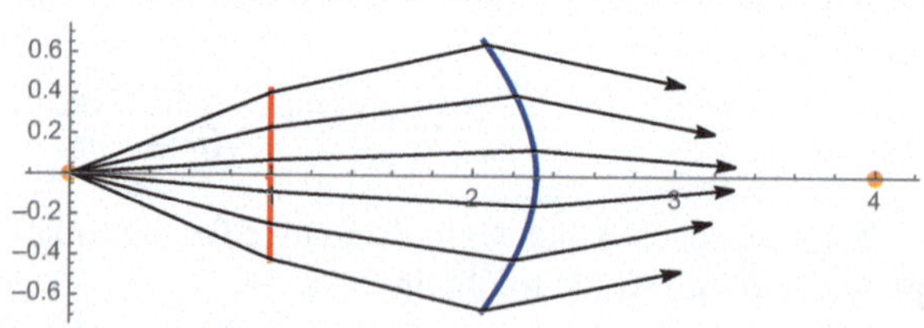

Fig. 6.11 Sample lens with $n_o = 1.83, n_e = 1.9, b = 4, a = 1, c = 5.47, \theta \approx \pi/4$

$$r - \sqrt{n_o^2 \cos^2(\theta) + n_e^2 \sin^2(\theta)}\mathbf{m} = \lambda_1 \boldsymbol{v}_1 + \frac{(n_o^2 - n_e^2)\cos(\theta)}{\sqrt{n_o^2 \cos^2(\theta) + n_e^2 \sin^2(\theta)}}\mathbf{d} \tag{6.27}$$

and

$$\sqrt{n_o^2 \cos^2(\theta) + n_e^2 \sin^2(\theta)}\mathbf{m} - \frac{P - P_2(\boldsymbol{x})}{|P - P_2(\boldsymbol{x})|} = \lambda_2 \boldsymbol{v}_2 - \frac{(n_o^2 - n_e^2)\cos(\theta)}{\sqrt{n_o^2 \cos^2(\theta) + n_e^2 \sin^2(\theta)}}\mathbf{d} \tag{6.28}$$

The constant λ_1 is determined from

$$\lambda_1 = (\boldsymbol{r} - \nabla_{\mathbf{v}} n_{\text{eff}}(\mathbf{m})) \cdot \boldsymbol{v} - \sqrt{n_{\text{eff}}(\mathbf{m})^2 - \left[|\boldsymbol{r} - \nabla_{\mathbf{v}} n_{\text{eff}}(\mathbf{m})|^2 - (\boldsymbol{r} - \nabla_{\mathbf{v}} n_{\text{eff}}(\mathbf{m})) \cdot \boldsymbol{v})^2\right]} \tag{6.29}$$

In this case, we find that

$$\lambda_1 = \cos(u) + \frac{(n_e^2 - n_o^2)\cos(\theta)}{\sqrt{n_o^2 \cos^2(\theta) + n_e^2 \sin^2(\theta)}}$$
$$- \sqrt{n_o^2 \cos^2(\theta) + n_e^2 \sin^2(\theta) - \left(\frac{(n_e^2 - n_o^2)\cos(\theta)}{\sqrt{n_o^2 \cos^2(\theta) + n_e^2 \sin^2(\theta)}} + \sin(u)\right)^2}$$

Note again that the integrand above is positive for u small enough. Now, if we know the director on the surface, then (6.27) allows to solve for $\mathbf{m}$. Thus as long as we are provided a boundary condition for the director, we may proceed as before.

Finally, we mention that the extension of the previous result to 3D is not a problem. One can define $x(u, v) = (\cos(u)\sin(v), \sin(u)\sin(v), \cos(v)$ using spherical coordinates and define

$$g(u, v) = \frac{a}{\cos(u)\sin(v)}$$

where a is such that the known surface is a plane passing through $(a, 0, 0)$. Then one can use the machinery from [169] and [154] to proceed.

6.3 Director Design Problem

With the additional flexibility of designing a director field within a liquid crystal, we are able to consider a new type of lens design problem: designing a lens to obtain a certain type of trajectory by choosing an appropriate director. We address this problem here for a linear trajectory.

In 2D we wish to design a director $\mathbf{d}$ such that the resulting trajectory $\mathbf{x}(t) = (x(t), y(t))$ is a line. In particular, we determine if we can obtain lines with nonconstant director. This would allow for more linear trajectories in nonhomogeneous media as discovered in [257], where one could obtain linear trajectories with refractive indices of the form

$$n(x, y) = \phi(x + ky)$$

provided the incident angle satisfies $\tan(\alpha) = k$. So, for a refractive index of the form $\phi(x + ky)$ with ϕ nonconstant, trajectories will be straight lines only if they have slope k.

Now, the Euler–Lagrange system for Q (recall Eq. 5.2) in 2D takes the form (assuming a director $\mathbf{d}$ which is C^1)

$$\begin{aligned} \frac{\partial n_{\text{eff}}^2}{\partial x} - \frac{d}{dt}\left(\frac{\partial n_{\text{eff}}^2}{\partial \dot{x}}\right) &= 0, \\ \frac{\partial n_{\text{eff}}^2}{\partial y} - \frac{d}{dt}\left(\frac{\partial n_{\text{eff}}^2}{\partial \dot{y}}\right) &= 0 \end{aligned} \tag{6.30}$$

Write $\mathbf{d} = (d_1(x, y), d_2(x, y))$; we obtain the equations

$$2(n_o^2 - n_e^2)\,(\dot{x}d_1 + \dot{y}d_2)\left(\dot{x}\frac{\partial d_1}{\partial x} + \dot{y}\frac{\partial d_2}{\partial x}\right) - \frac{d}{dt}\left(2n_e^2\dot{x} + 2(n_o^2 - n_e^2)(\dot{x}d_1 + \dot{y}d_2)d_1\right) = 0$$

$$2(n_o^2 - n_e^2)\,(\dot{x}d_1 + \dot{y}d_2)\left(\dot{x}\frac{\partial d_1}{\partial y} + \dot{y}\frac{\partial d_2}{\partial y}\right) - \frac{d}{dt}\left(2n_e^2\dot{y} + 2(n_o^2 - n_e^2)(\dot{x}d_1 + \dot{y}d_2)d_2\right) = 0$$

We find that (for the x equation)

$$\begin{aligned} &\frac{d}{dt}\left(2n_e^2\dot{x} + 2(n_o^2 - n_e^2)(\dot{x}d_1 + \dot{y}d_2)d_1\right) \\ &= 2n_e^2\ddot{x} + 2(n_o^2 - n_e^2)d_1\left\{\ddot{x}d_1 + \dot{x}\left(\frac{\partial d_1}{\partial x}\dot{x} + \frac{\partial d_1}{\partial y}\dot{y}\right) + \ddot{y}d_2 + \dot{y}\left(\frac{\partial d_2}{\partial x}\dot{x} + \frac{\partial d_2}{\partial y}\dot{y}\right)\right\} \\ &\quad + 2(n_o^2 - n_e^2)(\dot{x}d_1 + \dot{y}d_2)\left(\frac{\partial d_1}{\partial x}\dot{x} + \frac{\partial d_1}{\partial y}\dot{y}\right) \end{aligned}$$

and the y equation is obtained similarly. Thus the Euler–Lagrange equations become

$$
\begin{aligned}
(n_o^2 - n_e^2)(\dot{x}d_1 + \dot{y}d_2)\left(\dot{x}\frac{\partial d_1}{\partial x} + \dot{y}\frac{\partial d_2}{\partial x}\right) &= n_e^2\ddot{x} \\
&+ (n_o^2 - n_e^2)d_1\left\{\ddot{x}d_1 + \dot{x}\left(\frac{\partial d_1}{\partial x}\dot{x} + \frac{\partial d_1}{\partial y}\dot{y}\right)\right. \\
&\left.+\ddot{y}d_2 + \dot{y}\left(\frac{\partial d_2}{\partial x}\dot{x} + \frac{\partial d_2}{\partial y}\dot{y}\right)\right\} \\
&+ (n_o^2 - n_e^2)(\dot{x}d_1 + \dot{y}d_2)\left(\frac{\partial d_1}{\partial x}\dot{x} + \frac{\partial d_1}{\partial y}\dot{y}\right) \\
(n_o^2 - n_e^2)(\dot{x}d_1 + \dot{y}d_2)\left(\dot{x}\frac{\partial d_1}{\partial y} + \dot{y}\frac{\partial d_2}{\partial y}\right) &= n_e^2\ddot{y} \\
&+ (n_o^2 - n_e^2)d_2\left\{\ddot{x}d_1 + \dot{x}\left(\frac{\partial d_1}{\partial x}\dot{x} + \frac{\partial d_1}{\partial y}\dot{y}\right)\right. \\
&\left.+\ddot{y}d_2 + \dot{y}\left(\frac{\partial d_2}{\partial x}\dot{x} + \frac{\partial d_2}{\partial y}\dot{y}\right)\right\} \\
&+ (n_o^2 - n_e^2)(\dot{x}d_1 + \dot{y}d_2)\left(\frac{\partial d_2}{\partial x}\dot{x} + \frac{\partial d_2}{\partial y}\dot{y}\right)
\end{aligned}
\tag{6.31}
$$

which agree with (5.5). If we want a line (through the origin, say) of the form

$$x(t) = at, \qquad y(t) = bt$$

then clearly

$$\dot{x} = a, \quad \dot{y} = b, \quad \ddot{x} = 0, \quad \ddot{y} = 0$$

and the system (6.31) becomes

$$
\begin{aligned}
(ad_1 + bd_2)\left(a\frac{\partial d_1}{\partial x} + b\frac{\partial d_2}{\partial x}\right) &= d_1\left[a\left(\frac{\partial d_1}{\partial x}a + \frac{\partial d_1}{\partial y}b\right) + b\left(\frac{\partial d_2}{\partial x}a + \frac{\partial d_2}{\partial y}b\right)\right] + \\
&+ (ad_1 + bd_2)\left(a\frac{\partial d_1}{\partial x} + b\frac{\partial d_1}{\partial y}\right) \\
(ad_1 + bd_2)\left(a\frac{\partial d_1}{\partial y} + b\frac{\partial d_2}{\partial y}\right) &= d_2\left[a\left(\frac{\partial d_1}{\partial x}a + \frac{\partial d_1}{\partial y}b\right) + b\left(\frac{\partial d_2}{\partial x}a + \frac{\partial d_2}{\partial y}b\right)\right] \\
&+ (ad_1 + bd_2)\left(a\frac{\partial d_2}{\partial x} + b\frac{\partial d_2}{\partial y}\right)
\end{aligned}
\tag{6.32}
$$

Let us first assume the director only depends on x (say); then we obtain the system

$$(ad_1+bd_2)\left(a\frac{dd_1}{dx}+b\frac{dd_2}{dx}\right)=d_1\left[a\left(\frac{dd_1}{dx}a\right)+b\left(\frac{dd_2}{dx}a\right)\right]+(ad_1+bd_2)a\frac{dd_1}{dx}$$
$$0=d_2\left[a\left(\frac{dd_1}{dx}a\right)+b\left(\frac{dd_2}{dx}a\right)\right]+(ad_1+bd_2)a\frac{dd_2}{dx}$$

The only solution to the above system is $\mathbf{d}$ is constant; thus, if $\mathbf{d}$ only depends on one variable, it is necessarily constant.

Remark 6.3 It is interesting to include here the notions of bend, splay, and twist (recall from Chap. 3). In 2D it is known that there exists no director which has vanishing splay and constant nonzero bend [313]. Indeed, pre-compatibility conditions on the director in terms of bend and splay can be obtained both in 2D and 3D. Writing the director as $\mathbf{d} = (\cos(\theta(x, y)), \sin(\theta(x, y))$, the 2D compatibility condition in [313] was found to be

$$b^2+s^2+\mathbf{d}\cdot\nabla s-\mathbf{d}_\perp\cdot\nabla b=0 \tag{6.33}$$

where $\mathbf{d}_\perp = (-\sin(\theta(x, y)), \cos(\theta(x, y)))$ and b, s denote bend and splay, respectively. The bend and splay are related to the director by

$$\begin{aligned} s &= \nabla\cdot\mathbf{d} = \frac{\partial\theta}{\partial y}\cos(\theta)-\frac{\partial\theta}{\partial x}\sin(\theta)=\mathbf{d}_\perp\cdot\nabla\theta, \\ b &= |\mathbf{d}\cdot(\nabla\mathbf{d})^T| = |\frac{\partial\theta}{\partial x}\cos(\theta)+\frac{\partial\theta}{\partial y}\sin(\theta)| = |\mathbf{d}\cdot\nabla\theta| \end{aligned} \tag{6.34}$$

Equations (6.34) can be inverted to obtain

$$\nabla\theta = b\mathbf{d}+s\mathbf{d}_\perp = \begin{pmatrix}\cos(\theta) & -\sin(\theta)\\ \sin(\theta) & \cos(\theta)\end{pmatrix}\begin{pmatrix}b\\ s\end{pmatrix} \tag{6.35}$$

which can be integrated to find θ only if the mixed partial derivatives of θ commute; this in turn results in the necessary condition (6.33), i.e., that $\nabla\theta$ is conservative.

Considering the above remark, since we saw above that d_1, d_2 are constant, then the only admissible bend and splay in this case is $b = s = 0$.

Let us write in general

$$\mathbf{d}=\begin{pmatrix}d_1\\ d_2\end{pmatrix}=\begin{pmatrix}\cos(\theta(x, y))\\ \sin(\theta(x, y))\end{pmatrix}$$

Then (6.32) becomes

$$(a\cos(\theta(x, y))+b\sin(\theta(x, y)))\left(-a\sin(\theta(x, y))\frac{\partial\theta}{\partial x}+b\cos(\theta)\frac{\partial\theta}{\partial x}\right)=$$

$$
\cos(\theta)\left[a\left(-a\sin(\theta(x,y))\frac{\partial\theta}{\partial x}-b\sin(\theta(x,y))\frac{\partial\theta}{\partial y}\right)\right.
$$

$$
\left.+b\left(a\cos(\theta(x,y))\frac{\partial\theta}{\partial x}+b\cos(\theta(x,y))\frac{\partial\theta}{\partial y}\right)\right]
$$

$$
+\left(a\cos(\theta(x,y))+b\sin(\theta(x,y))\right)\left(-a\sin(\theta(x,y))\frac{\partial\theta}{\partial x}-b\sin(\theta)\frac{\partial\theta}{\partial y}\right),
$$

$$
\left(a\cos(\theta(x,y))+b\sin(\theta(x,y))\right)\left(-a\sin(\theta(x,y))\frac{\partial\theta}{\partial y}+b\cos(\theta)\frac{\partial\theta}{\partial y}\right)=
$$

$$
\sin(\theta)\left[a\left(-a\sin(\theta(x,y))\frac{\partial\theta}{\partial x}-b\sin(\theta(x,y))\frac{\partial\theta}{\partial y}\right)\right.
$$

$$
\left.+b\left(a\cos(\theta(x,y))\frac{\partial\theta}{\partial x}+b\cos(\theta(x,y))\frac{\partial\theta}{\partial y}\right)\right]
$$

$$
+\left(a\cos(\theta(x,y))+b\sin(\theta(x,y))\right)\left(a\cos(\theta(x,y))\frac{\partial\theta}{\partial x}+b\cos(\theta(x,y))\frac{\partial\theta}{\partial y}\right)
$$

i.e.,

$$
\begin{aligned}
(a^2+b^2)\cos(\theta)\sin(\theta)\frac{\partial\theta}{\partial x} &= b(b\cos(2\theta)-a\sin(2\theta))\frac{\partial\theta}{\partial y},\\
(a^2+b^2)\cos(\theta)\sin(\theta)\frac{\partial\theta}{\partial y} &= -a(a\cos(2\theta)+b\sin(2\theta))\frac{\partial\theta}{\partial x}
\end{aligned}
\tag{6.36}
$$

It can be seen that only constant θ will solve this system; perhaps unsurprisingly, we see linear ray trajectories in a constant $\mathbf{d}$ field.

Chapter 7
The Challenge of Absolute Instruments

In this chapter we illustrate the challenge of obtaining absolute instruments. In particular, if there is hope to use liquid crystals to construct such devices, a more comprehensive understanding of superintegrable systems with velocity dependent potentials is required. We will show that even obtaining linear integrals of motion for a system with general velocity dependent potential is rather difficult, in part due to the complicated nature of Hamilton's equations of motion. All this being said, there is an interesting connection to Riemannian geometry through the notion of cut locus which may provide an alternative approach to finding absolute instruments.

To make the connection between absolute instruments and the theory of super-integrable systems, we begin with an analogy between optics and mechanics which serves as the foundation for all that follows.

7.1 Optics–Mechanics Analogy

We begin by exploring the relation between optics and mechanics as discussed in [244], Chapter 2.3. The optical path length is defined in terms of the refractive index of a material

$$s = \int n d\ell = \int n\sqrt{dx^2 + dy^2 + dz^2} \tag{7.1}$$

where $d\ell$ denotes an increment of the geometrical path length. Suppose that the Cartesian coordinates are functions of a real parameter ξ. Then we can write

E. Stachura, *Mathematical Methods in Liquid Crystal Optics and Lens Design*, Springer Tracts in Modern Physics 294,
https://doi.org/10.1007/978-3-031-46614-4_7

$$s = \int n\sqrt{\left(\frac{dx}{d\xi}\right)^2 + \left(\frac{dy}{d\xi}\right)^2 + \left(\frac{dz}{d\xi}\right)^2} d\xi = \int \sqrt{n^2 v^2} d\xi, \qquad v = \frac{d\mathbf{r}}{d\xi}$$

where $\mathbf{r}$ denotes the ray trajectory. Forming the optical Lagrangian $\mathcal{L} = \sqrt{n^2 v^2}$, the corresponding Euler–Lagrange equation is

$$\frac{d}{d\xi}\frac{n^2 v}{\mathcal{L}} = \frac{v^2 \nabla n^2}{2\mathcal{L}} \tag{7.2}$$

If we choose ξ such that

$$d\xi = \frac{d\ell}{n}$$

we obtain the "speed"

$$\left|\frac{d\mathbf{r}}{d\xi}\right| = n\left|\frac{d\mathbf{r}}{d\ell}\right| = n$$

since $d\ell = \sqrt{dx^2 + dy^2 + dz^2}$. This parameterization leads to the following simplification of the Euler–Lagrange equation (7.2):

$$\frac{d^2\mathbf{r}}{d\xi^2} = \frac{\nabla n^2}{2} \tag{7.3}$$

But (7.3) is just Newton's Second Law for a mechanical particle moving in "time" ξ under the influence of the potential

$$V = -\frac{n^2}{2} + E, \qquad E \in \mathbb{R} \tag{7.4}$$

Note that we can then write $E = V + \frac{1}{2}n^2$, so this constant can be interpreted as the total energy of the particle (the sum of kinetic and potential energy).

Let us now move to 2D and write $\mathbf{x} = (x, y)$, so that $\dot{\mathbf{x}} = (\dot{x}, \dot{y})$. Starting from (7.4), we see that

$$n^2 = 2E - 2V \tag{7.5}$$

which means that the trajectories of a particle with Lagrangian

$$\mathcal{L}_m = \frac{v^2}{2} + V(\mathbf{r}) = \frac{\dot{x}^2 + \dot{y}^2}{2} + V(x, y) \tag{7.6}$$

and energy E have the same geometrical shapes as light rays in a medium with refractive index given by (7.5). Now, with $E = 0$ and the refractive index given by (3.11), the corresponding potential V becomes, writing $\mathbf{d}(\mathbf{x}) = (d_1(x, y), d_2(x, y))$,

$$V(x, y, \dot{x}, \dot{y}) = -\frac{1}{2}\left\{n_e^2\left(\dot{x}^2 + \dot{y}^2\right) + (n_o^2 - n_e^2)\left(\dot{x}d_1(x, y) + \dot{y}d_2(x, y)\right)^2\right\} \tag{7.7}$$

However, notice that V depends also on $\dot{x}$, $\dot{y}$ since the effective refractive index is anisotropic. Thus, we should have started with the *anisotropic* Lagrangian

$$\mathcal{L}_{m,a} = \frac{\dot{x}^2 + \dot{y}^2}{2} - U(x, y, \dot{x}, \dot{y}) \tag{7.8}$$

So, the starting point should be the anisotropic version of Maupertuis' principle [244]:

$$\delta \int_A^B \sqrt{2(E - U(\mathbf{r}, \dot{\mathbf{r}}))}dl = 0 \tag{7.9}$$

and proceed as above.

As an example, let us take the director to be the hedgehog:

$$\mathbf{d}(x, y) = \frac{1}{\sqrt{x^2 + y^2}}(x, y) \tag{7.10}$$

Then the potential takes the form

$$V(x, y, \dot{x}, \dot{y}) = -\frac{1}{2}\left\{n_e^2(\dot{x}^2 + \dot{y}^2) + \frac{(n_o^2 - n_e^2)(x\dot{x} + y\dot{y})^2}{x^2 + y^2}\right\} \tag{7.11}$$

which yields the following mechanical Lagrangian:

$$\begin{aligned}\mathcal{L}_h &= \frac{\dot{x}^2 + \dot{y}^2}{2} - \frac{1}{2}\left\{n_e^2(\dot{x}^2 + \dot{y}^2) + \frac{(n_o^2 - n_e^2)(x\dot{x} + y\dot{y})^2}{x^2 + y^2}\right\} \\ &= \frac{1}{2}\left\{(1 - n_e^2)(\dot{x}^2 + \dot{y}^2) + \frac{(n_e^2 - n_o^2)(x\dot{x} + y\dot{y})^2}{x^2 + y^2}\right\}\end{aligned} \tag{7.12}$$

Now, it turns out that it is also useful to formulate this problem in the Hamiltonian framework and understand the resulting Hamilton–Jacobi equation [413]. Indeed, this leads to the notion of superintegrable classical bound systems and separability of the Hamilton–Jacobi equation and related potential. But what is the connection to absolute instruments? Why work toward superintegrability? Well, the answer lies in

[413] and separability of the Hamilton–Jacobi equation. This is due to the Liouville–Arnold theorem, which says that a Hamiltonian system is maximally superintegrable if and only if the Hamiltonian is separable.

The method in [413] shows how, starting from a separable Hamiltonian, one can construct an absolute instrument. Indeed, generally the method will allow one to construct an appropriate mechanical potential V, which in turn will lead to the refractive index profile of the instrument through (7.5). This method has been used in [414] to construct absolute instruments without spherical symmetry as well.

Let us now investigate the possibility of quadratically maximal superintegrable systems, i.e., those maximally superintegrable systems with integrals of motion at most quadratic in the momenta. Why explore such systems? Well, for scalar potentials, separation of variables is characterized by the existence of first integrals which are quadratic or linear in the momenta. But what about velocity dependent potentials? This turns out to be a highly nontrivial problem [423] and has first been investigated in [112] for velocity dependent potentials which are linear in momenta.

The one result on separability for vector potentials in [288] that we are aware of essentially says that for a potential which is linear in momenta, all integrable systems in 2D with second order integrals of motion allow separation of variables in a coordinate system in which the corresponding Hamilton–Jacobi equation allows separation. This in turn would lead us to an absolute instrument; however, as we shall see below, this result is insufficient to handle the optical Hamiltonian for liquid crystals (see equation (7.28) below) which has a potential which is quadratic in momenta.

7.2 Separability and Superintegrability

In classical mechanics [277, 300], consider a Hamiltonian $H = T + V$. The system is described by position coordinates $q_j(t)$ and momentum coordinates $p_j(t)$. The dynamics of the system are given by Hamilton's equations

$$\frac{dq_j}{dt} = \frac{\partial H}{\partial p_j}, \qquad \frac{dp_j}{dt} = -\frac{\partial H}{\partial q_j}, \qquad j = 1, \ldots, n$$

Solutions of Hamilton's equations provide the trajectories of the system.

Let $q = \mathbf{x}(t)$. We consider a mechanical interpretation of a light ray as the trajectory of a particle with coordinates $\mathbf{x} = (x, y, z)$ and generalized momentum $\mathbf{p} = (p_x, p_y, p_z)$ which satisfy

$$\begin{aligned}\frac{dx}{dt} &= \frac{1}{2}\frac{\partial\mathcal{H}}{\partial p_x},\\ \frac{dy}{dt} &= \frac{1}{2}\frac{\partial\mathcal{H}}{\partial p_y},\\ \frac{dz}{dt} &= \frac{1}{2}\frac{\partial\mathcal{H}}{\partial p_z},\\ \frac{dp_x}{dt} &= -\frac{1}{2}\frac{\partial\mathcal{H}}{\partial x},\\ \frac{dp_y}{dt} &= -\frac{1}{2}\frac{\partial\mathcal{H}}{\partial y},\\ \frac{dp_z}{dt} &= -\frac{1}{2}\frac{\partial\mathcal{H}}{\partial z}\end{aligned} \tag{7.13}$$

where $\mathcal{H}(\mathbf{x},\mathbf{p})$ is the optical indicatrix given by (2.21). By taking the above mechanical interpretation of a light ray, we see that the optical indicatrix $\mathcal{H}(\mathbf{x},\mathbf{p})$ represents a Hamiltonian system

$$\begin{aligned}\frac{d\mathbf{x}}{dt} &= \frac{1}{2}\nabla_{\mathbf{p}}\mathcal{H}(\mathbf{x},\mathbf{p}),\\ \frac{d\mathbf{p}}{dt} &= -\frac{1}{2}\nabla_{\mathbf{x}}\mathcal{H}(\mathbf{x},\mathbf{p})\end{aligned} \tag{7.14}$$

We can now split $\mathcal{H}$ into the ordinary and extraordinary parts and obtain the Hamiltonian system for each type of ray.

7.2.1 Calculating Hamilton's Equations for the Extraordinary Ray

Continuing the previous example, we can find Hamilton's equations for the extraordinary ray starting from the Fresnel surface (2.21). Recall in particular Sect. 5.3 where we first saw this.

Consider in 2D the Hamiltonian for the extraordinary ray coming from (2.21), written in terms of the canonical momenta:

$$\mathcal{H}(x,y,p_1,p_2) = n_o^2|\mathbf{p}|^2 + (n_e^2 - n_o^2)(\mathbf{p}\cdot\mathbf{d})^2 - n_e^2 n_o^2 \tag{7.15}$$

Let us write the director as

$$\mathbf{d}(x,y) = \begin{pmatrix}\cos(\theta(x,y))\\ \sin(\theta(x,y))\end{pmatrix}$$

In this case Hamilton's equations (7.14) become

$$\begin{aligned}
\dot{x} &= n_o^2 p_1 + (n_e^2 - n_o^2)(\mathbf{p}\cdot\mathbf{d})\cos(\theta) \\
\dot{y} &= n_o^2 p_2 + (n_e^2 - n_o^2)(\mathbf{p}\cdot\mathbf{d})\sin(\theta) \\
\dot{p}_1 &= -(n_e^2 - n_o^2)(\mathbf{p}\cdot\mathbf{d})\left(p_2\cos(\theta)\frac{\partial\theta}{\partial x} - p_1\sin(\theta)\frac{\partial\theta}{\partial x}\right) \\
\dot{p}_2 &= -(n_e^2 - n_o^2)(\mathbf{p}\cdot\mathbf{d})\left(p_2\cos(\theta)\frac{\partial\theta}{\partial y} - p_1\sin(\theta)\frac{\partial\theta}{\partial y}\right)
\end{aligned} \tag{7.16}$$

We can solve the first two equations for p_1, p_2 yielding

$$\begin{aligned}
p_1 &= \frac{\dot{x}(n_e^2 + n_o^2) + (n_o^2 - n_e^2)\,(\dot{x}\cos(2\theta) + \dot{y}\sin(2\theta))}{2n_e^2 n_o^2} \\
p_2 &= \frac{\dot{y}(n_e^2 + n_o^2) - (n_o^2 - n_e^2)\,(\dot{y}\cos(2\theta) - \dot{x}\sin(2\theta))}{2n_e^2 n_o^2}
\end{aligned} \tag{7.17}$$

Now we differentiate the first two equations in (7.16) to obtain

$$\begin{aligned}
\ddot{x} &= n_o^2 \dot{p}_1 \\
&+ (n_e^2 - n_o^2)\left[\cos(\theta)\left(\dot{\mathbf{p}}\cdot\mathbf{d} - p_1\sin(\theta)\left(\frac{\partial\theta}{\partial x}\dot{x} + \frac{\partial\theta}{\partial y}\dot{y}\right)\right.\right. \\
&+ \left.\left. p_2\cos(\theta)\left(\frac{\partial\theta}{\partial x}\dot{x} + \frac{\partial\theta}{\partial y}\dot{y}\right)\right)\right. \\
&\left. -(\mathbf{p}\cdot\mathbf{d})\sin(\theta)\left(\frac{\partial\theta}{\partial x}\dot{x} + \frac{\partial\theta}{\partial y}\dot{y}\right)\right]
\end{aligned}$$

$$\begin{aligned}
\ddot{y} &= n_o^2 \dot{p}_2 \\
&+ (n_e^2 - n_o^2)\left[\sin(\theta)\left(\dot{\mathbf{p}}\cdot\mathbf{d} - p_1\sin(\theta)\left(\frac{\partial\theta}{\partial x}\dot{x} + \frac{\partial\theta}{\partial y}\dot{y}\right)\right.\right. \\
&+ \left.\left. p_2\cos(\theta)\left(\frac{\partial\theta}{\partial x}\dot{x} + \frac{\partial\theta}{\partial y}\dot{y}\right)\right)\right. \\
&\left. +(\mathbf{p}\cdot\mathbf{d})\cos(\theta)\left(\frac{\partial\theta}{\partial x}\dot{x} + \frac{\partial\theta}{\partial y}\dot{y}\right)\right]
\end{aligned}$$

Then, using (7.17), we can write

$$\dot{p}_1 = \frac{(n_e^2 - n_o^2)\left(-2\dot{x}\dot{y}\cos(2\theta) + (\dot{x}^2 - \dot{y}^2)\sin(2\theta)\right)\dfrac{\partial\theta}{\partial x}}{2n_e^2 n_o^2}$$

and similarly

$$\dot{p}_2 = \frac{(n_e^2 - n_o^2)\left(-2\dot{x}\dot{y}\cos(2\theta) + (\dot{x}^2 - \dot{y}^2)\sin(2\theta)\right)\dfrac{\partial\theta}{\partial y}}{2n_e^2 n_o^2}$$

Thus we arrive at

$$\begin{aligned}
\ddot{x} = \frac{(n_o^2 - n_e^2)}{8n_e^2 n_o^2} \Bigg\{ & 2\left(n_e^2 + n_o^2\right)\sin(2\theta)\left(\left(\dot{x}^2 + \dot{y}^2\right)\frac{\partial\theta}{\partial x} + 2\dot{x}\dot{y}\frac{\partial\theta}{\partial y}\right) \\
& -(n_e^2 - n_o^2)\left(\left(\dot{x}^2 + 3\dot{y}^2\right)\frac{\partial\theta}{\partial y} + 2\dot{x}\dot{y}\frac{\partial\theta}{\partial x}\right) \\
& + (n_e^2 - n_o^2)\sin(4\theta)\left(\left(\dot{y}^2 - \dot{x}^2\right)\frac{\partial\theta}{\partial x} + 2\dot{x}\dot{y}\frac{\partial\theta}{\partial y}\right) \\
& +(n_e^2 - n_o^2)\cos(4\theta)\left((\dot{x}^2 - \dot{y}^2)\frac{\partial\theta}{\partial y} + 2\dot{x}\dot{y}\frac{\partial\theta}{\partial x}\right) - 4\dot{y}^2\left(n_e^2 + n_o^2\right)\frac{\partial\theta}{\partial y}\cos(2\theta)\Bigg\}
\end{aligned} \tag{7.18}$$

and

$$\begin{aligned}
\ddot{y} = \frac{(n_e^2 - n_o^2)}{8n_e^2 n_o^2} \Bigg\{ & 2\left(n_e^2 + n_o^2\right)\sin(2\theta)\left(\left(\dot{x}^2 + \dot{y}^2\right)\frac{\partial\theta}{\partial y} + 2\dot{x}\dot{y}\frac{\partial\theta}{\partial x}\right) \\
& -(n_e^2 - n_o^2)\left(\left(3\dot{x}^2 + \dot{y}^2\right)\frac{\partial\theta}{\partial x} + 2\dot{x}\dot{y}\frac{\partial\theta}{\partial y}\right) + 4\dot{x}^2\left(n_e^2 + n_o^2\right)\frac{\partial\theta}{\partial x}\cos(2\theta) \\
& -(n_e^2 - n_o^2)\sin(4\theta)\left((\dot{x}^2 - \dot{y}^2)\frac{\partial\theta}{\partial y} + 2\dot{x}\dot{y}\frac{\partial\theta}{\partial x}\right) \\
& - (n_e^2 - n_o^2)\cos(4\theta)\left((\dot{x}^2 - \dot{y}^2)\frac{\partial\theta}{\partial x} - 2\dot{x}\dot{y}\frac{\partial\theta}{\partial y}\right)\Bigg\}
\end{aligned} \tag{7.19}$$

Note that in the limiting case $n_e = n_o = n$, we recover the equations

$$\ddot{x} = 0$$

$$\ddot{y} = 0$$

as expected for an isotropic medium.

Let us suppose now that θ just depends on y. Then the previous equations of motion simplify to

$$\ddot{x} = \frac{(n_o^2 - n_e^2)}{8n_e^2 n_o^2}\frac{\partial\theta}{\partial y}\Big\{4\left(n_e^2 + n_o^2\right)\sin(2\theta)\dot{x}\dot{y} - (n_e^2 - n_o^2)\left(\dot{x}^2 + 3\dot{y}^2\right)$$
$$+2(n_e^2 - n_o^2)\sin(4\theta)\dot{x}\dot{y} + (n_e^2 - n_o^2)\cos(4\theta)\left(\dot{x}^2 - \dot{y}^2\right) \tag{7.20}$$
$$- 4\dot{y}^2\left(n_e^2 + n_o^2\right)\cos(2\theta)\Big\}$$

and

$$\ddot{y} = \frac{(n_e^2 - n_o^2)}{8n_e^2 n_o^2}\frac{\partial\theta}{\partial y}\Big\{2\left(n_e^2 + n_o^2\right)\sin(2\theta)\left(\dot{x}^2 + \dot{y}^2\right) - 2(n_e^2 - n_o^2)\dot{x}\dot{y}$$
$$-(n_e^2 - n_o^2)\sin(4\theta)\left(\dot{x}^2 - \dot{y}^2\right) + 2(n_e^2 - n_o^2)\cos(4\theta)\dot{x}\dot{y}\Big\} \tag{7.21}$$

As an example, suppose now that $\theta(x, y) = y$.
Then the previous equations of motion simplify even further to

$$8n_e^2 n_o^2\ddot{x} = \left(\dot{y}^2 - \dot{x}^2\right)\left(n_e^2 - n_o^2\right)^2\cos(4y) + \left(\dot{x}^2 + 3\dot{y}^2\right)\left(n_e^2 - n_o^2\right)^2$$
$$+ 4\dot{x}\dot{y}\left(n_o^4 - n_e^4\right)\sin(2y) \tag{7.22}$$
$$- 2\dot{x}\dot{y}\left(n_e^2 - n_o^2\right)^2\sin(4y) + 4\dot{y}^2\left(n_e^4 - n_o^4\right)\cos(2y)$$

and

$$\ddot{y} = \frac{\begin{array}{r}(n_o^2 - n_e^2)\sin(2y)\left((n_e^2 - n_o^2)(2\dot{x}\dot{y}\sin(2y)\right.\\ \left.+(\dot{x}^2 - \dot{y}^2)\cos(2y)) - (\dot{x}^2 + \dot{y}^2)\left(n_e^2 + n_o^2\right)\right)\end{array}}{4n_e^2 n_o^2} \tag{7.23}$$

Let us take $n_e = 1.5$ and $n_o = 1.6$; then we can use a Runge–Kutta method to solve (7.22)–(7.23). Then under certain initial conditions, the resulting trajectory and director field are plotted below in Fig. 7.1.

7.2.2 *Superintegrability*

Next we recall some basic definitions regarding superintegrability. Define the Poisson bracket of two functions $R(\mathbf{p}, \mathbf{q})$ and $S(\mathbf{p}, \mathbf{q})$ as the function

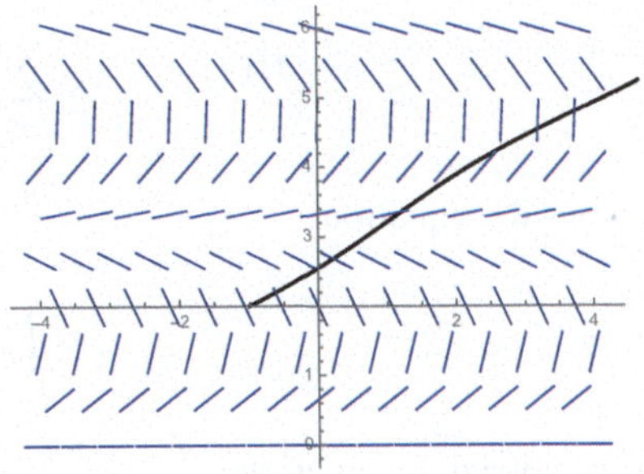

Fig. 7.1 The director field with $\theta(x, y) = y$ and the resulting trajectory from solving (7.22)–(7.23)

$$\{R, S\}(\mathbf{p}, \mathbf{q}) = \sum_{j=1}^{n} \left(\frac{\partial R}{\partial p_j}\frac{\partial S}{\partial q_j} - \frac{\partial R}{\partial q_j}\frac{\partial S}{\partial p_j} \right)$$

By definition, if $\{H, S\} = 0$, we say S is a constant of the motion. We call a system with Hamiltonian H integrable if it admits n constants of motion $P_1 = H, P_2, \ldots, P_n$ that are in involution:

$$\{P_j, P_k\} = 0, \quad 1 \le j, k \le n$$

and are functionally independent. By functionally independent, we mean the following: If $F = (f_1(\mathbf{q}, \mathbf{p}), \ldots, f_N(\mathbf{q}, \mathbf{p}))$ is a set of functions defined and locally analytic in some region in phase space, then F is called functionally independent if the $N \times 2n$ matrix $\left(\frac{\partial f_l}{\partial q_j}, \frac{\partial f_l}{\partial p_k}\right)$ has rank N throughout the region.

Next we discuss superintegrability. A Hamiltonian system is polynomially integrable if it is integrable and the constants of motion are each of the polynomials in the momenta globally defined. A classical system in n dimensions is polynomially superintegrable if it admits $n + k$ with $k = 1, \ldots, n - 1$ functionally independent constants of motion that are polynomial in the momenta and are globally defined except possibly for singularities on a lower dimensional manifold. It is maximally polynomially superintegrable if $k = n - 1$. For more details we refer to the review article [300].

Of particular interest is quadratic superintegrability, where all integrals of motion are at most quadratic polynomials in the momenta. In this case, integrability is related to separation of variables in the Hamilton–Jacobi equation. To be precise about separation of variables, let us fix a classical Hamiltonian of the form

$$H = T + V = \frac{1}{2}\sum_{j=1}^{n} p_j^2 + V(q), \qquad q = (q_1, ..., q_n)$$

Then, separation of variables amounts to finding curvilinear coordinates $x = (x_1, .., x_n)$ with canonical momenta coordinates $y = (y_1, \ldots, y_n)$ such that the transformed Hamiltonian

$$H(x, y) = \frac{1}{2} \sum_{i,j=1}^{n} g^{ij}(x) y_i y_j + V(q(x))$$

allows the equation

$$H\left(x, \frac{\partial u}{\partial x}\right) = E$$

to be solved via an ansatz

$$u(x; \alpha) = \sum_i u_i(x_i, \alpha) \tag{7.24}$$

which depends on a set of separation constants $\alpha = (\alpha_1, \ldots, \alpha_n)$, see, e.g., [423]. Finally, let us comment on the presence of the g^{ij} above. The coordinates (p, q) are canonical coordinates on the cotangent bundle $T^*\mathbb{R}^n$. Changing coordinates $q \mapsto x$ implies a change of coordinates $p \mapsto y$ which requires (x, y) to be canonical; as such, the quadratic form $\frac{1}{2} \sum p_j^2$ transforms to $\frac{1}{2} \sum g^{ij} y_i y_j$ when the metric transforms to $ds^2 = \sum g_{ij} dx_i dx_j$.

Consider the generalized Hénon–Heiles potential

$$V = \frac{1}{2}\left(q_1^2 + q_2^2\right) + q_1^2 q_2 + 2q_2^3$$

considered in [438]. By introducing translated parabolic coordinates (x_1, x_2) by

$$q_1 = \sqrt{x_1 x_2}, \qquad q_2 = \frac{1}{2}(x_1 - x_2) + \frac{3}{4}$$

one uses the so-called Bertrand–Darboux equation as in [423] to find that

$$V = \frac{f_1(x_1) + f_2(x_2)}{x_1 + x_2}$$

for some functions $f_1(x_1)$ and $f_2(x_2)$. This comes from the solution of Bertrand–Darboux equation

$$(x_1 + x_2) \frac{\partial^2}{\partial x_1 \partial x_2} ((x_1 + x_2) V) = 0$$

In these parabolic coordinates, the Hamilton–Jacobi equation is

$$\frac{2}{x_1 + x_2} \left[x_1 \left(\frac{\partial u}{\partial x_1} \right)^2 + x_2 \left(\frac{\partial u}{\partial x_2} \right)^2 \right] + V = E$$

Then, inserting the ansatz $u = u_1(x_1) + u_2(x_2)$ yields

$$2x_1(u_1')^2 + f_1(x_1) = Ex_1 + \alpha_2, \qquad 2x_2(u_2')^2 + f_2(x_2) = Ex_2 - \alpha_2$$

where α_2 is a separation constant (and $E = \alpha_1$ can also be realized as a separation constant). Thus the potential V is indeed separable in parabolic coordinates (x_1, x_2).

The presence of a velocity dependent potential causes further difficulties. Indeed, if we are trying to understand when separation of variables applies, then in general the existence of second order integrals of motion does not imply the separation of variables (unlike for scalar potentials) [42].

Remark 7.1 The case of the ordinary ray is straightforward, which is why we focus on the extraordinary ray. Indeed, starting from the Hamiltonian for the ordinary ray

$$\mathcal{H}_o = |\mathbf{p}|^2 - n_o^2$$

we see that this Hamiltonian is separable, and as such, known methods (e.g., from [414]) can be used to construct absolute instruments.

Consider a 3D Hamiltonian of the form

$$H(x, y, z, \dot{x}, \dot{y}, \dot{z}) = \frac{\dot{x}^2 + \dot{y}^2 + \dot{z}^2}{2} + A_1(x, y, z)\dot{x} + A_2(x, y, z)\dot{y} + A_3(x, y, z)\dot{z} + W(x, y, z) \tag{7.25}$$

which was first considered in 2D in [112]. There are only two cases [36, 371] in which this Hamiltonian separates in Cartesian coordinates in 3D:

1. $W(x, y, z) = W_1(x) + W_2(y), \qquad \mathbf{A}(x, y, z) = (0, 0, u_1(y) - u_2(x)).$
2. $W(x, y, z) = W_1(x), \qquad \mathbf{A}(x, y, z) = (0, u_3(x), -u_2(x)).$

In both cases the system admits two integrals [276]:

$$X_1 = \dot{x}^2 - 2(u_2(x)\dot{z} - W_1(x)), \quad X_2 = \dot{y}^2 + 2(u_1(y)\dot{z} + W_2(y)) \quad \text{for Case (1)} \tag{7.26}$$

and

$$X_2 = \dot{y}, \quad X_2 = \dot{z} \qquad \text{for Case (2)} \tag{7.27}$$

Since we know the refractive index we are interested in for the optics problem, our starting point is actually the potential (7.7). At first glance, the corresponding Hamiltonian

$$\mathcal{H} = \frac{1}{2}\left(\dot{x}^2 + \dot{y}^2\right) - \frac{1}{2}\left\{n_e^2\left(\dot{x}^2 + \dot{y}^2\right) + (n_o^2 - n_e^2)\left(\dot{x}d_1(x, y) + \dot{y}d_2(x, y)\right)^2\right\} \tag{7.28}$$

does not fit the form that was considered in [112]:

$$\mathcal{H} = \frac{1}{2}\left(\dot{x}^2 + \dot{y}^2\right) + A(x, y)\dot{x} + B(x, y)\dot{y} + W(x, y) \tag{7.29}$$

since the potential (7.7) is quadratic in the momenta. Thus, the Hamiltonian type that we need to study is the following:

$$\mathcal{H} = \frac{1}{2}\left(p_x^2 + p_y^2\right) + A(x, y)p_x^2 + B(x, y)p_y^2 + C(x, y)p_x p_y + D(x, y)p_x + E(x, y)p_y + W(x, y) \tag{7.30}$$

We shall follow [112] to illustrate the challenge in determining integrals of motion. First we determine the Hamilton equations of motion:

$$\begin{aligned}
\dot{x} &= \frac{\partial \mathcal{H}}{\partial p_x} = p_x + 2A(x, y)p_x + C(x, y)p_y + D(x, y)\\
\dot{y} &= \frac{\partial \mathcal{H}}{\partial p_y} = p_y + 2B(x, y)p_y + C(x, y)p_x + E(x, y)\\
\dot{p_x} &= -\frac{\partial \mathcal{H}}{\partial x} = -\left(A_x p_x^2 + B_x p_y^2 + C_x p_x p_y + D_x p_x + E_x p_y + W_x\right)\\
\dot{p_y} &= -\frac{\partial \mathcal{H}}{\partial y} = -\left(A_y p_y^2 + B_y p_y^2 + C_y p_x p_y + D_y p_x + E_y p_y + W_y\right)
\end{aligned} \tag{7.31}$$

Solving the first two equations for p_x, p_y yields

$$\begin{aligned}
p_x &= \frac{-((1 + 2B(x, y))D(x, y)) + \dot{x}(1 + 2B(x, y)) - \dot{y}C(x, y) + C(x, y)E(x, y)}{1 + 2B(x, y) + 2A(x, y)(1 + 2B(x, y)) - C(x, y)^2},\\
p_y &= \frac{-C(x, y)\dot{x} + C(x, y)D(x, y) + (1 + 2A(x, y))\dot{y} - E(x, y)(1 + 2A(x, y))}{1 + 2B(x, y) + 2A(x, y)(1 + 2B(x, y)) - C(x, y)^2}
\end{aligned}$$

Now we eliminate the momenta and obtain the x equation

$$\begin{aligned}
\ddot{x} &= \dot{p_x} + 2\left(\left(A_x\dot{x} + A_y\dot{y}\right)p_x + A\dot{p_x}\right) + \left(C_x\dot{x} + C_y\dot{y}\right)p_y + C(x, y)\dot{p_y}\\
&\quad + D_x\dot{x} + D_y\dot{y}\\
&= P_1\dot{x}^2 + P_2\dot{y}^2 + P_3\dot{x} + P_4\dot{y} + P_5\dot{x}\dot{y} + P_6
\end{aligned} \tag{7.32}$$

for some P_j. These coefficients can be found explicitly; a Mathematica script is included in Appendix A.3 to find them.

We similarly obtain the y equation

$$\begin{aligned}\ddot{y} &= \dot{p}_y + 2\left(\left(B_x\dot{x} + B_y\dot{y}\right)p_y + B\dot{p}_y\right) + \left(C_x\dot{x} + C_y\dot{y}\right)p_x + C(x,y)\dot{p}_x \\ &\quad + E_x\dot{x} + E_y\dot{y} \\ &= Q_1\dot{x}^2 + Q_2\dot{y}^2 + Q_3\dot{x} + Q_4\dot{y} + Q_5\dot{x}\dot{y} + Q_6\end{aligned} \tag{7.33}$$

where the Q_j can also be found in Appendix A.3. Suppose a classical first (linear) integral of motion takes the form

$$C = f_0(x,y)\dot{x} + f_1(x,y)\dot{y} + h(x,y) \tag{7.34}$$

for some f_0, f_1, h independent of time. More generally we would also like to consider the integral of motion

$$C = g_0(x,y)\dot{x}^2 + g_1(x,y)\dot{x}\dot{y} + g_2(x,y)\dot{y}^2 + k_0(x,y)\dot{x} + k_1(x,y)\dot{y} + h(x,y) \tag{7.35}$$

The determining equations for f_0, f_1, h in (7.34) come from the requirement that

$$\{\mathcal{H}, C\} = \frac{dC}{dt} = 0 \tag{7.36}$$

when (7.32)–(7.33) are satisfied, i.e., C is a constant on the solutions of the equations of motion. Then $dC/dt = 0$ becomes

$$\frac{dC}{dt} = \ddot{x}\, f_0 + \dot{x}\left(\frac{\partial f_0}{\partial x}\dot{x} + \frac{\partial f_0}{\partial y}\dot{y}\right) + \ddot{y} f_1 + \dot{y}\left(\frac{\partial f_1}{\partial x}\dot{x} + \frac{\partial f_1}{\partial y}\dot{y}\right) + h_x\dot{x} + h_y\dot{y} = 0$$

then use the equations of motion (7.32)–(7.33), and equate the coefficients of each term in $\dot{x}$ and $\dot{y}$ to zero. For the $\dot{x}^2$ term, we have

$$f_0 P_1 + \frac{\partial f_0}{\partial x} + f_1 Q_1 = 0 \tag{7.37}$$

For the $\dot{y}^2$ term, we have

$$f_0 P_2 + \frac{\partial f_1}{\partial y} + f_1 Q_2 = 0 \tag{7.38}$$

For the $\dot{x}\dot{y}$ term, we have

$$f_0 P_5 + \left(\frac{\partial f_0}{\partial y} + \frac{\partial f_1}{\partial x}\right) + f_1 Q_5 = 0 \tag{7.39}$$

For the $\dot{x}$ term, we have

$$f_0 P_3 + h_x + f_1 Q_3 = 0 \tag{7.40}$$

For the $\dot{y}$ term, we have

$$f_0 P_4 + h_y + f_1 Q_4 = 0 \tag{7.41}$$

Finally the constant term gives

$$f_0 P_6 + f_1 Q_6 = 0 \tag{7.42}$$

Thus it is clear that even finding a linear integral of motion is a highly nontrivial endeavor. The search for quadratic integrals of motion would proceed in a similar manner, starting from (7.35).

7.3 Stigmatic Systems Via Homotopy

As can be seen above, the search for absolute instruments is generally difficult. Recall that absolute instruments require stigmatic imaging, i.e., an AI is a device that stigmatically images a 3D domain of space.

Another way one can view a stigmatic system is with a homotopy, as was considered in [146]. Recall that a homotopy between elements $f_0, f_1 : X \to Y$, where X and Y are topological spaces, is a continuous function $F : X \times [0, 1] \to Y$ such that $F(0, x) = f_0(x)$ and $F(1, x) = f_1(x)$. Now, one can consider a stigmatic system as a homotopy that obeys Fermat's principle. The extreme fixed points of the homotopy are the object and image points. That is, the relevant homotopy of a stigmatic system is the one such that its ray paths obey Fermat's principle.

In particular, in [146] it is shown that the homotopy $H(t, s)$ that follows the Fermat principle in a stigmatic system is the solution of the Eikonal equation

$$|\nabla S(s)|^2 = n(s)^2 \tag{7.43}$$

where $S(s) = H(\cdot, s)$.

The key elements in obtaining the homotopy in [146], see also [145], were a vector form of Snell's Law and Fermat's Principle. The first step is actually to determine the shape of a second surface given a first surface in order to correct spherical aberration. This is reminiscent of the lens design problem considered in [169] and in Sect. 6.2. In [146] the first surface was assumed to have axial symmetry and to be smooth.

Ultimately a natural extension is to liquid crystal lens design and in particular the design of liquid crystal lenses to construct absolute instruments.

7.4 Connection to Riemannian Geometry

There is an interesting connection between absolute instruments and Riemannian geometry, and it comes through the so-called Wiedersehenmannigfaltigkeit theorem. To this end, let us first review the classical Maxwell fisheye [48]. This is a particularly prominent example since the Maxwell fisheye surrounded by a mirror [243] is only one of the two devices that are known to be absolute instruments in *both* the wave and ray sense (the other being Pendry's slab [334]).

Maxwell's fisheye is characterized by a spherical medium with radial refractive index

$$n(r) = \frac{1}{1+(r/a)^2} n_0, \qquad n_0,\ a \in \mathbb{R} \tag{7.44}$$

To determine the equation of the rays with this refractive index, we use the fact that if (r, θ) are the polar form of a plane curve, so that $r = f(\theta)$, the angle ϕ between the radial vector to a point P on the curve and the tangent at P is given by

$$\tan(\phi) = \frac{f(\theta)}{f'(\theta)}$$

Indeed, if we think of the curve as $y = F(x)$ and let θ be a parameter so that $\dot{r} = f'(\theta)$, we have

$$y'(x) = \frac{\dot{y}}{\dot{x}} = \frac{\dot{r}\tan(\theta) + r}{\dot{r} - r\tan(\theta)}$$

which results in

$$\tan(\phi) = \frac{y' - \tan(\theta)}{1 + y'\tan(\theta)} = \frac{r + r\tan^2(\theta)}{\dot{r} + \dot{r}\tan^2(\theta)} = \frac{r}{\dot{r}}$$

We have used the fact that

$$\dot{x} = \dot{r}\cos(\theta) - r\sin(\theta)\dot{\theta},$$
$$\dot{y} = \dot{r}\sin(\theta) + r\cos(\theta)\dot{\theta}$$

This in particular implies that $\sin(\phi)$ is given by

$$\sin(\phi) = \frac{r(\theta)}{\sqrt{r(\theta)^2 + \left(\frac{dr}{d\theta}\right)^2}} \tag{7.45}$$

Now, consider the ray given by $\mathbf{r}(s)$ so that it is parameterized by arclength. We find that (see [48, Section 3.2])

$$\frac{d}{ds}(\mathbf{r} \times n(\mathbf{r})\langle \mathbf{S} \rangle) = \dot{\mathbf{r}} \times n(\mathbf{r})\langle \mathbf{S} \rangle + \mathbf{r} \times \frac{d}{ds}(n(\mathbf{r})\langle \mathbf{S} \rangle) \tag{7.46}$$

where $\langle \mathbf{S} \rangle$ denotes the (normalized) time averaged Poynting vector. The first term is zero since $\dot{\mathbf{r}}(s) = \langle \mathbf{S} \rangle$. This follows from the Eikonal equation, so that if $\mathcal{S} =$ constant denote the wavefronts, then $\nabla \mathcal{S}/n$ is a unit vector, which is $\langle \mathbf{S} \rangle$, and the ray equation can be written as

$$n\frac{d\mathbf{r}}{ds} = \nabla \mathcal{S}$$

We can write the second term in (7.46) as $\mathbf{r} \times \nabla n$ using the ray equation (4.7). Since the refractive index here is radial, we take a derivative and see that

$$\nabla n = \frac{\mathbf{r}}{|\mathbf{r}|} n'(r)$$

which means that the right hand side of (7.46) vanishes:

$$\frac{d}{ds}(\mathbf{r} \times n(\mathbf{r})\langle \mathbf{S} \rangle) = 0 \implies \mathbf{r} \times n\langle \mathbf{S} \rangle = \text{constant}$$

Thus all the rays are plane curves, and along each ray there holds

$$n|\mathbf{r}|\sin(\phi) = \text{constant} \tag{7.47}$$

where ϕ as previous is the angle between $\mathbf{r}$ and the tangent on the ray. Then, from (7.45) and (7.47), we see that

$$\frac{dr}{d\theta} = \frac{r}{c}\sqrt{n(r)^2 r^2 - c^2}, \qquad c \text{ constant}$$

and so the equation of the rays can be written as

$$\theta = c\int_0^r \frac{dr}{r\sqrt{n(r)^2 r^2 - c^2}}, \qquad c \in \mathbb{R}$$

Now, plugging the refractive index (7.44) into this equation, it can be shown that the equation of the rays becomes

$$\sin(\theta - \alpha) = \frac{c}{\sqrt{a^2 n_0^2 - 4c^2}} \frac{r^2 - a^2}{ar}, \qquad \alpha \in \mathbb{R} \tag{7.48}$$

The one-parameter family of rays through a given fixed point $P_0 = (r_0, \theta_0)$ is thus given by

$$\frac{r^2 - a^2}{r \sin(\theta - \alpha)} = \frac{r_0^2 - a^2}{r_0 \sin(\theta_0 - \alpha)}$$

and whatever α may be, this equation is satisfied by $r = a^2/r_0^2, \theta = \theta_0 + \pi$. Thus, all the rays from the point P_0 meet in a point on the line joining the origin to P_0, so the fisheye is an absolute instrument. Finally, by looking at the equation of the rays in Cartesian coordinates, it can easily be seen that (7.48) becomes

$$(x + b\sin(\alpha))^2 + (y - b\cos(\alpha))^2 = a^2 + b^2, \qquad b = \frac{a}{2c}\sqrt{a^2 n_0^2 - 4c^2}$$

showing that each ray traces out a circular trajectory.

The connection to Riemannian geometry comes through the search for manifolds whose geodesics are closed. To this end, we look on a sphere at a metric such that the cut locus of any point is a single point (called the antipode). Recall that, given a Riemannian manifold M and a fixed point P on M, the cut locus of P is the set of cut points of the geodesics emanating from P [37]. The cut point can be defined through the following:

Theorem 7.1 (Theorem 86 [37]) *Let M be a complete Riemannian manifold. Let $m = \gamma(0)$ be the initial point of a geodesic γ. Then, there is a number $t > 0$, called the cut value of γ, such that γ is a segment from m to $\gamma(s)$ for every $s \leq t$ and for $s > t$ is never again a segment from m to any $\gamma(s)$ for any $s > t$. The point $\gamma(t)$ is called the cut point of γ.*

For any point on a simply connected Riemannian manifold with nonpositive curvature, the cut locus is always empty. This essentially follows from the von Mangoldt–Hadamard–Cartan Theorem (see, e.g., [37], Theorem 72). Indeed, if M is a simply connected Riemannian manifold with nonpositive curvature, then the exponential map is one-to-one, M is diffeomorphic to $\mathbb{R}^n$, and any two points can be joined by a unique segment. For a sphere, the cut locus of every point is a single point: its antipode.

Now, at any given point m on a Riemannian manifold M, the infimum of the cut values of the geodesics emanating from m is positive, is called the injectivity radius of M at m, and is denoted $\mathrm{Inj}(m)$. If M is compact, then the injectivity radius $\mathrm{Inj}(M)$ is the positive minimum of the continuous function $\mathrm{Inj}(m)$. A natural question is the following: What are the compact Riemannian manifolds M such that $\mathrm{Inj}(M)$ is equal to the diameter of M? To answer this (so-called Blaschke's conjecture), we finally recall a pair of points x, y on a compact Riemannian manifold such that every geodesic through x also passes through y and conversely is called a Wiedersehen pair. Then Green in [150] answered:

Theorem 7.2 ("Wiedersehenmannigfaltigkeit") *On a sphere S^n, only the metric of constant curvature can be Wiedersehen, i.e., the metric such that the cut locus of any point is a single point.*

Let us call a complete, connected Riemannian manifold such that the cut locus of any point is a single point, a *Wiedersehen manifold*. Summing up we have the following:

Theorem 7.3 ([43]) *Let M be a simply connected Wiedersehen manifold. Then:*

1. *M is isometric to a sphere.*
2. *$Inj(M)$ is equal to the diameter of M.*
3. *All unit speed geodesics of M are periodic with period 2 $diam(M)$.*

Thus we see that the only Wiedersehen manifolds are the standard spheres. In the language of absolute instruments, these results say that the standard sphere is an absolute instrument: There is everywhere perfect focusing with any angle up to 2π. There is no way to deform the sphere and preserve this property.

Chapter 8
Electromagnetics of Helical Medial

8.1 Introduction and Background

We briefly previously encountered the cholesteric phase of a liquid crystal: This occurs when the constituent molecules are chiral (not equal to their mirror image). If this happens and the nematic phase twists,[1] this phase is called cholesteric. In fact, the cholesteric phase was the first observed liquid crystal phase [348]; the material was cholesteryl benzoate (chemical formula $C_{34}H_{50}O_2$), thus the name cholesteric. There are really only two distinctions between a cholesteric and a nematic, which we outline below.

First, in contract to the nematic phase, the director **d** turns about one spatial direction **c** which is orthogonal to **d**. We will denote **c** the cholesteric axis, as seen in Fig. 8.1.

For example, in a coordinate system (as in Fig. 8.1) where the z-axis is parallel to **c**, we can write the components of the director **d** as

$$\mathbf{d}(x, y, z) = \left(\cos\left(\frac{2\pi}{p}z\right), \sin\left(\frac{2\pi}{p}z\right), 0\right) \tag{8.1}$$

where p is called the helix pitch. A rotation by some angle α around **c** is equivalent to a translation by $p\alpha/2\pi$. Since **d** and $-\mathbf{d}$ are equivalent in a nematic, a chiral nematic liquid crystal is periodic with period $p/2$.

We can quantify the amount of "chirality" by introducing a parameter c^* which measures the excess concentration in an enantiomer mixture [323], which, when racemic (i.e., $c^* = 0$), becomes a uniaxial nematic. Roughly speaking, an enantiomer is a molecule which does not contain a mirror plane; that is, a pair

[1] That is, a spontaneous twist about an axis occurs due to the optical activity of the molecules. Optical activity will be discussed in Sect. 8.4.

E. Stachura, *Mathematical Methods in Liquid Crystal Optics and Lens Design*,
Springer Tracts in Modern Physics 294,
https://doi.org/10.1007/978-3-031-46614-4_8

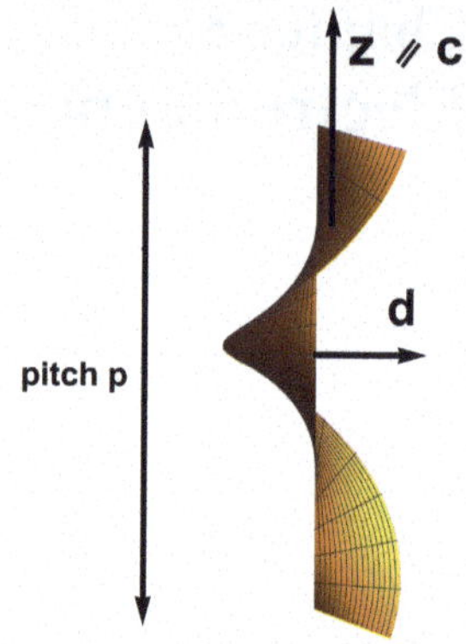

Fig. 8.1 Here the cholesteric axis **c** is parallel to the z-axis, and **d** is orthogonal to **c**. Mathematically, a helix given parametrically by $(a\cos(t), a\sin(t), bt)$ for $0 \leq t \leq 2\pi$ has radius a and pitch $p = 2\pi b$

of such molecules can be mirror images of each other but not superimposed on each other. The mixture is racemic when two enantiomers are present in equal proportions; for this reason usually one write $c^* = c_+ - c_-$, where $c_\pm$ denotes the concentration of an enantiomer.

Remark 8.1 We have put the word "chirality" in quotations for the following reason: Strictly speaking, chirality is a *geometric* concept; an object is either chiral or achiral, i.e., it either has lacks or has mirror symmetry. Chirality is not a state of matter. More precisely, in geometry, an object is chiral if it does not coincide with its image under a rotation or translation. There is also the notion of "electromagnetic chirality" which appears in electromagnetic scattering when the scatterer handles incoming waves differently depending on their handedness (see in particular Section 5 of [400]). This notion of chirality was introduced in [126], and there are some recent mathematical results regarding scattering of time-harmonic electromagnetic waves for electromagnetically chiral media [16, 173]. Throughout this chapter, though, we stick with the more standard "geometric" notion of chirality.

In the case of low chirality, i.e., such that $c^* \ll 1$, the wave vector $q := 2\pi/p$ is actually proportional to c^* [323], so in the limit that $c^* \to 0$ the helix pitch $p \to \infty$.

The second distinction from the nematic phase is that there are actually three directors which characterize this phase: the director **d** along the molecular axes, the chiral axis **c**, and $\mathbf{r} = \mathbf{d} \times \mathbf{c}$, which is orthogonal to both. Thus we actually have biaxial orientational order in the cholesteric phase.

Besides the usual nematic and smectic phases, it is also worth mentioning here there are *columnar phases*, which is a long-range two dimensional positional order with translational symmetry. Here, the molecules stack up in columns and are free to slide with respect to each other. The molecules can be normal to the column axis or have some tilt angle. The columns themselves are parallel and form a 2D lattice, so essentially columnar phases are two dimensional crystals [220, 323]. Typically here molecules are considered as discs (discotic) rather than rods.

Remark 8.2 In discotic liquid crystals [75, 76, 437], there are different kinds of nematic phases, still described by a director **d** which describes the average orientation of the direction normal to the discotic molecules. Besides the anticipated (discotic) nematic mesophase and chiral mesophase, there are also columnar nematic mesophases and lateral nematic mesophases, both formed via appropriate configurations of discotic mesogens. The nematic phase for rod-like molecules and the discotic nematic phase have the same symmetry and order parameters but could have different signs of birefringence [185]. Biaxial nematic discotic liquid crystals also occur; for instance, they are expected in mixtures of rod-like and discotic mesogens [412]. Moreover, selective reflection (discussed in Sect. 8.4.1 below) has also been observed in a discotic chiral liquid crystal [231].

The point of mentioning such a phase here is that there is a columnar phase made of chiral molecules: the so-called *blue phases* (I, II, II). In a blue phase (which is between the isotropic and cholesteric phases, in a small temperature range), chiral molecules organize in an inhomogeneous manner. If **d** is a director parallel to an axis (say), then in a state with lowest energy, the chiral molecules near **d** tend to rotate along a helix along all directions perpendicular to **d**, not just along the helix axis **c** like the typical cholesteric phase. This geometry is called a double-twist [220], and as one goes further from the axis which is parallel to **d**, this double-twist smoothly disappears.

Typically the blue phases of types I and II are modeled as networks of disclination lines with a certain periodicity (related to the pitch p). Let us explore this a bit more in the context of the Oseen–Frank free energy . It is convenient here to write the Oseen–Frank free energy $W(\mathbf{d}, \nabla\mathbf{d})$ in a coordinate system such that a point P of interest, the director **d** coincides with the z-axis. Then we may define the local deformations (we have already seen these, just written differently), with $\mathbf{d} = (d_1, d_2, d_3)$,

$$\begin{aligned}
\text{splay}: \quad & s_1 := \frac{\partial d_1}{\partial x}(P), \quad s_2 := \frac{\partial d_2}{dy}(P), \\
\text{twist}: \quad & t_1 := -\frac{\partial d_2}{\partial x}(P), \quad t_2 := \frac{\partial d_1}{\partial y}(P), \\
\text{bend}: \quad & b_1 := \frac{\partial d_1}{\partial z}(P), \quad b_2 := \frac{\partial d_2}{\partial z}(P)
\end{aligned} \tag{8.2}$$

In the Oseen–Frank formalism (recall from Sect. 3.5) of cholesterics, there is an additional term [220] which is linear in $\nabla\mathbf{d}$, so that the energy W from (3.23) becomes

$$\begin{aligned}
W_c(\mathbf{d}, \nabla\mathbf{d}) = {} & K_1(\nabla \cdot \mathbf{d})^2 + K_2(\mathbf{d} \cdot \nabla \times \mathbf{d} + \mathrm{q})^2 + K_3|\mathbf{d} \times \nabla \times \mathbf{d}|^2 \\
& + (K_2 + K_4)\left(\mathrm{Tr}(\nabla\mathbf{d})^2 - (\nabla \cdot \mathbf{d})^2\right)
\end{aligned} \tag{8.3}$$

where $\mathfrak{q}$ is a pseudoscalar (recall this is a scalar which changes sign under spatial inversion). But $\mathfrak{q}$ is also a material constant, which thus implies there is some kind of chirality (arising from molecular chirality). The ground state of the energy functional associated with (8.3) is given by configurations of the form (8.1), which is why we use the letter $\mathfrak{q}$.

Since we may write (recall $\mathbf{d}$ is a unit vector)

$$\mathbf{d} = (d_1, d_2, 1 - d_1^2 - d_2^2)$$

we can write the first three terms (neglecting second order saddle-splay for now) of the Oseen–Frank free energy as

$$\begin{aligned} w(s_1, s_2, t_1, t_2, b_1, b_2) &= \frac{1}{2}K_1(s_1 + s_2)^2 + \frac{1}{2}K_2(t_1 + t_2)^2 \\ &+ \frac{1}{2}K_3(b_1^2 + b_2^2) + K_4(s_1 s_2 + t_1 t_2) \end{aligned}$$

Let us turn our attention to the twist term, so consider the free energy contribution

$$f_{c,t} = \frac{1}{2}K_2(t + \mathfrak{q})^2$$

Clearly this energy is minimized when $t = -\mathfrak{q}$, which produces $f_{c,t} = 0$. Now, if we consider the local energy for a double-twist, we need the K_4 term since $t_t = t_2 = t$:

$$f_{dt}(t) = \frac{1}{2}K_2(2t + \mathfrak{q})^2 + K_4 t^2$$

We can minimize this as a function of t, and we find that the minimum occurs when

$$t = \frac{-K_2 \mathfrak{q}}{K_4 + 2K_2}$$

and this produces

$$f_{dt} = \frac{K_2 K_4 \mathfrak{q}^2}{2(K_4 + 2K_2)}$$

Thus, a double-twist is (locally) more favorable than a cholesteric twist if $-2K_2 < K_4 < 0$. This is a necessary condition for the existence of a blue phase but in fact is not sufficient [323]. This is because a double-twist cannot be satisfied everywhere in $\mathbb{R}^3$.

In an effort to obtain a configuration which is (energetically) preferable to the cholesteric helix, one can assemble so-called double-twist cylinders [323], where the director takes the form (taking $q = 1$ for simplicity)

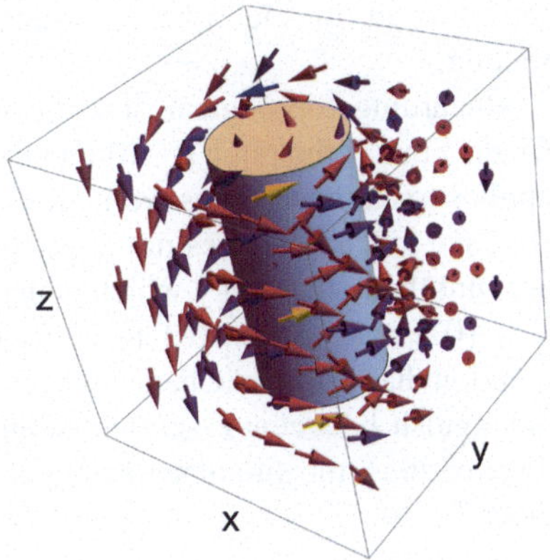

Fig. 8.2 The director for a double-twist cylinder

$$\mathbf{d}(x, y, z) = \left(-\frac{y \sin\left(\sqrt{x^2 + y^2}\right)}{\sqrt{x^2 + y^2}}, \frac{x \sin\left(\sqrt{x^2 + y^2}\right)}{\sqrt{x^2 + y^2}}, \cos\left(\sqrt{x^2 + y^2}\right) \right)$$

This vector field is the Cartesian form of the cylindrical field

$$\mathbf{d}(r, \theta, z) = (0, \sin(r), \cos(r))$$

via the transformation

$$\begin{pmatrix} d_1(x, y, z) \\ d_2(x, y, z) \\ d_3(x, y, z) \end{pmatrix} = \begin{pmatrix} \cos(\theta) & -\sin(\theta) & 0 \\ \sin(\theta) & \cos(\theta) & 0 \\ 0 & 0 & 1 \end{pmatrix} \begin{pmatrix} 0 \\ \sin(r) \\ \cos(r) \end{pmatrix}$$

which transfers between cylindrical and Cartesian coordinates. This is seen in Fig. 8.2.

Note for the director above, we find that

$$\mathbf{d} \cdot \nabla \times \mathbf{d} = 2$$

or more generally, it is equal to 2q, hence the name "double-twist." The point here is that this double-twist cylinder is locally preferred to the cholesteric helix, at least assuming the necessary condition $-2K_2 < K_4 < 0$ is satisfied. Finally, to connect back to disclinations, one can construct double-twist cylinders oriented in such a way (with the right symmetry) so that **d** is continuous everywhere in the space occupied by the cylinders. To finish this model, it is necessary to fill the gaps between the cylinders, and this introduces wedge disclinations [289, 323].

Interestingly enough, under a polarized microscope, the blue phases appear blue. The third type of blue phase is less understood. Nonetheless, there is Bragg reflection in the blue phases, which will be discussed in more detail in the next section.

Regarding topological classification of defects , we previously mentioned in Chap. 3 that if the pitch is sufficiently small, then the discussion on biaxial nematics applies to cholesterics as well. However, in general, the brief discussion on defects we gave in Chap. 3 does not apply to a general cholesteric, which should take into account the high energy which is necessary to change the pitch p. See in particular [74, 107] and section 6.4 in [219] for more on defects in cholesterics. We only mention that, while in Chap. 3 we discussed χ-disclinations, for cholesterics one can distinguish between χ-screw disclinations and χ-edge disclinations. For χ-screw disclinations, the singular line is parallel to the cholesteric axis, and the director is given by

$$\mathbf{d}_{sd} = \left(\cos\left(s\arctan(y/x) + \frac{2\pi}{p}z\right), \sin\left(s\arctan(y/x) + \frac{2\pi}{p}z\right), 0\right), \quad s \in \pm\frac{1}{2}\mathbb{N}$$

where the notation above means that $s = \pm n/2$ for some integer n and p again denotes the helix pitch. For χ-edge disclinations, the singular line is orthogonal to the cholesteric axis, so the director takes the form [70, 363]

$$\mathbf{d}_{ed} = \left(\cos\left(\frac{N}{2}\arctan(z/x) + \frac{2\pi}{p}z\right), \sin\left(\frac{N}{2}\arctan(z/x) + \frac{2\pi}{p}z\right), 0\right)$$

where N is not necessarily an integer. Note that for a χ-edge disclination, the pitch is changed around the singular line where $N \in \mathbb{N}$.

Next we enumerate a few rigorous mathematics results for cholesteric liquid crystals. In [405] the Oseen–Frank free energy for cholesterics above (W_c) can be obtained as the Γ-limit[2] of a chiral doped nematic. This particular model is of Maier–Saupe type [270] for a two-species system: the host H (which is the achiral nematic) and a chiral dopant D. Long-range, attractive interactions are considered. The energy is described as a competition between pairwise interactions and entropy in the two-species system. There is a small parameter associated with the energy which is directly proportional to the dopant concentration and inversely proportional to the sample size; this parameter is particularly relevant as forces a relationship between the density of the dopant and the sample size.

In [410] defects around a spherical colloidal particle in a cholesteric were investigated using a spectral method [372], which allows to capture the fine structure of defects without substantial computational costs. The goal is to find a minimizer of the relevant free energy. Here the authors work in the Landau–de Gennes framework and include a surface term to impose boundary conditions on the surface of the

[2] A sequence of functionals $\{F_n\}$ is said to Γ-converge to the Γ-limit F if (1) for every strongly convergent sequence $x_n \to x$ there holds that $F(x) \leq \liminf_{n\to\infty} F_n(x_n)$ and (2) for every x there exists a strongly convergent sequence $x_n \to x$ such that $F(x) \geq \limsup_{n\to\infty} F_n(x_n)$. See, e.g., [95].

colloidal particles:

$$\mathcal{F}_s = \int_{\partial\Omega} |\mathbf{Q}(\mathbf{x}) - \mathbf{Q}_b(\mathbf{x})|^2 d\sigma$$

where $d\sigma$ denotes surface measure, $\mathbf{Q}$ is the usual Q-tensor from the Landau–de Gennes framework, and $\mathbf{Q}_b$ is a Q-tensor involving the given director in the surface $\partial\Omega$. The above surface energy contribution models strong homeotropic anchoring. For this anchoring, it is shown that a cholesteric dipole can be seen as a nematic dipole[3] with the cholesteric structure.

In one dimension, a simplified Oseen–Frank model for cholesterics has been used with disparate elastic constants based on elastic deformations of the director [142]. Such a problem can be realized as a relaxation of the twist (K_2 term) on W_c; in 2D the variational problem takes the form

$$\inf_{\mathcal{A}} E_\epsilon(u), \qquad u : \Omega \to \mathbb{R}^3$$

where Ω denotes the sample domain and

$$E_\epsilon(u) = \int_\Omega \frac{\epsilon}{2}|\nabla u|^2 + \frac{1}{4\epsilon}\left(|u|^2 - 1\right)^2 + \frac{L}{2}(u \cdot \nabla \times u - 2\pi N)^2 \, dx$$

with

$$\mathcal{A} = \left\{u \in H^1(\Omega; \mathbb{R}^3) : u|_{\partial\Omega} = u_0\right\}$$

for some boundary condition u_0 and some integer N. Here L can be thought of as playing the role of the constant K_2. It is interesting to note that simulations for the dynamics of this problem lead to textures which resemble so-called cholesteric fingerprint textures[4] observed in experiments [324]. In particular, in [142] it is shown that local minimizers of E_ϵ exist for every positive integer value of twist, and in certain parameter regimes, the local minimizers of E_ϵ turn out to be global.

We have only listed a few rigorous results in the literature, which as usual is by no means exhaustive. In general, though, there is a less rich mathematical literature on cholesterics as opposed to (uniaxial) nematics.

Next up, before proceeding to the geometric optics regime, we address how chirality can be seen rigorously in electromagnetism and provide a new mathematical result along the way.

---

[3] This so-called metastable configuration is well known [260, 341, 342].

[4] A random stripe pattern or periodic bright and dark lines can occur in cholesterics [106] and can be seen in transmission electron microscope images.

8.2 Mathematics of Chiral Media: Maxwell's Equations

We follow the framework of [350] and start with the constitutive relation (2.6). By taking Fourier transform of the constitutive relations, we obtain a multiplication operator

$$\mathcal{F}\begin{pmatrix}\mathbf{D}\\ \mathbf{B}\end{pmatrix} = \begin{pmatrix}\varepsilon & \xi\\ \eta & \mu\end{pmatrix}\begin{pmatrix}\mathbf{D}\\ \mathbf{B}\end{pmatrix} + C\begin{pmatrix}\mathbf{D}\\ \mathbf{B}\end{pmatrix} = C\begin{pmatrix}\mathbf{D}\\ \mathbf{B}\end{pmatrix}$$

where $\mathcal{F}$ denotes Fourier transform and

$$C = \mathcal{F}\begin{pmatrix}\varepsilon_d & \xi_d\\ \eta_d & \mu_d\end{pmatrix}$$

so that

$$C = \begin{pmatrix}\varepsilon + \mathcal{F}(\varepsilon_d) & \xi + \mathcal{F}(\xi_d)\\ \eta + \mathcal{F}(\eta_d) & \mu + \mathcal{F}(\mu_d)\end{pmatrix} := \begin{pmatrix}\widehat{\varepsilon} & \widehat{\xi}\\ \widehat{\eta} & \widehat{\mu}\end{pmatrix}$$

Defining

$$\widehat{\xi} = \kappa + i\chi, \qquad \widehat{\eta} = \kappa - i\chi \tag{8.4}$$

the parameter χ is a chirality parameter which measures the handedness of the material, and κ describes the magnetoelectric effect;[5] materials with $\kappa \neq 0$ are called nonreciprocal. We can then further classify media as follows:

Definition 8.1 A material is called:

1. **Isotropic** if $\kappa = 0$ and $\chi = 0$.
2. **Tellegen** if $\kappa \neq 0$ and $\chi = 0$.
3. **Reciprocal chiral/Pasteur** if $\kappa = 0$ and $\chi \neq 0$.
4. **nonreciprocal chiral** if $\kappa \neq 0$ and $\chi \neq 0$.

An important example of reciprocal chiral media is the Drude–Born–Fedorov constitutive relation [233, 234, 250], which take the form (writing tilde for Fourier transform)

$$\widetilde{\mathbf{D}} = \varepsilon_{DBF}\left(\widetilde{\mathbf{E}} + \beta_{DBF}\nabla \times \widetilde{\mathbf{E}}\right), \quad \widetilde{\mathbf{B}} = \mu_{DBF}\left(\widetilde{\mathbf{H}} + \beta_{DBF}\nabla \times \widetilde{\mathbf{H}}\right) \tag{8.5}$$

where the medium is characterized now by three material parameters, where β is a chirality measure. To simplify notation from here on, let us drop tildes and the DBF

---

[5] The magnetoelectric effect was first reported by Wilhelm Röntgen [351] who discovered that a dielectric material moving through an electric field would become magnetized.

subscript. By taking Fourier transform of the Maxwell system (2.1)–(2.4), we obtain the system

$$\begin{aligned}
\nabla \times \mathbf{H} &= i\omega \mathbf{D} + \mathbf{J}, \\
\nabla \times \mathbf{E} &= -i\omega \mathbf{B}, \\
\nabla \cdot \mathbf{D} &= \rho, \\
\nabla \cdot \mathbf{B} &= 0
\end{aligned} \tag{8.6}$$

with (say) the boundary condition which prescribes the tangential component of the electric field on the boundary $\partial\Omega$ of the region Ω

$$\boldsymbol{\nu}(\mathbf{x}) \times \mathbf{E}(\mathbf{x}) = f(\mathbf{x}), \qquad \mathbf{x} \in \partial\Omega$$

When $f = 0$, the boundary condition is that of a perfect conductor. Above, ω denotes the angular frequency of the fields. Using the Drude–Born–Fedorov constitutive relations (8.5) in (8.6) yields the system

$$\begin{aligned}
\nabla \times \mathbf{H} &= i\omega\left(\varepsilon\left(\mathbf{E} + \beta\nabla \times \mathbf{E}\right)\right) + \mathbf{J}, \\
\nabla \times \mathbf{E} &= -i\omega\left(\mu\left(\mathbf{H} + \beta\nabla \times \mathbf{H}\right)\right)
\end{aligned} \tag{8.7}$$

where $i^2 = -1$. After simplification, the above can be written in the form (see [350] and [10])

$$\begin{aligned}
\nabla \times \mathbf{E} &= \frac{\omega^2\varepsilon\mu\beta}{1-\omega^2\beta^2\varepsilon\mu}\mathbf{E} + i\frac{\omega\mu}{1-\omega^2\beta^2\varepsilon\mu}\mathbf{H}, \\
\nabla \times \mathbf{H} &= \frac{\omega^2\varepsilon\mu\beta}{1-\omega^2\beta^2\varepsilon\mu}\mathbf{H} - i\frac{\omega\mu}{1-\omega^2\beta^2\varepsilon\mu}\mathbf{E}
\end{aligned} \tag{8.8}$$

A typical solution of interest is the so-called fundamental solution, so let us recall here the result from [13] which gives the fundamental solution of the isotropic Drude–Born–Fedorov system.

Denote by $\mathcal{L}_0$ the isotropic Drude–Born–Fedorov operator

$$\mathcal{L}_0 = \begin{pmatrix} 0 & \dfrac{-i\gamma^2\beta_0}{\varepsilon_0} + \dfrac{i}{\beta_0}\nabla \times \cdot \\ i\dfrac{\gamma^2\beta_0}{\mu_0} - \dfrac{i}{\mu_0}\nabla \times \cdot & 0 \end{pmatrix} \tag{8.9}$$

where the subscript zero in the previous material parameters denotes their constant value in $\mathbb{R}^3 \setminus \overline{\Omega}$. Here we have introduced the parameter

$$\gamma^2 = \frac{\omega^2 \varepsilon \mu}{1 - \omega^2 \beta^2 \varepsilon \mu}$$

Now, define the kernel G by

$$G = \frac{1}{2}(G_1 + G_2) \tag{8.10}$$

where

$$G_1 = \mathcal{G}^1 \begin{pmatrix} 1 & i\sqrt{\frac{\mu_0}{\varepsilon_0}} \\ -i\sqrt{\frac{\varepsilon_0}{\mu_0}} & 1 \end{pmatrix}, \qquad G_2 = \mathcal{G}^2 \begin{pmatrix} 1 & -i\sqrt{\frac{\mu_0}{\varepsilon_0}} \\ i\sqrt{\frac{\varepsilon_0}{\mu_0}} & 1 \end{pmatrix}$$

and the $\mathcal{G}^j$'s form the fundamental outgoing solution of the isotropic Maxwell system in the exterior domain:

$$\begin{aligned} \nabla \times \mathcal{E}_j &= i\omega_j \mu_0 \mathcal{H}_j, \quad \mathbf{x} \in \mathbb{R}^3 \setminus \overline{\Omega}, \\ \nabla \times \mathcal{H}_j &= -i\omega_j \varepsilon_0 \mathcal{E}_j, \quad \mathbf{x} \in \mathbb{R}^3 \setminus \overline{\Omega} \end{aligned} \tag{8.11}$$

Above, $\overline{\Omega}$ denotes the closure of the domain $\Omega \subset \mathbb{R}^3$. To recall the explicit form of $\mathcal{G}^j$, denote by $g_j(x, y)$ the fundamental outgoing solution of the Helmholtz equation $-(\Delta + m_j^2)$

$$g_j(x, y) = \frac{e^{im_j|x-y|}}{|x - y|}, \quad x \neq y$$

Finally, we have

$$\mathcal{G}^j = \frac{m_j^2}{\omega_j} \begin{pmatrix} 1 + \frac{\nabla\nabla\cdot}{m_j^2} & i\sqrt{\frac{\mu_0}{\varepsilon_0}} \frac{1}{m_j} \nabla\times \\ -i\sqrt{\frac{\mu_0}{\varepsilon_0}} \frac{1}{m_j} \nabla\times & 1 + \frac{\nabla\nabla\cdot}{m_j^2} \end{pmatrix}$$

where, here and above,

$$\omega_1 = \frac{\omega}{1 - \omega\beta_0\sqrt{\varepsilon_0\mu_0}}, \qquad \omega_1 = \frac{\omega}{1 + \omega\beta_0\sqrt{\varepsilon_0\mu_0}} \tag{8.12}$$

and

$$m_1 = \frac{\omega\sqrt{\varepsilon_0\mu_0}}{1 - \omega\beta_0\sqrt{\varepsilon_0\mu_0}}, \qquad m_2 = \frac{\omega\sqrt{\varepsilon_0\mu_0}}{1 + \omega\beta_0\sqrt{\varepsilon_0\mu_0}}$$

The point here is that the kernel G given by (8.10) is a fundamental solution of the isotropic Born-Drude-Fedorov system, so that

$$(\mathcal{L}_0 - \omega_0)\, G = \delta I$$

where δ is the Dirac Delta, I is the identity matrix, and

$$\omega_0 = \frac{\omega}{1 - \omega^2 \beta_0^2 \varepsilon_0 \mu_0}$$

Remark 8.3 We mention that general nonreciprocal chiral media have also been studied, see [49, 181, 182, 251, 252, 357], e.g., in the context of the inverse problem of given scattering data to determine information about the material parameters of the nonreciprocal chiral medium. In fact, a nonreciprocal (isotropic) medium was introduced by Tellegen [406] to realize a new nonreciprocal circuit element; for this reason, a nonchiral bi-isotropic medium is called Tellegen. The point is that, by offering extra material parameters, it is possible to satisfy additional constraints beyond just isotropic media.

Let us now move back to the time domain, where chirality and dispersiveness of the medium is captured by the convolution term in the general constitutive relations (2.6). Here we present some new solvability results.

8.3 Sobolev Regularity of Solutions to a General Maxwell System

As an example, we show solvability of a general Maxwell system with convolution type constitutive relations (which again capture the chirality and dispersiveness of the material) on a Lipschitz domain $\Omega \subset \mathbb{R}^3$. Before formulating the problem, we provide the necessary definitions and function spaces relevant for this problem.

8.3.1 Definitions and Function Spaces

Definition 8.2 A bounded domain $\Omega \subset \mathbb{R}^3$ is called a *Lipschitz domain* if for each point $p \in \partial\Omega$, there exist an open set $O \subset \mathbb{R}^3$ such that $p \in O$ and an orthogonal coordinate system with coordinates $\xi = (\xi_1, \xi_2, \xi_3)$ having the following property:

There exist a vector $\mathbf{b} \in \mathbb{R}^3$ so that

$$O = \{\xi : -b_j < \xi_j < b_j,\ 1 \leq j \leq 3\}$$

and a Lipschitz continuous function ϕ defined on the set

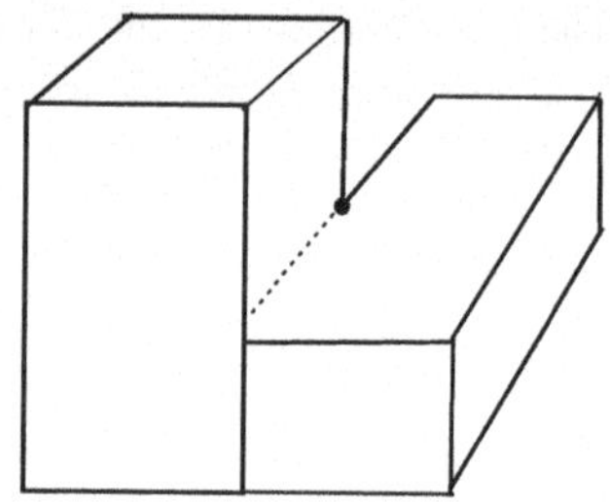

Fig. 8.3 A classical example of a non-Lipschitz domain. At the point in the figure, the boundary of the region cannot be written as the graph of a Lipschitz function

$$\mathcal{O}' = \{\xi' \in \mathbb{R}^2 : -b_j < \xi_j < b_j,\ 1 \leq j \leq 2\}$$

such that

$$\Omega \cap \mathcal{O} = \{\xi : \xi_3 < \phi(\xi'),\ \xi' \in \mathcal{O}'\} \quad \text{and} \quad \partial\Omega \cap \mathcal{O} = \{\xi : \xi_3 = \phi(\xi'),\ \xi' \in \mathcal{O}'\}$$

Lipschitz domains are classical domains to study partial differential equations. Indeed, for numerical purposes, they can be covered by a mesh of tetrahedra, which makes finite element implementation easier. That being said, on such a domain (which can have corners and reentrant edges), the unit normal is only defined at almost every point of the boundary, so any theoretical analysis of differential equations with such geometry requires more delicate arguments. Simple examples of Lipschitz domains include polyhedra. An example of a non-Lipschitz domain can be seen in Fig. 8.3.

For the following, let $\Omega \subseteq \mathbb{R}^3$ be an open connected set (possibly all of $\mathbb{R}^3$). Recall that $L^2(\Omega)$ denotes the Lebesgue space of square integrable functions on Ω, and $\left(L^2(\Omega)\right)^3$ denotes the space of square integrable vectors $f = (f_1, f_2, f_3)$. The main Hilbert space we work in will be $\mathcal{H} := \left(L^2(\Omega)\right)^3 \times \left(L^2(\Omega)\right)^3$.

For an element $U \in \left(L^2(\Omega)\right)^3$, any derivative of U here will be understood in the sense of distributions. Recall next the standard Sobolev spaces $W^{s,p}(\Omega)$ for $s \in \mathbb{Z}_+$ and $1 \leq p < \infty$ are defined by

$$W^{s,p}(\Omega) = \left\{\phi \in L^p(\Omega) : D^\alpha \phi \in L^p(\Omega)\ \forall\ |\alpha| \leq s\right\}$$

One way of defining the spaces $H^s(\Omega)$ is by restriction

$$H^s(\Omega) = \left\{u \in \mathcal{D}'(\Omega) : u = U|_\Omega \text{ for some } U \in W^{s,2}(\mathbb{R}^3)\right\}$$

where $\mathcal{D}'(\Omega)$ denotes the space of distributions on Ω.

We next define the fractional Sobolev spaces on Ω. Let $s \in \mathbb{R}_+$ such that $s = m + \sigma$, with $m \in \mathbb{Z}_+$ and $0 < \sigma < 1$. Then $H^s(\Omega)$ is defined to be the space of distributions $U \in \mathcal{D}'(\Omega)$ such that $U \in W^{m,2}(\Omega)$ and

$$\int_{\Omega}\int_{\Omega}\frac{|D^{\alpha}U(x)-D^{\alpha}U(y)|^2}{|x-y|^{3+2\sigma}}dxdy<+\infty,\quad \forall\ |\alpha|=m$$

For $s \in \mathbb{R}_+$, the space $H^s(\Omega)$ is a Hilbert space. Then, for $s > 1$, define the normed space

$$H^s(\partial\Omega)=\left\{u\in L^2(\partial\Omega):u=U|_{\partial\Omega}\ \text{ for some }\ U\in H^{s+1/2}(\Omega)\right\}$$

with norm

$$||u||_{H^s(\partial\Omega)}=\inf_{U\in H^{s+1/2}(\Omega),u=U|_{\partial\Omega}}||U||_{H^{s+1/2}(\Omega)}$$

For $0 \leq s < 1$, we define the space $H^s(\partial\Omega)$ on a Lipschitz domain as follows, see, e.g., [283].

Definition 8.3 Let $\Omega \subset \mathbb{R}^3$ be a bounded, Lipschitz domain. A distribution U defined on $\partial\Omega$ is said to belong to the space $W^{s,p}(\partial\Omega)$ for $0 < s < 1$ if

$$U\circ\phi\in W^{s,p}\left(\mathcal{O}'\cap\phi^{-1}(\partial\Omega\cap\mathcal{O})\right)$$

for all possible $\mathcal{O}, \phi$ as in Definition 8.2.

We next define the spaces

$$\mathbf{H}(\text{div},\Omega):=\left\{U\in\left(L^2(\Omega)\right)^3:\nabla\cdot U\in L^2(\Omega)\right\},$$

$$\mathbf{H}(\text{curl},\Omega):=\left\{U\in\left(L^2(\Omega)\right)^3:\nabla\times U\in\left(L^2(\Omega)\right)^3\right\}$$

and

$$\mathbf{H}^0(\text{div},\Omega):=\ \text{completion of }\left(C_0^{\infty}(\Omega)\right)^3\ \text{in the }\mathbf{H}(\text{div},\Omega)\text{ norm},$$

$$\mathbf{H}^0(\text{curl},\Omega):=\ \text{completion of }\left(C_0^{\infty}(\Omega)\right)^3\ \text{in the }\mathbf{H}(\text{curl},\Omega)\text{ norm}$$

As usual, the space $C_0^{\infty}(\Omega)$ denotes the space of infinitely differentiable compactly supported functions on Ω. For more details and discussion of relevant Sobolev spaces, see, e.g., [140, 283, 304].

8.3.2 Problem Formulation

We use the constitutive relations (2.6) and define $\mathbf{U} = (\mathbf{E}, \mathbf{H})^T$ and $\mathbf{d} = (\mathbf{D}, \mathbf{B})^T$, where T denotes transpose. Furthermore, consider the matrices

$$M := \begin{bmatrix} \varepsilon & \xi \\ \eta & \mu \end{bmatrix} \tag{8.13}$$

and

$$N(t) := \begin{bmatrix} \varepsilon_d(t) & \xi_d(t) \\ \eta_d(t) & \mu_d(t) \end{bmatrix} \tag{8.14}$$

The matrix M is sometimes called the optical response, and N sometimes is referred to as the susceptibility kernel (which takes into account memory effects) [194].

Define the unbounded[6] Maxwell operator $A : \mathscr{D}(A) \to \mathscr{H}$ with domain $\mathscr{D}(A) := \mathbf{H}^0(\text{curl}, \Omega) \times \mathbf{H}(\text{curl}, \Omega) \subset \mathscr{H}$ by $A := \begin{bmatrix} 0 & -\nabla\times \\ \nabla\times & 0 \end{bmatrix}$. Consider on $\mathscr{H}$ the inner product $(\mathbf{U}, \mathbf{V})_{\mathscr{H}} := (\mathbf{U}, M\mathbf{V})_{L^2}$. This will indeed be an inner product on $\mathscr{H}$ provided that M is almost everywhere coercive, i.e., if

$$M(x) \cdot \mathbf{y} \cdot \mathbf{y} \geq C_0 |\mathbf{y}|^2 \tag{8.15}$$

for a.e. $x \in \Omega$ and for all $\mathbf{y} \in \mathbb{R}^3 \times \mathbb{R}^3$; see, e.g., Proposition 8.3 in [194]. Thus it remains to check under what conditions on the material parameters $\varepsilon, \mu, \eta, \xi$ are required in order for (8.15) to hold. Thus we assume that:

(H1) There exists $C > 0$ such that for each $1 \leq j, k \leq 6$, for a.e. $\mathbf{x} \in \Omega$, and for each $\mathbf{w} \in \mathbb{R}^6$, there holds

$$w_j M_{jk}(x) w_k \geq C|w|^2$$

Regarding differentiability of each of the matrices $\varepsilon, \mu, \eta, \xi$, we will only assume each element of each matrix is measurable (in addition to the above mentioned boundedness). We also require:

(H2) $M(x)\mathbf{y} \cdot \mathbf{y} > 0$ for almost every $x \in \Omega$ and all nonzero $\mathbf{y} \in \mathbb{R}^3 \times \mathbb{R}^3$.

Now we define $\kappa(t) := -M^{-1}N(t)$. We will further assume:

(H3) $\kappa(t) \in C^1([0, T); \mathcal{B}(\mathscr{H}))$

[6] Given normed linear spaces X, Y and a linear operator $A : X \to Y$, we define the operator norm of A by $||A||_{\text{op}} = \sup\left\{\frac{||Ax||}{||x||} : x \neq 0\right\}$. If $||A||_{\text{op}} < \infty$, then A is called bounded; otherwise it is unbounded. It is well known that A is bounded if and only if A is continuous at zero [455].

where $\mathcal{B}(X)$ denotes the Banach algebra of bounded operators on X, endowed with the strong operator norm topology. Thus, in terms of the material parameters $\varepsilon_d, \xi_d, \eta_d, \mu_d$, we require each to be at least continuously differentiable in time. We could have also imposed a spatial regularity constraint on κ, but this would have required further restrictions on the material parameters within the matrix M.

Let us turn to the boundary condition, as was similarly done in [396, 397]. Since Ω is Lipschitz, we define the trace space [404] by

$$\mathbf{T}(\partial\Omega) = \Big\{\xi \in \left(H^{-1/2}(\partial\Omega)\right)^3 : \exists\, \eta \in H^{-1/2}(\partial\Omega)\ s.t.\ \forall\, \phi \in H^2(\Omega),$$

$$\langle \xi, \gamma(\nabla\phi)\rangle_{H^{1/2}(\partial\Omega)} = \langle \eta, \gamma(\phi)\rangle_{H^{1/2}(\partial\Omega)}\Big\}$$

where γ denotes the scalar trace mapping $\gamma : f \mapsto f|_{\partial\Omega}$. When Ω is smooth (e.g., the graph of a continuously differentiable function), this space becomes the more familiar space of tangential vector fields of order $-1/2$, $H^{-1/2}(\text{div}, \partial\Omega)$, see, e.g., [86]. In [59] it was shown that $\mathbf{T}(\partial\Omega)$ is isomorphic to the space

$$H^{-1/2}\left(\text{div}_{\partial\Omega}, \partial\Omega\right) = \left\{u \in V_\pi' : \text{div}_{\partial\Omega} u \in H^{-1/2}(\partial\Omega)\right\}$$

where V_π' is the dual of the space V_π, and V_π is the image of $H^{1/2}(\partial\Omega)$ under the map $\pi(\mathbf{w}) = \boldsymbol{\nu} \times (\mathbf{w} \times \boldsymbol{\nu})|_{\partial\Omega}$. Finally, the operator $\text{div}_{\partial\Omega} : L_t^2(\partial\Omega) \to H^{-1}(\partial\Omega)$ is the adjoint of the surface gradient $\nabla_{\partial\Omega} : H^1(\Omega) \to L_t^2(\partial\Omega)$, where $L_t^2(\partial\Omega)$ denotes the space of square integrable w on $\partial\Omega$ such that $\boldsymbol{\nu} \cdot \mathbf{w} = 0$. For more details on these trace operators and function spaces, see [59, 397].

We impose now the boundary condition that prescribes the tangential component[7] of the electric field on the boundary:

$$\nu(\mathbf{x}) \times \mathbf{E}(\mathbf{x}, t)|_{\partial\Omega} = \mathbf{f}(\mathbf{x}) \in \mathbf{T}(\partial\Omega), \quad \mathbf{x} \in \partial\Omega,\ t \geq 0 \tag{8.16}$$

Now, to instead deal with homogeneous boundary data, we will lift $\mathbf{f}$ [73], which requires surjectivity of the tangential trace map. Tartar [404] showed that the tangential trace map which maps $\mathbf{H}(\text{curl}, \Omega)$ to $\mathbf{T}(\partial\Omega)$ is surjective. Hence there exists a right inverse, call it $R : \mathbf{T}(\partial\Omega) \to \mathbf{H}(\text{curl}, \Omega)$. Then we define the adjusted fields $\widetilde{\mathbf{E}}$ and $\widetilde{\mathbf{H}}$ as follows: $\widetilde{\mathbf{E}} := \mathbf{E} - R\mathbf{f}$ and $\widetilde{\mathbf{H}} := \mathbf{H}$, and this results in a homogeneous (perfect conductor) boundary condition for $\widetilde{\mathbf{E}}$. Thus, from here on we will drop the tilde and just consider that $\mathbf{E}$ satisfies a homogeneous boundary condition.

In terms of the Cauchy problem, we wish to solve

[7] To be precise, we should really write $\gamma_t(\mathbf{E}) = \mathbf{f}$, where γ_t is the tangential trace operator. We have skipped this to avoid any additional unnecessary technicalities, but the precise trace operator and function spaces where it is defined, as well as its properties, can be found in [59, 397]. The point is that the geometry affects how one defines the boundary conditions.

$$\frac{d}{dt}\left(M\mathbf{U}+\int_0^t N(t-s)\mathbf{U}(s)ds\right) = -A\mathbf{U}(t)+\mathbf{G}(t)$$
$$\mathbf{U}(0) = \mathbf{U}_0 \tag{8.17}$$
$$\boldsymbol{\nu}\times\mathbf{E}(t) = 0$$

where

$$\mathbf{G}(t) = M^{-1}\begin{pmatrix}\mathbf{J}(t)\\ \nabla\times R\mathbf{f}\end{pmatrix}$$

Now, by hypothesis (H2), we first see that (8.17) becomes an integrodifferential equation of Volterra type:

$$\mathbf{U}'(t) = \left(-\kappa(0)-M^{-1}A\right)\mathbf{U}(t)+\int_0^t \kappa'(t-\tau)\mathbf{U}(\tau)d\tau$$
$$\mathbf{U}(0) = \mathbf{U}_0 \tag{8.18}$$
$$\nu\times\mathbf{E}(t) = 0$$

This follows immediately from the differentiability properties of convolution, see, e.g., [194, Lemma B.2].

Lemma 8.1 *Suppose* $F \in W^1_{loc}([0,T);\mathcal{B}(\mathcal{H}))$. *Then the convolution* $F*g$ *is differentiable and*

$$(F*g)'(t) = F(0)g(t)+(F'*g)(t)$$

Now, the operator $A_\kappa := -A+\kappa(0)$ with domain $\mathcal{D}(A_\kappa) = \mathcal{D}(A)$ generates a C_0-semigroup $T_\kappa(t)$ which also satisfies a Volterra equation [120]

$$T_\kappa(t)\mathbf{h} = T(t)\mathbf{h}+\int_0^t T(t-\tau)\kappa(0)T_\kappa(\tau)\mathbf{h}d\tau \tag{8.19}$$

for each $t\geq 0$ and $\mathbf{h}\in\mathcal{H}$. Above, $T(t)$ is the semigroup generated by $-M^{-1}A$.

Problems of the form (8.18) have been solved in [278]. Indeed, provided $\mathbf{U}_0 \in \mathcal{D}(-M^{-1}A) = \mathcal{D}(A)$ and $\mathbf{G}(t)$ is continuously differentiable, a classical solution $\mathbf{U}$ of (8.18) exists and in fact is given explicitly by the representation

$$\mathbf{U}(t) = \sum_{n=0}^{\infty} V_n(t)\mathbf{U}_0+\sum_{n=0}^{\infty} g_n(t)$$

where $V_0(t) = T_\kappa(t)$, $V_n(t) = \int_0^t T_\kappa(t-\tau)\int_0^\tau \kappa'(\tau-s)V_{n-1}(s)dsd\tau$, $g_0(t) = \int_0^t T_\kappa(t-\tau)\mathbf{G}(\tau)d\tau$, and $g_n(t) = \int_0^t T_\kappa(t-\tau)\int_0^\tau \kappa'(\tau-s)g_{n-1}(s)dsd\tau$. The above series converge absolutely and uniformly in each finite interval of $[0, \infty)$. Thus we have the following, which was essentially shown in [194]. We have just clarified how to impose nonzero boundary data (via a nontrivial lifting) for the electric field and reduce it to the homogeneous case.

Theorem 8.1 *Suppose that hypotheses (H1), (H2), and (H3) hold. Furthermore, suppose that* $\mathbf{J} \in C^1([0, T); \mathscr{H})$ *and that* $\mathbf{U}_0 \in \mathbf{H}(curl, \Omega) \times \mathbf{H}(curl, \Omega)$ *and* $\mathbf{f} \in \mathbf{T}(\partial\Omega)$. *Then there exists a unique classical solution of (8.18)* $\mathbf{U}$, *such that*

$$\mathbf{U} \in C^1([0, \infty); \mathbf{H}(curl, \Omega) \times \mathbf{H}(curl, \Omega))$$

Let us clarify what is meant by classical solution. A function $\mathbf{U}$ is a classical solution of (8.18) if $\mathbf{U}$ is continuously differentiable, $\mathbf{U}(t) \in \mathscr{D}(A)$ for $t > 0$, and $\mathbf{U}$ satisfies (8.18) pointwise for $t > 0$. This contrasts with the notion of mild solution, which is only continuous, satisfies $\int_0^t \mathbf{U}(\tau)d\tau \in \mathscr{D}(A)$ for $t > 0$ and $\mathbf{U}$ satisfies an integrated version of (8.18).

We next explore a higher regularity result for the same problem (which appears to be new).

Using the spaces $H^s(\Omega)$, we can define the higher order $\mathbf{H}$(curl) spaces for $s \geq 0$ by

$$\mathbf{H}^s(\text{curl}, \Omega) := \left\{U \in \left(H^s(\Omega)\right)^3 : \nabla \times U \in \left(H^s(\Omega)\right)^3\right\} \tag{8.20}$$

The norm on $\mathbf{H}^s(\text{curl}, \Omega)$ is given by $||U||^2_{\mathbf{H}^s(\text{curl},\Omega)} := ||U||^2_{H^s(\Omega)} + ||\nabla \times U||^2_{H^s(\Omega)}$. It is clear what is meant by this space if $s \in \mathbb{Z}_+$. Now, if $s \in \mathbb{R}_+$, write $s = m + \sigma$, where $m \in \mathbb{Z}_+$ and $0 < \sigma < 1$. The norm on $\mathbf{H}^s(\text{curl}, \Omega)$ then becomes

$$\begin{aligned} ||U||^2_{\mathbf{H}^s(\text{curl},\Omega)} &= ||U||^2_{H^m(\Omega)} + ||\nabla \times U||^2_{H^m(\Omega)} + \\ \sum_{|\alpha|=m} \Bigg(\int_\Omega \int_\Omega & \frac{|D^\alpha U(x) - D^\alpha U(y)|^2}{|x-y|^{3+2\sigma}} dxdy + \\ & \int_\Omega \int_\Omega \frac{|D^\alpha (\nabla \times U)(x) - D^\alpha (\nabla \times U)(y)|^2}{|x-y|^{3+2\sigma}} dxdy \Bigg) \end{aligned}$$

From now on we will assume $s \geq 1$ is an integer. Then we can define the space

$$\mathbf{H}^s_0(\text{curl}, \Omega) = \text{ closure of } \left(C_0^\infty(\Omega)\right)^3 \text{ in } \mathbf{H}^s(\text{curl}, \Omega)$$

We have (see [396]) the following lemma:

Lemma 8.2 *Let* $\Omega \subset \mathbb{R}^3$ *be a bounded, Lipschitz domain with unit normal* $\boldsymbol{\nu}$, *and let* $s \in \mathbb{Z}_+$.

1. $\mathbf{H}^s(curl, \Omega) = \overline{\left(C^\infty(\overline{\Omega})\right)^3}^{||\cdot||_{\mathbf{H}^s(curl,\Omega)}}$.

2. *The tangential trace map for* $W \in \left(C^\infty(\overline{\Omega})\right)^3$ *given by* $\boldsymbol{\nu} \times W|_{\partial\Omega}$ *extends to a continuous linear map from* $\mathbf{H}^s(curl, \Omega)$ *into* $\left(H^{-(s-1/2)}(\partial\Omega)\right)^3$.
3. $\mathbf{H}_0^s(curl, \Omega) = \{U \in \mathbf{H}^s(curl, \Omega) : (U, \nabla\times\phi)_{H^s} = (\nabla\times U, \phi)_{H^s} \forall \phi \in \left(C^\infty(\overline{\Omega})\right)^3\}$.
4. $\mathbf{H}_0^s(curl, \Omega) = \{U \in \mathbf{H}^s(curl, \Omega) : \boldsymbol{\nu} \times U = 0 \ on \ \partial\Omega\}$.

Let $s \geq 1$ be an integer, and we consider $\mathcal{H} = (H^s(\Omega))^3 \times (H^s(\Omega))^3$ with the bilinear form $(\mathbf{U}, \mathbf{V})_{\mathcal{H}} = (\mathbf{U}, M\mathbf{V})_{H^s}$.

The Maxwell operator A is then defined on

$$\mathcal{D}(A) = \mathbf{H}_0^s(\text{curl}, \Omega) \times \mathbf{H}^s(\text{curl}, \Omega) \tag{8.21}$$

To understand what conditions on M are required in order to obtain classical solutions in this setting, we first require that the above bilinear form defines an inner product on $\mathcal{H}$. In fact, we assume:

(H4) $M(x)$ has enough smoothness so that $(\mathbf{U}, \mathbf{V})_{\mathcal{H}}$ is an inner product on $\mathcal{H}$.

We additionally suppose that hypothesis (H2) holds, in order that the operator $-M^{-1}A$ is the infinitesimal generator[8] of a C_0-semigroup.[9] Finally we will assume that hypothesis (H3) holds for $\kappa(t)$. Then, provided we impose the homogeneous boundary condition

$$\boldsymbol{\nu}(\mathbf{x}) \times \mathbf{E}(\mathbf{x}, t)|_{\partial\Omega} = 0, \quad \mathbf{x} \in \partial\Omega, \ t \geq 0 \tag{8.22}$$

then the techniques discussed previously can be used to obtain classical solutions of (8.17) with $\mathbf{f} \equiv 0$, provided now the initial data $\mathbf{U}_0 \in \mathbf{H}_0^s(\text{curl}, \Omega) \times \mathbf{H}^s(\text{curl}, \Omega)$. This in particular improves Theorem 6.4 in [396] as we do not require merely constant material parameters. However, the difficulty of treating nonzero boundary data remains, as this requires showing the tangential trace map on the higher order Sobolev spaces is surjective, which we have not done.

Now, suppose that hypotheses (H1) and (H2) hold with the boundary condition (8.22) and that $\varepsilon, \mu, \eta, \xi \in L^\infty\left(\Omega, \mathbb{R}^{3\times3}\right)$. We will rid ourselves of any differentiability requirements of the material parameters in hypothesis (H4). When $N(t) \equiv 0$ and $\xi = \eta \equiv 0$, the following H^s regularity result has been shown in [198].

[8] A one-parameter semigroup on a Banach space X is a function $T : [0, \infty) \to \mathcal{L}(X)$, taking values in the space of bounded, linear operators on X, such that $T(t+s) = T(t)T(s)$ for all $t, s \geq 0$. If also $\lim_{t\to0^+} T(t)x = x$ for all $x \in X$, then T is called a C_0 semigroup. Given a C_0 semigroup $T(t)$, the infinitesimal generator of $T(t)$ is given by an operator I such that $Ix = \lim_{t\to0} \frac{1}{t}(T(t) - \text{Id})x$ whenever the limit exists, where Id denotes the identity operator [120].

[9] Note that $\mathcal{D}(-M^{-1}A) = \mathcal{D}(A)$ with $\mathcal{D}(A)$ given by (8.21), as one might expect.

Theorem 8.2 *There exists $s^* \in (0, s_0)$ that depends on Ω, ε, μ such that for all $s \in [0, s^*]$ and initial data $\mathbf{U}_0 \in H^s(\Omega)$, one has $\mathbf{U}(t) \in C([0, \infty); H^s(\Omega))$.*

It is further assumed in the above result that ε, μ satisfy the following multiplier property:

$$\varepsilon \mathbf{G}, \mu \mathbf{G} \in H^{s_0}(\Omega) \text{ for all } \mathbf{G} \in H^{s_0}(\Omega) \text{ for some } s_0 \in (0, 1/2) \tag{8.23}$$

Notice that this property holds in particular for piecewise constant material parameters.

Let $\mathcal{H}_0^s(\text{curl}, \Omega)$ and $\mathcal{H}_0^s(\text{div}, \Omega)$ denote for $0 \le s \le 1$ the interpolation spaces $\left((L^2(\Omega))^3, \mathbf{H}_0(\text{curl}, \Omega)\right)_s$ and $\left((L^2(\Omega))^3, \mathbf{H}_0(\text{div}, \Omega)\right)_s$, respectively. Let us briefly recall how to define interpolation spaces in general; for this we use the notation from Appendix B in [283].

Let X_0, X_1 be normed subspaces of a vector space. Then X_0, X_1 form what is called a compatible pair $X = (X_0, X_1)$. Given a compatible pair X, then, it is possible (via the so-called K and J methods, see, e.g., Appendix B in [283]) to construct a family of normed spaces

$$X_{\theta,p} = (X_0, X_1)_{\theta,p}, \qquad 0 < \theta < 1, \ 1 \le p \le \infty$$

each of which is between X_0 and X_1 in the sense that

$$X_0 \cap X_1 \subseteq X_{\theta,p} \subseteq X_0 + X_1$$

and has the following interpolation property: If $Y = (Y_0, Y_1)$ is another compatible pair and $A_0 : X_0 \to Y_0$ and $A_1 : X_1 \to Y_1$ are bounded, linear operators, then, if

$$A_0 w = A_1 w, \qquad w \in X_0 \cap X_1$$

then A_0 and A_1 are called compatible, and there exists a unique bounded linear operator $A_\theta : X_{\theta,p} \to Y_{\theta,p}$ such that

$$A_\theta w = A_0 w = A_1 w, \qquad w \in X_0 \cap X_1$$

The interpolation spaces mentioned above, namely, the spaces

$$\left((L^2(\Omega))^3, \mathbf{H}_0(\text{curl}, \Omega)\right)_s, \qquad \left((L^2(\Omega))^3, \mathbf{H}_0(\text{div}, \Omega)\right)_s$$

correspond to taking $\theta = s$ and $p = 2$. It turns out that these spaces actually coincide with $\mathbf{H}_0^s(\text{curl}, \Omega)$ and $\mathbf{H}_0^s(\text{div}, \Omega)$; see, e.g., [253].

Now, suppose that $\Omega = R = (0, a) \times (0, b) \times (0, c)$ is a rectangle. Let $A \in L^\infty(R, \mathbb{R}^{3\times3})$ be uniformly positive definite, so there exists $K > 0$ such that

$$\xi A(y)\xi \geq K|\xi|^2, \quad \forall\, y \in R, \xi \in \mathbb{R}^3$$

Suppose further that A satisfies the multiplier property:

$$Af \in H^{s_0}(R) \quad \forall\, f \in H^{s_0}(R) \quad \text{for some } s_0 \in (0, 1/2) \tag{8.24}$$

By [198, Theorem 1], there exist $\bar{s} \in (0, s_0)$ and a constant $c_0 > 0$ which depends only on A, such that for all $s \in [0, \bar{s}]$ and $\mathbf{E} \in \mathcal{H}_0^s(\text{curl}, R)$ such that $A\mathbf{E} \in \mathcal{H}_0^s(\text{div}, R)$, we have $\mathbf{E} \in (H^s(R))^3$ and

$$||\mathbf{E}||_{H^s(R)} \leq c_0 \left(||A\mathbf{E}||_{\mathcal{H}_0^s(\text{div}, R)} + ||\mathbf{E}||_{\mathcal{H}_0^s(\text{curl}, R)} \right)$$

Now we let Ω be a bounded Lipschitz domain, with A satisfying, for some $0 < s_0 < 1/2$,

$$Af \in H^s(\Omega) \quad \forall\, s \in [0, s_0] \quad \text{and } f \in H^s(\Omega) \tag{8.25}$$

An H^s regularity result of the Maxwell system with $\eta = \xi = \varepsilon_d = \xi_d = \eta_d = \mu_d = 0$ is proved in [198, Theorem 3]. We can show an analogous result here provided we impose certain conditions on the material parameters.

Theorem 8.3 *Suppose that $\varepsilon, \xi, \eta, \mu$ satisfy the H^s multiplier property (8.25) for some $s_0 \in (0, 1/2)$. Furthermore, suppose that hypotheses (H1)–(H3) hold. Then there exists a classical solution of (8.18) for each $\mathbf{U}_0 \in \mathbf{H}_0^s(curl, \Omega) \times \mathbf{H}^s(curl, \Omega)$, provided $\mathbf{J} \in C^1([0, T); \mathcal{H})$.*

Proof According to the technique of Marti [278], we need:

1. $-\kappa(0) - M^{-1}A$ generates a C_0-semigroup.
2. $\kappa(t)$ is continuously differentiable in time.
3. $\mathbf{G}(t)$ is continuously differentiable in time.
4. $\mathbf{U}_0 \in D(-\kappa(0) - M^{-1}A)$.

Item (2) holds by assumption (H3), and item (3) holds if we assume the current density $\mathbf{J}(t)$ is continuously differentiable in time. Since M satisfies the H^s multiplier property, we have that $-M^{-1}A$ generates a C_0-group $T(t)$ on $H^s(\Omega)$ for all $s \in [0, \bar{s})$, where $\bar{s}$ is as in the paragraph following (8.24). This follows directly from the proof of Theorem 3 in [198]. Indeed, the group can be seen to be

$$T(t) = \exp(-tM^{-1}A)$$

Now, by (H3) we know that $-\kappa(0) \in \mathcal{B}(\mathcal{H})$, so we consider the operator

$$A_P = -M^{-1}A - \kappa(0)$$

with domain $D(A_P) = D(-A) = \mathbf{H}_0^s(\text{curl}, \Omega) \times \mathbf{H}^s(\text{curl}, \Omega)$. Then (see, e.g., [120]), A_p also generates a C_0 semigroup $T_p(t)$ which satisfies

$$T_p(t)x = T(t)x + \int_0^t T(t-s)\kappa(0)T_p(s)ds$$

for each $t \geq 0, x \in \mathcal{H}$. Thus Marti's method implies there exists a unique C^1 solution $\mathbf{U}(t) \in C^1\left([0, \infty); \mathbf{H}_0^s(\text{curl}, \Omega) \times \mathbf{H}^s(\text{curl}, \Omega)\right)$, provided the initial data $\mathbf{U}_0 \in \mathbf{H}_0^s(\text{curl}, \Omega) \times \mathbf{H}^s(\text{curl}, \Omega)$.

In particular, we have replaced any differentiability requirements from **(H4)** by merely requiring the material parameters satisfy the H^s multiplier property (8.25).

Finally, we return to the geometric optics regime to study the optics of helical media.

8.4 Optics of Chiral Media

In plane-polarized light, the direction in which the electric field oscillates is contained in a single plane. Similarly, the magnetic field oscillates within another plane, and these planes are orthogonal to each other. Optical activity is the ability of a chiral molecule to rotate the plane of plane-polarized propagating light. Thus, an optically active material can change the polarization state, and indeed this is the case with chiral liquid crystals [106].

The periodic, helical structure of a chiral nematic liquid crystal results in a periodic dielectric ε; this results in various optical phenomena, such as selective reflection (discussed below), Bragg reflection (discussed in Sect. 8.4.4), and the Borrmann effect [74].

Remark 8.4 We will not discuss the Borrmann effect here but suffice it to say that it is a classical effect in X-ray radiation where there is a dramatic increase in transmitted X-ray intensity through a crystal subject to Bragg's Law of reflection. A similar optical effect has been observed in *absorbing* cholesteric liquid crystals[10] [21, 77, 402]. See in particular [312] for light propagating at normal incidence (i.e., propagating along the cholesteric axis) and [34] for arbitrary incidence angle (see also [119]). In the aforementioned references, it is shown how to suppress absorption by having an ideal ordering of molecules, and the following relationship holds between the first two of the principal dielectric parameters:

---

[10] We are generally considering non-absorbing cholesteric liquid crystals in what follows; that is, we assume that cholesteric liquid crystals do not absorb light. In fact, there are interesting optical phenomena when one considers absorbing cholesterics, such as the Borrmann effect, and this absorption is related to diffraction via the absorption of individual molecules [35].

$$\Im(\varepsilon_1 - \varepsilon_2) = \pm 2\Im\left(\frac{\varepsilon_1 + \varepsilon_2}{2}\right)$$

where $\Im$ denotes the imaginary part. More precisely, by suppression of absorption, it is meant that for a given polarization of light and incident angle, the intensity of the reflected light from the liquid crystal is equal to the incident intensity [35].

8.4.1 Selective Reflection

One of the most striking features of cholesterics is *selective reflection*. The helical structure of a cholesteric can cause circularly polarized light of a particular wavelength to be reflected. This remarkable phenomenon is of keen interest for various optical device development, from reflective displays [25, 297, 337, 379, 446, 447, 461] to lasers [91, 193, 249, 393, 418, 448, 459], as well as the ability to control light [51, 460]. It has also been reported that certain cholesteric liquid crystals can control the phase of light using selective reflection [81, 82, 224, 356, 435].

It is interesting to see how selective reflection arises mathematically, so we recall here the argument from [115] and [114] in order to arrive at a version of Snell's Law, which has been one of the main focuses of this book.

Consider the cholesteric director (8.1). We will consider two coordinate systems: the "lab" frame with coordinates (x, y, z) with the z-axis parallel to the chiral axis $\mathbf{c}$ and a local coordinate system (ξ, η, χ) with the χ-axis parallel to the z-axis; see Fig. 8.4 and Figure 1 in [114]. The later system is chosen so that the permittivity ε is diagonal in this system; write $\varepsilon = \mathrm{diag}(\varepsilon_1, \varepsilon_2, \varepsilon_3)$ then. In particular, $\varepsilon = \varepsilon(q)$ where as before $q = (2\pi/p)z$, with p denoting the pitch of the helix. In the lab frame, ε can be written [114] as

$$\varepsilon = \begin{pmatrix} \epsilon(1 + \beta\cos(2q)) & \epsilon\beta\sin(2q) & 0 \\ \epsilon\beta\sin(2q) & \epsilon(1 - \beta\cos(2q)) & 0 \\ 0 & 0 & \varepsilon_3 \end{pmatrix} \tag{8.26}$$

where $\epsilon = (1/2)(\varepsilon_1 + \varepsilon_2)$ and $\beta = (\varepsilon_1 - \varepsilon_2)/(\varepsilon_1 + \varepsilon_2)$ is the so-called dielectric anisotropy. In the system (ξ, η, χ), the diagonal ε can be written in terms of ϵ and β as follows:

$$\varepsilon_1 = \epsilon(1 + \beta), \qquad \varepsilon_2 = \epsilon(1 - \beta), \qquad \varepsilon_3 = \varepsilon_3$$

As usual, $\mathbf{D}$ is related to $\mathbf{E}$ through the relation $\mathbf{D} = \varepsilon\mathbf{E}$. Thus, to study waves propagating from an isotropic medium to a cholesteric liquid crystal, one needs to consider the wave equation

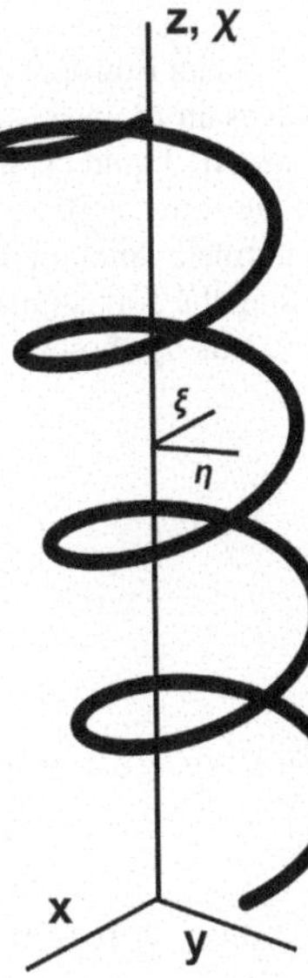

Fig. 8.4 The two coordinate systems for selective reflection. The z-axis is also parallel to the chiral axis **c**

$$\frac{1}{c^2}\frac{\partial^2 \mathbf{D}}{\partial t^2} = \Delta \mathbf{E} - \nabla (\nabla \cdot \mathbf{E}) \tag{8.27}$$

This is of course obtained from Maxwell's equations (2.1)-(2.2) in the standard way:

$$\frac{\partial^2 \mathbf{D}}{\partial t^2} = \nabla \times \frac{\partial \mathbf{H}}{\partial t} = \nabla \times \frac{1}{\mu}\frac{\partial \mathbf{B}}{\partial t} = -\frac{1}{\mu}\nabla \times \nabla \times \mathbf{E}$$

assuming μ to be constant and all fields to be sufficiently differentiable. Then simply apply the vector calculus identity

$$\nabla \times \nabla \times \mathbf{E} = \nabla (\nabla \cdot \mathbf{E}) - \Delta \mathbf{E}$$

where

$$\Delta \mathbf{E} = (\Delta E_1, \Delta E_2, \Delta E_3), \qquad \mathbf{E} = (E_1, E_2, E_3), \quad \Delta = \frac{\partial^2}{\partial x^2} + \frac{\partial^2}{\partial y^2} + \frac{\partial^2}{\partial z^2}$$

Assume now that light passes through an isotropic material with refractive index n_1 and strikes the cholesteric with some incident angle θ_1 as in Fig. 8.5.

Since the cholesteric is periodic, waves generally will not simply be plane waves [30]. Thus, we can look for solutions to (8.27) which have a plane wave form in the direction normal to the cholesteric axis (for instance, the x-axis here since **c** is parallel to the z-axis) but is otherwise arbitrary in z:

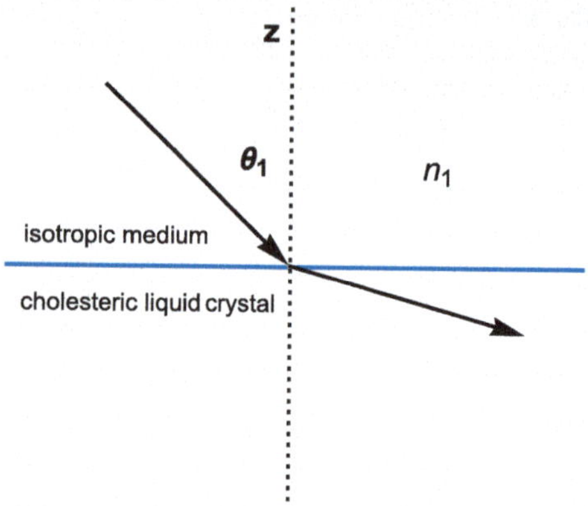

Fig. 8.5 An incident light ray from an isotropic medium (such as air or glass) strikes a cholesteric liquid crystal, making some angle θ_1 with the normal, which in this case is along the z-axis (and along the χ-axis by choice)

$$\mathbf{E}(z) = \mathbf{F}(z)e^{i\omega(t-(m/c)x)} \tag{8.28}$$

for some constant m. Suppose the components of $\mathbf{F}$ in lab frame are given by $\mathbf{F}_{\text{lab}}(z) = (F_1(z), F_2(z), F_3(z))$ and in the local frame by $\mathbf{F}_{\text{local}}(z) = (f_1(z), f_2(z), F_3(z))$. By using (8.28) in (8.27), we obtain a pair of second order equations with periodic coefficients for f_1, f_2:

$$\begin{aligned} \frac{d^2 f_1}{dq^2} - 2\frac{df_2}{dq} + (a_1 + a_2\cos(2q))f_1 = 0, \\ \frac{d^2 f_2}{dq^2} + 2\frac{df_1}{dq} + b_1 f_2 - a_2\sin(2q)f_1 = 0 \end{aligned} \tag{8.29}$$

and for the third component we obtain

$$F_3 - i\frac{\lambda}{p}\frac{m}{\varepsilon_2 - m^2}\frac{dF_1}{dq} = 0 \tag{8.30}$$

where $\lambda = 2\pi/\omega$ and where

$$\begin{aligned} a_1 &= \left(\frac{p}{\lambda}\right)^2\left(\varepsilon_1 - \frac{m^2}{2}\left(\frac{\varepsilon_1}{\varepsilon_2} + 1\right)\right) - 1, \\ a_2 &= \left(\frac{p}{\lambda}\right)^2\frac{m^2}{2}\left(1 - \frac{\varepsilon_1}{\varepsilon_2}\right), \\ b_1 &= \left(\frac{p}{\lambda}\right)^2\left(\varepsilon_2 - m^2\right) - 1 \end{aligned}$$

Remark 8.5 We have implicitly assumed above that the cholesteric is locally uniaxial, so that the ξ-axis is parallel to **d**, which implies due to symmetry of the nematic that $\varepsilon_2 = \varepsilon_3$.

In lieu of (8.29), one can consider the equivalent fourth order equation

$$\frac{d^4 f_1}{dq^4} + (a + a_2\cos(2q))\frac{d^2 f_1}{dq^2} - 6a_2\sin(2q)\frac{df_1}{dq} + (b + da_2\cos(2q))f_1 = 0 \tag{8.31}$$

where $a = 4 + a_1 + b_1$, $b = a_1 b_1$, and $d = b_1 - 8$. Floquet's Theorem implies there exist solutions of the form

$$f_1(q) = e^{i\mu q}P_1(q), \qquad f_2(q) = e^{i\mu q}P_2(q) \tag{8.32}$$

where each P_j is a periodic function with period π and μ is the so-called Floquet exponent. This follows from determination of the characteristic equation of (8.31), which should vanish. Let us provide a few more details here. First, note that the system (8.29) can be written in the form

$$\frac{dX}{dq} = AX$$

where Id denotes the identity matrix and

$$X = \begin{pmatrix} f_1 \\ f_2 \\ \frac{df_1}{dq} \\ \frac{df_2}{dq} \end{pmatrix}, \qquad A = \begin{pmatrix} 0 & \text{Id} \\ -C & -Q \end{pmatrix}$$

with

$$Q = \begin{pmatrix} 0 & -2 \\ 2 & 0 \end{pmatrix}, \quad C = \begin{pmatrix} a_1 + a_2\cos(2q) & 0 \\ -a_2\sin(2q) & b_1 \end{pmatrix}$$

Since the coefficients of A have period $T = \pi$, we may build the so-called monodromy matrix

$$M = \widetilde{X}(\pi)$$

where $\widetilde{X}$ is the fundamental matrix, defined by its satisfaction of $X'(q) = AX$ with the initial condition $X(0) = \text{Id}$. That is, the monodromy matrix is the value at the "end of the period" of the fundamental matrix. The characteristic equation is then

$$\det(M - \lambda \mathrm{Id}) = 0$$

Eigenvalues of M are called characteristic multipliers, and a complex number μ is a Floquet exponent if ρ is a characteristic multiplier and $e^{\pi\mu} = \rho$ [407].

The point is that this exponent μ will dictate whether wave propagation is allowed or not. Now, the solutions (8.32) then lead to an electric field of the form [114]

$$\mathbf{E} = \begin{pmatrix} \widetilde{f}_1(z) \\ \widetilde{f}_2(z) \\ \widetilde{f}_3(z) \end{pmatrix} e^{i\omega\left(t - \frac{m}{c}x + \frac{\mu}{c}\frac{\lambda}{p}z\right)} \tag{8.33}$$

for some p-periodic functions $\widetilde{f}_j(z)$, $j = 1, 2, 3$, where λ denotes the vacuum wavelength and m is a constant. There are four characteristic exponents μ, two conjugate pairs with opposite signs. Now, by identifying the wave vector $\mathbf{k}$ in (8.33) and using the fact that a refractive index can be identified by $n = c|\mathbf{k}|/\omega$, we see that if μ is real (so that the wave is *stable*[11]), then such a wave propagates like a standard plane wave in a material with refractive index

$$n = \sqrt{g} := \sqrt{m^2 + \mu^2 \left(\frac{\lambda}{p}\right)^2} \tag{8.34}$$

and making an angle $\theta = \arcsin(m/\sqrt{g})$ with the cholesteric axis.

8.4.2 Snell's Law for Stable Waves

Consider again Fig. 8.5. We will derive a classical version of Snell's Law now. Indeed, consider a plane wave originating in the isotropic material (having refractive index $n_1 = \sqrt{\varepsilon_{\text{iso}}}$), and suppose it comes in with an incident angle θ_1.

By imposing phase matching at along the interface $z = 0$, i.e., equating components of the corresponding wave vectors, we find that

$$m = n_1 \sin(\theta_1) \tag{8.35}$$

which gives a meaning to the constant m. But if we use the definition of $\sqrt{g}$ and θ, this yields the scalar Snell's Law at the interface $z = 0$:

[11] When the characteristic exponent is real, in the language of Floquet theory, corresponding solutions are called stable since they will be uniformly bounded.

$$n_1 \sin(\theta_1) = \sqrt{g} \sin(\theta) \tag{8.36}$$

with effective refractive index $\sqrt{g}$. As is the case classically, the largest the angle θ can be is $\pi/2$ (we explore this in more detail below). This implies that

$$n_1 \sin(\theta_{\text{crit}}) = \sqrt{g} \tag{8.37}$$

for θ_1 the critical angle for refraction. But from (8.34) this implies that we must have $\mu = 0$. Recall, though, that μ is a Floquet exponent, which means that μ is only defined up to an (additive) integer since if $e^{\pi\mu}$ is a characteristic multiplier, then so is $e^{\pi(\mu+2\pi ij/\pi)}$ for some integer j (here $i^2 = -1$ is the imaginary unit). Ultimately, the point is that we obtain the following limit for wave propagation (i.e., the boundary between stable and unstable solutions)

$$\mu = j, \quad j \in \mathbb{Z} \tag{8.38}$$

That is, the curves (8.38) give conditions for selective reflection; the challenge of course is determining μ explicitly. We will not go into this here; we only mention that one can obtain an implicit equation for μ to determine so-called reflection bands [114]. These bands, provided by characteristic curves, determine where reflection is limited. The curves separate regions of stable waves (solutions) from those of unstable solutions.

For obliquely incident light (approaching at some angle θ_1 as before), a typical method for determining the reflection spectrum of cholesterics is the Berreman 4×4 method [38, 40, 441]. The idea of this method is as follows. Considering an electric field of the form (8.28) and a magnetic field of a similar form, i.e.,

$$\mathbf{H}(z) = \mathbf{G}(z)e^{i\omega(t-(m/c)x)}$$

then, writing $\psi = (E_x, E_y, H_x, H_y)^T$, Maxwell's equations lead to the matrix equation

$$\frac{\partial \psi}{\partial z} = Q(z) \cdot \psi(z) \tag{8.39}$$

where $Q(z)$ is the so-called Berreman matrix, taking the form

$$Q(z) = \begin{pmatrix} 0 & 0 & 0 & Q_{14} \\ 0 & 0 & Q_{23} & 0 \\ Q_{31} & Q_{32} & 0 & 0 \\ Q_{41} & Q_{42} & 0 & 0 \end{pmatrix}$$

where each Q_{ij} depends on the components of ε in the lab frame, the wave number $\omega m/c$, and the wave number of free space; see, for instance, equation (8.11) in [441].

Now, for small h, the solution of (8.39) can be written as a relation between ψ at $z+h$ and $\psi(z)$ via the matrix exponential:

$$\psi(z+h) = e^{Q(z)h} \cdot \psi(z)$$

where

$$e^{Q(z)h} = \sum_{k=0}^{\infty} \frac{1}{k!} (Q(z)h)^k$$

Now, by the Cayley–Hamilton Theorem, the characteristic polynomial $p(\lambda) = \det(\lambda \mathrm{Id}_4 - Q(z)h) = 0$ is a linear combination of the first three powers of $Q(z)h$:

$$Q(z)h = a_1 \mathrm{Id}_4 + a_2 Q(z)h + a_3 (Q(z)h)^2 + a_4 (Q(z)h)^3$$

where Id_4 denotes the 4×4 identity matrix. The constants a_j are determined from the eigenvalues of $Q(z)h$. The reflectance and transmittance can be calculated in a layered cholesteric cell using numerical approximations of the matrix exponential; see equations (8.27) and (8.28) in [441].

The case of normal incidence is simpler and can be studied using this 4×4 matrix method. In this case, it was shown in [38] using a simple symmetry argument that ψ should take the form

$$\psi(z) = e^{i\phi z} \begin{pmatrix} A_1 e^{i\beta z} + B_1 e^{-i\beta z} \\ A_2 e^{i\beta z} + B_2 e^{-i\beta z} \\ A_1 e^{i\beta(z-\frac{\pi}{2\beta})} + B_1 e^{-i\beta(z-\frac{\pi}{2\beta})} \\ A_2 e^{i\beta(z-\frac{\pi}{2\beta})} + B_2 e^{i\beta(z-\frac{\pi}{2\beta})} \end{pmatrix}$$

for some A_1, A_2, B_1, B_2 and phase ϕ to be determined. Here β is the same as in (8.26). By plugging this ψ into (8.39), one gets four equations, from which one can obtain A_2, B_2 if A_1, B_1 are prescribed, and the phase ϕ can be found explicitly. In this normal incidence case, other simpler methods can be used, see in particular [103, 275, 322].

Let us now return to Snell's Law (8.36) and develop this in vector form, as this is one of our main focuses for lens design applications. We will do this in a standard way, see, for instance, [141]. Keeping with the notation of Chap. 4, let $\mathbf{a}$ denote the direction of the incident ray upon the cholesteric sample. We will compute the refracted direction $\mathbf{m}$. Keeping with the setup of Fig. 8.5, let $\boldsymbol{\nu}$ denote the unit normal to the planar interface, so that $\boldsymbol{\nu}$ is parallel to the z-axis. Assume that $\mathbf{a}$ and $\mathbf{m}$ are unit vectors, and we decompose $\mathbf{a} = \mathbf{a}_{\parallel} + \mathbf{a}_{\perp}$ and similar for $\mathbf{m}$. Now, we see that

$$\mathbf{a}_{\parallel} = \mathbf{a} + (\mathbf{a} \cdot \boldsymbol{\nu})\boldsymbol{\nu}$$

which gives

$$\mathbf{m}_{\|} = \frac{n_1}{\sqrt{g}}\mathbf{a}_{\|} = \frac{n_1}{\sqrt{g}}(\mathbf{a} + (\mathbf{a}\cdot\boldsymbol{\nu})\boldsymbol{\nu})$$

If θ again denotes the angle that the refracted ray makes with the normal $\boldsymbol{\nu}$, we find from (8.36) that

$$\cos(\theta) = \sqrt{1-\sin^2(\theta)} = \sqrt{1-\left(\frac{n_1}{\sqrt{g}}\right)^2 \sin^2(\theta_1)} = \sqrt{1-\left(\frac{n_1}{\sqrt{g}}\right)^2 (1-(\mathbf{a}\cdot\boldsymbol{\nu})^2)}$$

This means that

$$\mathbf{m}_{\perp} = -\sqrt{1-\left(\frac{n_1}{\sqrt{g}}\right)^2 (1-(\mathbf{a}\cdot\boldsymbol{\nu})^2)}$$

where the minus sign indicates the correct choice of unit normal (pointing into the cholesteric). Thus we find that

$$\mathbf{m} = \mathbf{m}_{\|} + \mathbf{m}_{\perp} = \frac{n_1}{\sqrt{g}}(\mathbf{a} + (\mathbf{a}\cdot\boldsymbol{\nu})\boldsymbol{\nu}) - \sqrt{1-\left(\frac{n_1}{\sqrt{g}}\right)^2 (1-(\mathbf{a}\cdot\boldsymbol{\nu})^2)} \tag{8.40}$$

We can write this to match Snell's Law forms we have seen in Chap. 4. Indeed, rearranging terms, we find that (8.40) can be written as

$$\mathbf{a} - \frac{\sqrt{g}}{n_1}\mathbf{m} = \alpha\boldsymbol{\nu}, \qquad \alpha \in \mathbb{R} \tag{8.41}$$

where

$$\alpha = \mathbf{a}\cdot\boldsymbol{\nu} - \frac{\sqrt{g}}{n_1}\sqrt{1-\left(\frac{n_1}{\sqrt{g}}\right)^2 (1-(\mathbf{a}\cdot\boldsymbol{\nu})^2)} \tag{8.42}$$

Note that we have said that $\alpha \in \mathbb{R}$, but since $\alpha = \alpha(\mu)$, this statement is true provided that $\mu \in \mathbb{R}$ so that we have a stable wave. Thus (8.41) provides a vector form of Snell's Law for stable waves at the interface of a cholesteric liquid crystal with pitch p whose chiral axis $\mathbf{c}$ is parallel to the z-axis.

The fact that $\mu \in \mathbb{R}$ means that μ itself is assumed to be a Lyapunov exponent (recall that the real part of a Floquet exponent is a Lyapunov exponent [87]). Note that if $\mu < 0$, then $|\rho| = |e^{\pi\mu}| < 1$, and if $\mu > 0$, then $|\rho| = |e^{\pi\mu}| > 1$, so the sign of μ dictates whether the modulus of ρ is larger or smaller than 1 (this is used to classify stability of solutions; see, e.g., Theorem 2.89 in [87]).

From the vector form of Snell's Law, one can start to study the various reflection and refraction problems (at least for stable/normal waves) discussed in Chap. 1, such as determining surfaces of uniform refraction [155] and determination of Fresnel formulas [168], among many others. For instance, a typical starting point for such problems after the development of a vector form of Snell's Law is to determine a physical constraint for refraction. We have seen this earlier in (8.37) as it was shown in [114]. Let us now use the vector form of Snell's Law (8.41) to determine a physical constraint for refraction so that there is no total internal reflection.

Let us consider two cases: when $\sqrt{g} < n_1$ and when $\sqrt{g} > n_1$. Consider first the case when $\sqrt{g} < n_1$. Then, from (8.36), we have that

$$\sin(\theta_1) = \frac{\sqrt{g}}{n_1} \sin(\theta) < \sin(\theta)$$

and so $\theta_1 < \theta$. Thus, the maximum angle of refraction is $\theta = \pi/2$ and, again from (8.36), occurs when

$$n_1 \sin(\theta_1) = \sqrt{g}$$

i.e., when $\mu = 0$. Thus for $\sqrt{g} < n_1$, there is no refraction, i.e., the total internal reflection occurs.

Next consider the case when $\sqrt{g} > n_1$; we aim for a physical constraint for refraction so that the total internal reflection does not occur. Indeed, in this case the maximum angle of refraction $\theta_{\max}$ is achieved when $\theta_1 = \pi/2$ and thus

$$\theta_{\max} = \arcsin\left(\frac{n_1}{\sqrt{g}}\right)$$

Notice the argument of the previous arcsine is well-defined since we have assumed that $\sqrt{g} > n_1$. Again from (8.36), we have that

$$\theta_1 - \theta = \arcsin\left(\frac{\sqrt{g}}{n_1} \sin(\theta)\right) - \theta$$

which is an increasing function of θ on the interval $[0, \theta_{\max}]$. Then we have $0 \le \theta_1 - \theta \le \frac{\pi}{2} - \theta_{\max}$, which implies that

$$\cos(\theta - \theta_1) \ge \cos\left(\frac{\pi}{2} - \theta_{\max}\right) = \sin(\theta_{\max}) = \frac{n_1}{\sqrt{g}}$$

Thus we obtain the following physical constraint for refraction in the case $\sqrt{g} \ge n_1$:

$$\mathbf{a} \cdot \mathbf{m} = \cos(\theta - \theta_1) \ge \frac{n_1}{\sqrt{g}} \tag{8.43}$$

Remark 8.6 Notice that the condition $\sqrt{g} \ge n_1$ reduces to

$$\mu \ge \frac{p^2}{\lambda^2} n_1^2 \cos^2(\theta_1) \tag{8.44}$$

which certainly implies that $\mu > 0$. Recall that all the above analysis was for stable waves (we assumed that $\mu \in \mathbb{R}$). A more detailed Lyapunov stability analysis is beyond the scope of this book but seems to be a worthwhile pursuit since such optical refraction problems in cholesterics have, until now, largely not been considered.

8.4.3 Free Energy of a Twist Pattern

A natural question to ask is what is the free energy of an arbitrary twist with specified boundary conditions [375]? How should the director look like? To this end, let us generalize (8.1) and consider a director of the form

$$\mathbf{d}(z) = (\cos(\theta(z)), \sin(\theta(z)), 0) \tag{8.45}$$

so that $\mathbf{d}$ lies in the $x - z$ plane (such θ is sometimes called the twist angle). Clearly with $\mathbf{d}$ given by (8.45), we have $\nabla \cdot \mathbf{d} = 0$, and we also compute that $\nabla \times \mathbf{d}(z) = \theta'(z)(\cos(\theta(z)), \sin(\theta(z)), 0)$. Then since $\mathbf{d} \times \nabla \times \mathbf{d} = \mathbf{0}$ and $\mathrm{Tr}(\nabla \mathbf{d}) = 0$, the Oseen–Frank energy functional becomes

$$I[\theta] = K_2 \int_\Omega \left| \frac{d\theta}{dz} - \mathrm{q} \right|^2 d\mathbf{x} = K_2 A \int_0^L \left| \frac{d\theta}{dz} - \mathrm{q} \right|^2 dz$$

where A denotes the area of the cholesteric liquid crystal cell in the $x - z$ plane (which we have assumed to be constant for simplicity). We also assume that the thickness of the liquid crystal cell is L. The Euler–Lagrange equation in this case is particularly simple and yields $\theta''(z) = 0$, which yields $\theta(z) = c_1 z + c_2$ for some integration constants c_1, c_2. In [375] a twisted nematic cell is considered; that is, the directors at the two walls of the cell are perpendicular to each other. In this case, the integration constant c_1 is determined from the equation $c_1 L = \pm\pi/2$, and so the twist is uniformly distributed across the cell. Note the presence of the plus/minus sign is because these angles are indistinguishable for a nematic and implies that $\theta'(z)$ can be positive or negative: This means the twist can have either handedness.

One can also include in the free energy the magnetic field contribution

$$W_{\mathrm{mag}} = -\frac{\Delta\chi}{2} (\mathbf{H} \cdot \mathbf{d})^2$$

where $\mathbf{H}$ is the external magnetic field and χ is a measure of the local anisotropy called the diamagnetic susceptibility. Similarly the electric field contribution can be added to the total free energy density

$$W_{\mathrm{elec}} = -\frac{\Delta\varepsilon}{8\pi} (\mathbf{E} \cdot \mathbf{d})^2$$

where $\Delta\varepsilon = n_o^2 - n_e^2$. Clearly these contributions will only be nonzero if the applied fields are in the $x - y$ plane. Indeed, depending on which plane the fields live, the additional contributions to the integral I will take the form (ignoring the constants for now)

$$\cos(\theta(z))^2 \qquad \text{or} \qquad \sin(\theta(z))^2$$

leading to an Euler–Lagrange equation of the form

$$\frac{d^2\theta}{dz^2} \pm \sin(\theta)\cos(\theta) = 0 \tag{8.46}$$

This arises, for instance, in the model of a Fréedericksz transition[12] or cholesteric–nematic transition. For such problems one can impose boundary conditions to find the solution which has the lowest energy, and the maximum of θ is attained in the middle of the cell, such that

$$\left.\frac{d\theta}{dz}\right|_{z=L} = -\left.\frac{d\theta}{dz}\right|_{z=0} \qquad \text{and } \theta(L) = \theta(0),$$

see, e.g., [310]. The solutions of such a problem can be found in terms of elliptic integrals. Indeed, making the substitution $\sin(\theta) = \sin(\theta(1/2))\sin(\beta) = r\sin(\beta)$ and using the above assumptions, after integrating equation (8.46), one obtains

$$\pm\left(\frac{1}{2} - z\right) = F\left(\frac{\pi}{2}, r\right) - F(\beta, r)$$

where F is the elliptic integral of the first kind:

$$F(a, b) = \int_0^a (1 - b^2\sin^2(\gamma))^{-1/2} d\gamma$$

Inverting $F(a, b)$ yields the solution $\theta(z)$ by undoing the previous substitution. For example, taking say the minus sign in (8.46) yields the general solution

$$\theta(z) = \mathrm{JA}\left(\sqrt{(c_1 - 1)(z + c_2)^2}; \frac{1}{1 - c_1}\right)$$

where JA denotes the Jacobi amplitude, which can be defined in terms of an incomplete elliptic integral of the first kind. Indeed, by definition of the Jacobi

[12] In 1927 Fréedericksz observed [132] that a nematic liquid crystal cell would see an abrupt change in optical properties under the influence of an external magnetic field when the magnetic field strength is sufficiently large.

amplitude, $\theta(z)$ satisfies

$$\sqrt{(c_1 - 1)(z + c_2)^2} = \int_0^{\theta} \frac{d\phi}{\sqrt{1 - \left(\frac{1}{1 - c_1}\right) \sin^2(\phi)}}$$

where c_1 and c_2 are integration constants determined by the boundary conditions. That is, the Jacobi elliptic function $\mathrm{JA}(x; k)$ can be written as

$$\mathrm{JA}(x, k) = \phi, \qquad \phi = f(\phi, k)$$

where f is a so-called Legendre integral:

$$f(\phi, k) = \int_0^{\phi} \frac{d\gamma}{1 - k^2 \sin^2(\gamma)}$$

Then, we use the fact that derivatives of the Jacobi amplitude satisfy

$$\frac{d}{ds}\mathrm{JA}(s, m) = \mathrm{dn}(s, m)$$

where dn is the so-called delta amplitude, which can be written in the form

$$\mathrm{dn}(s, m) = \sqrt{1 - m \sin^2(\mathrm{JA}(s, m))}$$

if $s \in \mathbb{R}$ and $m \in [0, 1]$. See, e.g., equation (9) in [310] for an example solution under a particular choice of surface energy (so that the field θ minimizes the total free energy which is the volume plus surface contributions) and [1] for more details on the various special functions considered above.

Several numerical studies have been conducted for modeling twisted nematics, see in particular [24, 39] among others, which rely on various approximations (such as the one-constant approximation for the K_j's). It is clear that the resulting Euler–Lagrange equations for systems with external fields are particularly challenging to solve analytically. One interesting approach to handle this challenge is the use of density functional theoretic methods. Such methods have been previously used to study various nematic liquid crystals, see, e.g., [33, 392, 454] among many others.

For example, in [102] density functional theory was used in lieu of the previous minimization approach. Here the idea is to replace the Oseen–Frank energy by a free energy which is a functional of local one-particle density. Still, a number of simplifying assumptions are typically required, such as making a particular coefficient approximation for the excess part of the free energy density functional. One benefit of such an approximation (as used by Onsager in his study of the bulk nematic [321]) is that the Euler–Lagrange equation for the corresponding one-body density can be formed as an integral equation:

$$\rho(\mathbf{x}) = C \exp\left(\beta V_{\text{ext}}(\mathbf{x}_1) - \int \rho(\mathbf{x}_2) f(\mathbf{x}_1, \mathbf{x}_2) d\mathbf{x}_2\right)$$

where C is a normalization factor, $\beta = 1/k_B T$ with k_B Boltzmann's constant, V_{ext} is a confining external potential, and

$$f(\mathbf{x}_1, \mathbf{x}_2) = 1 - e^{-\beta V(\mathbf{x}_1, \mathbf{x}_2)}$$

where $V(\mathbf{x}_1, \mathbf{x}_2)$ is an intramolecular potential. Once one finds ρ, then it is possible to construct the angular distribution h in terms of ρ. From here, one can construct the local Q-tensor

$$Q(\mathbf{x}) = \int_{S^2} h(\mathbf{x}, \mathbf{a}) \left(\mathbf{a} \otimes \mathbf{a} - \frac{1}{3}\mathbf{I}\right) d\mathbf{a}$$

recall equation (3.16). Finally, determine the local eigenvalues and eigenvectors of Q, and from here the local director $\mathbf{d}$ can be identified with the eigenvector of the largest eigenvalue of Q (see [98]).

8.4.4 Bragg Reflection

The Bragg–Snell Law is well known in the theory of photonic crystals [197]. This was first studied for X-rays which are diffracted by atomic crystals [195] and later for waves in layered media [355, 452]. Before proceeding, we mention that a thorough mathematical analysis of photonic crystals is well beyond the scope of this book (especially as it relates to the spectral theory of periodic differential operators and Floquet theory) and instead refer the interested reader to [11, 12, 200, 232], among many others.

Over a small range of frequencies, photonic crystals prevent light from passing through in certain directions (via destructive interference). Thus, the light is reflected back through diffraction. The Bragg–Snell Law arises from the classical Snell Law and Bragg's Law of Diffraction, which says

$$m\lambda = 2d \sin(\theta) \tag{8.47}$$

where $m \in \mathbb{N}$ is the order of diffraction, d is the interplanar spacing in the crystal, and θ is the angle of incident of propagating light (as with the Snell Law, this is measured with respect to the normal). That is, Bragg reflection occurs if the incident angle θ is such that the path length difference ($2d \sin(\theta)$) between reflections of successive layers is an integer number (m) of wavelengths λ; see Fig. 8.6.

The Bragg Law (8.47) is then coupled with the classical Snell's Law

$$n_1 \sin(\theta) = n_2 \sin(\phi) \tag{8.48}$$

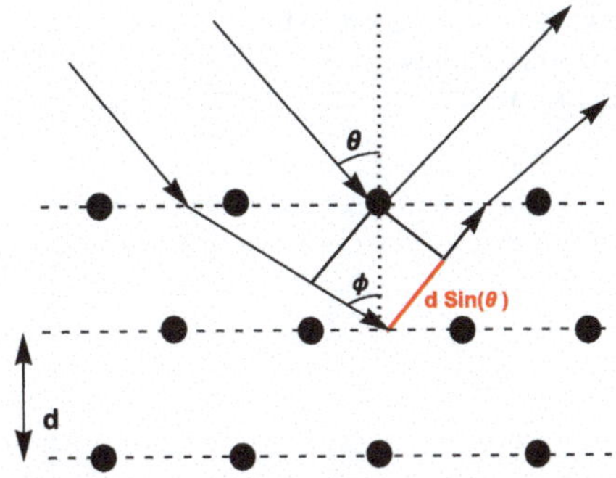

Fig. 8.6 A classical setup for Bragg's Law. We see from the figure that for constructive interference, the path difference should be an integer multiple of the wavelength. The incident angle θ is measured with respect to the normal (illustrated by the vertical dashed line)

where ϕ is the angle of the refracted ray, n_1 is the refractive index of the interstitial medium, and n_2 is the refractive index of the microstructures [281]. To obtain the Bragg–Snell Law (see also [370]), suppose the Bragg condition is satisfied at wavelength λ_B:

$$m\lambda_B = 2d\sin(\theta)$$

Then we see from Fig. 8.6 that this condition can be written in terms of the angle ϕ as $m\lambda = 2d\cos(\phi)$. But then Snell's Law implies that

$$\frac{\lambda_B}{\lambda} = \frac{\sin(\theta)}{\sin(\phi)} = \frac{n_{\text{eff}}}{n_{\text{air}}} = n_{\text{eff}}$$

where we have assumed that the incoming wave material is air for simplicity. The Bragg condition then becomes

$$m\lambda_B = m\lambda n_{\text{eff}} = 2dn_{\text{eff}}\cos(\phi)$$

Again from Snell's Law, we find that $\sin(\phi) = \sin(\theta)/n_{\text{eff}}$, and so

$$\cos^2(\phi) = 1 - \sin^2(\phi) = 1 - \frac{\sin^2(\theta)}{n_{\text{eff}}}$$

which all together yields the Bragg–Snell Law for photonic crystals:

$$m\lambda = 2d\sqrt{n_{\text{eff}}^2 - \sin^2(\theta)} \tag{8.49}$$

where n_{eff} is an average effective refractive index. Note this is not the same effective refractive index we have been working with in previous chapters; here, typically, for a dielectric medium with small voids, one can take

$$n_{\text{eff}} = \sqrt{fn_2^2 + (1-f)n_1^2}$$

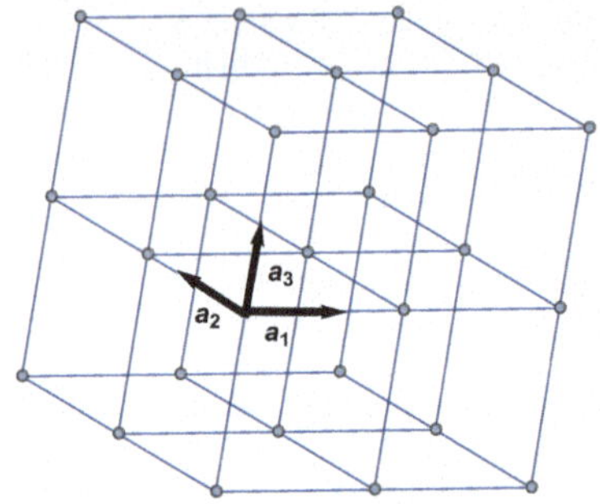

Fig. 8.7 A space lattice in 3D with primitive axes $\mathbf{a}_1, \mathbf{a}_2, \mathbf{a}_3$

f is the void volume filling fraction. Sometimes this is known as the Drude model for a two-material system [258]. In line with the notation of Chap. 4, if $\mathbf{a}$ denotes the incident direction unit vector and $\boldsymbol{\nu}$ the normal to the crystal planes, then we may write (8.49) in terms of the incident direction $\mathbf{a}$

$$m\lambda = 2d\sqrt{n_{\text{eff}}^2 - ||\mathbf{a} \times \boldsymbol{\nu}||^2} \tag{8.50}$$

Now, as we have largely been interested in the vector formulation of Snell's Law and the like, let us recall how to write the Bragg Law in vector form.[13] This will again illustrate the differences from "typical" optical lens design problems as discussed in Chap. 1 where the starting point is typically a vector version of Snell's Law; here, the periodic nature of a photonic crystal means we need to consider reciprocal lattice vectors.

The following derivation will be for X-rays. Now, consider a lattice: an array of points that are related by a translation operator $\mathbf{T} = u_1\mathbf{a}_1 + u_2\mathbf{a}_2 + u_3\mathbf{a}_3$, where $u_1, u_2, u_3 \in \mathbb{Z}$ and $\mathbf{a}_1, \mathbf{a}_2, \mathbf{a}_3$ are the crystal axes (the lattice in 3D can be defined using these three translation vectors). The lattice is called primitive if the axes $\mathbf{a}_1, \mathbf{a}_2, \mathbf{a}_3$ span a cell of the smallest volume for a particular lattice; see Fig. 8.7.

That is, no cell of volume smaller than $\mathbf{a}_1 \cdot \mathbf{a}_2 \times \mathbf{a}_3$ (the parallelepiped with axes $\mathbf{a}_1, \mathbf{a}_2, \mathbf{a}_3$) can serve as a unit cell for the structure.

Any local physical property of a crystal is invariant under $\mathbf{T}$. The position of the center of an atom in the basis $\{\mathbf{a}_1, \mathbf{a}_2, \mathbf{a}_3\}$ is given by $\mathbf{r} = x\mathbf{a}_1 + y\mathbf{a}_2 + z\mathbf{a}_3$. To introduce the reciprocal lattice, we consider the Fourier expansion of function invariant under $\mathbf{T}$ in 3D:

$$f(\mathbf{r}) = \sum_{\mathbf{G}} f_{\mathbf{G}} e^{i\mathbf{G}\cdot\mathbf{r}}$$

where the Fourier coefficients $f_{\mathbf{G}}$ are given by

[13] This is a classical exercise in solid state physics, see, e.g., Chapter 2 in [216], but we think it relevant enough to include here.

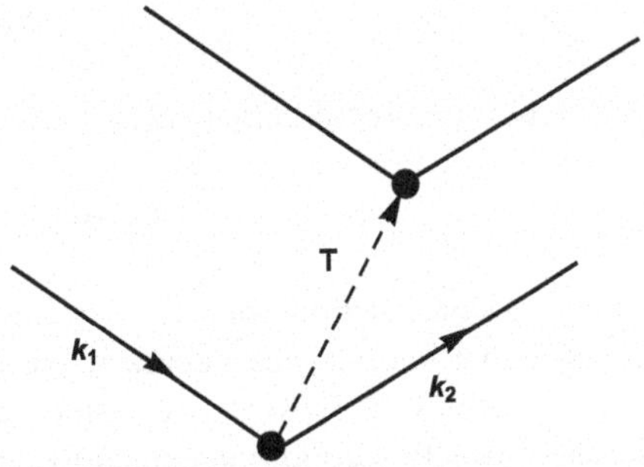

Fig. 8.8 An incident X-ray has direction $\mathbf{k}_1$ and is elastically scattering into the direction $\mathbf{k}_2$

$$f_{\mathbf{G}} = \frac{1}{V(\text{cell})} \int_{\text{cell}} f(\mathbf{r}) e^{-i\mathbf{G}\cdot\mathbf{r}} d\mathbf{r}$$

where integration over a cell of the crystal in $V(\text{cell})$ denotes its volume. The determination of the vectors $\mathbf{G}$ above leads to the reciprocal lattice. To this end, define the vectors

$$\mathbf{b}_1 = 2\pi \frac{\mathbf{a}_2 \times \mathbf{a}_3}{\mathbf{a}_1 \cdot \mathbf{a}_2 \times \mathbf{a}_3}, \quad \mathbf{b}_2 = 2\pi \frac{\mathbf{a}_3 \times \mathbf{a}_1}{\mathbf{a}_1 \cdot \mathbf{a}_2 \times \mathbf{a}_3}, \quad \mathbf{b}_3 = 2\pi \frac{\mathbf{a}_1 \times \mathbf{a}_2}{\mathbf{a}_1 \cdot \mathbf{a}_2 \times \mathbf{a}_3}$$

Note that the denominators above $\mathbf{a}_1 \cdot \mathbf{a}_2 \times \mathbf{a}_3$ give precisely the volume of a cell. The **b**'s form the axis vectors of the reciprocal lattice, and a vector $\mathbf{G}$ in the form

$$\mathbf{G} = v_1\mathbf{b}_1 + v_2\mathbf{b}_2 + v_3\mathbf{b}_3$$

is a reciprocal lattice vector. The point here is that the vectors $\mathbf{G}$ in the Fourier series above are precisely the reciprocal lattice vectors, and moreover, the set of reciprocal lattice vectors actually determines the possible X-ray reflections [216].

Now, consider an incident X-ray with direction $\mathbf{k}_1$ and corresponding wave vector $\mathbf{k} = 2\pi\mathbf{k}_1/\lambda$, where λ denotes the wavelength of the X-ray (Fig. 8.8). The X-ray is scattered (elastically) in direction $\mathbf{k}_2$ with the corresponding wave vector $\mathbf{k}' = 2\pi\mathbf{k}_2/\lambda$. Then, the condition for constructive interference is that the path difference between the scattered X-ray and the incident ray should be an integer number of wavelengths; in terms of wave vectors, this condition reads

$$(\mathbf{k}' - \mathbf{k}) \cdot \mathbf{T} = 2\pi m, \quad m \in \mathbb{N}$$

In terms of direction vectors, this condition reads

$$\mathbf{T} \cdot (\mathbf{k}_1 - \mathbf{k}_2) = m\lambda, \quad m \in \mathbb{N} \tag{8.51}$$

Defining the scattering wave vector $\Delta\mathbf{k} = \mathbf{k}' - \mathbf{k}$, the condition for diffraction becomes

$$\Delta \mathbf{k} = \mathbf{G}$$

which is a reciprocal lattice vector; this is because such vector **G** satisfies

$$\mathbf{G} \cdot \mathbf{T} = 2\pi n, \quad n \in \mathbb{Z}$$

since by construction, $\mathbf{b}_i \cdot \mathbf{a}_j = 2\pi \delta_{ij}$ for $i, j = 1, 2, 3$. Thus we see that constructive interference occurs if the scattering wave vector $\Delta \mathbf{k}$ is a reciprocal lattice vector. Finally, since we assume elastic scattering and we can write $\mathbf{k} + \mathbf{G} = \mathbf{k}'$, the diffraction condition can equivalently be written as

$$2\mathbf{k} \cdot \mathbf{G} + |\mathbf{G}|^2 = 0$$

Since if **G** is a reciprocal lattice vector, then so is $-\mathbf{G}$, we finally arrive at the vector form of Bragg's Law

$$2\mathbf{k} \cdot \mathbf{G} = |\mathbf{G}|^2 \tag{8.52}$$

To see why this is equivalent to the scalar Bragg condition (8.47), suppose that the incident and scattered wave vectors $\mathbf{k}$ and $\mathbf{k}'$ satisfy (8.52) where **G** is a reciprocal lattice vector. Since we have assumed elastic scattering, it holds that $|\mathbf{k}| = |\mathbf{k}'|$ because photon energy is conserved. Thus, the wave vectors $\mathbf{k}$ and $\mathbf{k}'$ make the same angle θ with the plane perpendicular to **G**. The scattering can thus be viewed as a Bragg reflection with angle θ from the family of planes orthogonal to **G**. It is straightforward to see that the distance between any two successive planes in this family satisfies $|\mathbf{G}_{\text{short}}| = 2\pi/d$, where $\mathbf{G}_{\text{short}}$ is the shortest wave vector which is parallel to **G**. But **G** must be an integer multiple of $\mathbf{G}_{\text{short}}$ (recall that **G** is a reciprocal lattice vector); thus there exists $m \in \mathbb{Z}$ such that $\mathbf{G} = m\mathbf{G}_{\text{short}}$. But simple trigonometry implies that $|\mathbf{G}| = 2|\mathbf{k}| \sin(\theta)$, we find that $|\mathbf{k}| \sin(\theta) = m\pi/d$, and since $|\mathbf{k}| = 2\pi/\lambda$, we arrive at (8.47).

Returning now to cholesteric liquid crystals, it turns out that, due to the periodic and helical nature of cholesteric liquid crystals, these Bragg reflections can occur when the helix pitch p is on the order of the wavelength of incident light propagating along the cholesteric axis [46, 74]. For normally incident light, the liquid crystal exhibits Bragg reflection at wavelength $\lambda_0 = nP$, where $n = n_e, n_o$. That is, the wavelengths $n_e P$ and $n_o P$ provide the long and short edges of the so-called stop band (we will see why this is the case shortly). The reflection peak thus has a bandwidth $P(n_e - n_o)$. The Bragg condition in this case becomes [98, Equation (6.4)]

$$p = m\lambda, \quad m \in \mathbb{Z} \tag{8.53}$$

where recall p is the pitch. One experimentally observes Bragg reflection when $m = 1$. To see why this is the case, we use the fact that locally a cholesteric behaves like a (uniaxial) nematic; as such, we can use Eq. (2.15) to write the permittivity ε at a point $\mathbf{x}$ in terms of $\mathbf{d}$, with $\mathbf{d}$ given by (8.1). We follow [98] in the derivation below but find it worthwhile to include more details and make it slightly more mathematically precise. Now, assuming ε has a Fourier transform (to be made more precise below), we will Fourier transform ε; we take the following as our definition of Fourier transform:

$$\widehat{\varepsilon}(\mathbf{q}) := \int_{\mathbb{R}^3} \varepsilon(\mathbf{x}) e^{-i\mathbf{x}\cdot\mathbf{q}} d\mathbf{x} \tag{8.54}$$

We will use the Fourier transform in the context of the so-called scattering amplitude to see under what conditions $\widehat{\varepsilon}$ does not vanish. Indeed, given an electric field radiated by a dipole, the scattering amplitude is known to be proportional to $\widehat{\varepsilon}$, and the initial and final polarization directions; see, e.g., equation (3.81) in [98]. Thus, determining where $\widehat{\varepsilon} \neq 0$ will reveal where the scattering amplitude is nonzero.

Now, to simplify notation, let us write $q_0 = 2\pi/p$. Then we have

$$\begin{aligned}
\varepsilon_{11}(\mathbf{x}) &= n_o^2 + (n_e^2 - n_o^2)\cos^2(q_0 z),\\
\varepsilon_{12}(\mathbf{x}) &= (n_e^2 - n_o^2)\cos(q_0 z)\sin(q_0 z),\\
\varepsilon_{13}(\mathbf{x}) &= 0,\\
\varepsilon_{21}(\mathbf{x}) &= \varepsilon_{12}(\mathbf{x}),\\
\varepsilon_{22}(\mathbf{x}) &= n_o^2 + (n_e^2 - n_o^2)\sin^2(q_0 z),\\
\varepsilon_{23}(\mathbf{x}) &= 0,\\
\varepsilon_{31}(\mathbf{x}) &= 0,\\
\varepsilon_{32}(\mathbf{x}) &= 0,\\
\varepsilon_{33}(\mathbf{x}) &= n_o^2
\end{aligned}$$

While the Fourier transform of ε is defined as an integral over all of $\mathbb{R}^3$, we assume that ε is supported on a compact subset $K_{xy} \subset \mathbb{R}^2$ of the xy plane so that the integral is really over $K_{xy} \times \mathbb{R}$ (recall we have assumed the cholesteric axis to be parallel to the z-axis).

In fact, seeing the explicit form of ε, we realize that the Fourier transform that we need is that on the space of *tempered distributions*. Indeed, since clearly one can find constants c, N such that

$$\int_{|z|\leq R} |\varepsilon(z)| dz \leq c R^N, \qquad \text{as } R \to \infty$$

then ε defines a tempered distribution [401].

Remark 8.7 We briefly recall the construction of tempered distributions here and refer, for instance, to [401] for more details. Tempered distributions are particularly useful as they give a rigorous meaning to the Dirac Delta function, for example.

The Schwartz space $\mathcal{S}(\mathbb{R}^n)$ of rapidly decreasing functions on $\mathbb{R}^n$ is the collection of infinitely differentiable functions $\phi : \mathbb{R}^n \to \mathbb{C}$ such that for every multi-index $\alpha = (\alpha_1, \alpha_2, \ldots, \alpha_n)$ and $\beta = (\beta_1, \beta_2, \ldots, \beta_n) \in \mathbb{N}_0^n$, there holds

$$\sup_{\mathbf{x}\in\mathbb{R}^n} |\mathbf{x}^\beta D^\alpha \phi(\mathbf{x})| < \infty$$

where $\mathbf{x}^\beta = x_1^{\beta_1} x_2^{\beta_2} \ldots x_n^{\beta_n}$ and $D^\alpha = \partial_{x_1}^{\alpha_1} \partial_{x_2}^{\alpha_2} \ldots \partial_{x_n}^{\alpha_n}$. The space of tempered distributions $\mathcal{S}'(\mathbb{R}^n)$ is the dual space of $\mathcal{S}(\mathbb{R}^n)$, that is, the set of all continuous, linear functionals on $\mathcal{S}(\mathbb{R}^n)$.

Note that the Fourier transform formula (8.54) still holds for Schwartz functions. The Fourier transform of a tempered distribution $f \in \mathcal{S}'(\mathbb{R}^n)$ is defined to be the tempered distribution

$$\langle \widehat{f}, \phi \rangle = \langle f, \widehat{\phi} \rangle, \quad \forall \phi \in \mathcal{S}(\mathbb{R}^n)$$

where

$$\langle g, h \rangle = \int_{\mathbb{R}^n} g(\mathbf{x}) h(\mathbf{x}) d\mathbf{x}$$

That is, the Fourier transform of a tempered distribution f is the functional on $\mathcal{S}$ which assigns to ϕ the value $\langle f, \widehat{\phi} \rangle$. Note that if f is actually a Schwartz function, then $\widehat{f}$ is the tempered distribution which is identified with the function $\widehat{f}$.

Since we are interested in determining nonzero contributions to $\widehat{\varepsilon}$, any constant terms (such as the various n_o^2) for $q \neq 0$ will contribute nothing, as claimed in [98]. But, let us see explicitly why this is the case. The Fourier transform of, say, n_o^2, viewed as a tempered distribution, is

$$\langle \widehat{n_o^2}, \phi \rangle = \langle n_o^2, \widehat{\phi} \rangle = \int_{\mathbb{R}} \widehat{\phi}(q) dq = C(n_o)\delta(q)$$

That is, the Fourier transform of a constant is the Dirac Delta function (the constant C depends on a factor of 2π). As such, if $q \neq 0$, there is indeed no contribution from the constant terms. Now it essentially remains to compute three Fourier transforms: the remaining nonzero terms in ε.

Regarding the $\cos^2(q_0 z)$ term, we may write $\cos^2(q_0 z) = \frac{1}{2} + \frac{1}{4}\left(e^{2iq_0 z} + e^{-2iq_0 z}\right)$. Ignoring the $1/2$ term for the reason mentioned above, it is enough to transform the sum of the exponential terms. A straightforward calculation shows that the transform takes the form

$$C\left(\delta(q-2q_0)+\delta(q+2q_0)\right)$$

for some absolute constant C, and this transform of course vanishes unless $q = \pm 2q_0$. The case $q = -2q_0$ is ruled out for physical considerations (this would lead to the wave vector $k_1 > k_0$ which is forbidden), and so the only nonzero contribution here is when $q = 2q_0$. In this case it means that $k_0 = -k_1 = q_0 = 2\pi/p$. But since $k_0 = 2\pi/\lambda$, this means that $p = \lambda$, which is precisely the Bragg condition (8.53) with $m = 1$. Now, the $\sin^2(q_0 z)$ term can be reduced to this case by a simple trigonometric identity. Finally, one can write

$$\sin(q_0 z)\cos(q_0 z) = \frac{i}{4}\left(e^{-2iq_0 z} - e^{2iq_0 z}\right)$$

which as above only remain when $q = 2q_0$. Thus one does indeed obtain Bragg reflection for normal incidence when $m = 1$.

Let us return now to the stop band mentioned previously. For an electromagnetic wave of frequency ω propagating along the z-axis, the following dispersion relation can be obtained [98]:

$$k_0^2 - q_0^2 = \pm k_1^2 \tag{8.55}$$

where

$$k_0^2 = \left(\frac{\omega}{c}\right)^2 \frac{n_e^2 + n_o^2}{2}, \quad k_1^2 = \left(\frac{\omega}{c}\right)^2 \frac{(n_e^2 - n_o^2)}{2}$$

This implies that

$$\omega_o = \frac{cq_0}{n_o}, \qquad \omega_e = \frac{cq_0}{n_e}$$

are the upper and lower bounds of the so-called frequency gap. This leads to the familiar formulas for the short and long wavelength edges of the reflection band:

$$\lambda_o = n_o p, \qquad \lambda_e = n_e p$$

Now, what happens at oblique incidence? In this case, Bragg reflections of all orders are observed under the law

$$p\cos(\phi) = m\lambda, \quad m \in \mathbb{N} \tag{8.56}$$

where ϕ is the refracted angle (given by Snell's Law), just as in the case of a photonic crystal. In this case, the effective refractive index is just the average refractive index

$$\overline{n} = \frac{n_o + n_e}{2}$$

so the scalar form of Snell's Law which provides ϕ takes the form (assuming $n_1 = 1$ so the incoming ray is coming from air for simplicity)

$$\sin(\theta) = \overline{n} \sin(\phi)$$

where θ is as usual the incident angle. Thus

$$\phi = \arcsin\left(\frac{1}{\overline{n}} \sin(\theta)\right)$$

and so

$$\cos(\phi) = \sqrt{1 - \frac{1}{\overline{n}^2} \sin^2(\theta)} = \frac{1}{\overline{n}} \sqrt{\overline{n}^2 - \sin^2(\theta)}$$

This gives the Bragg–Snell law for cholesteric liquid crystals (compare with (8.49))

$$m\lambda = \frac{p}{\overline{n}} \sqrt{\overline{n}^2 - \sin^2(\theta)} = \frac{p}{\overline{n}} \sqrt{\overline{n}^2 - ||\mathbf{a} \times \boldsymbol{\nu}||^2}, \quad m \in \mathbb{N} \tag{8.57}$$

where as usual $\mathbf{a}$ is the incident direction unit vector and $\boldsymbol{\nu}$ the unit normal vector to the sample interface.

Now, consider oblique reflections in the following scenario (as briefly discussed in Section 6.1.5 in [98]), which is in practice useful for so-called polydomain samples. To clarify the terminology, in this context a *domain* is not to be understood in the mathematical sense (i.e., an open connected set), but rather refers to regions between disclinations. A single domain or monodomain is a region which can be characterized by a single director **d**, but when directors characterize domains that are somewhat random with respect to each other, the region is referred to as a polydomain [79, 227]. In the context of so-called liquid crystal elastomers, a monodomain is the case when the director is the same at every point in the relaxed elastomer, and a polydomain is the case if the director points in different directions at different points in the relaxed elastomer [45].

Remark 8.8 Liquid crystal elastomers are a very rich subject we have not really attempted to discuss [3, 50, 432, 433, 436]. For instance, this class of materials has potential applications in soft actuators [314, 368], e.g., for artificial muscle applications [376, 409] or for energy absorption [152, 196], among many other areas. Also worth mentioning is the ability to obtain lasing in cholesteric elastomers [129, 419] and related optical applications [47, 67, 431].

We will use a vector formulation rather than the classical Berreman 4×4 matrix method (see, e.g., section 4.1.5 in [74]); this vector formulation stays in the spirit of Chap. 4. Our goal will be a form of Bragg's law in what is sometimes called Fergason's equation [125], [227, Equation (2)], but written in terms of incident and

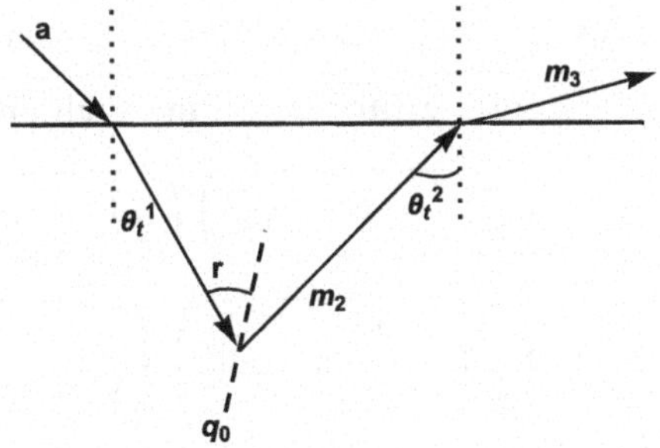

Fig. 8.9 Geometric setup of the polydomain example to determine λ_0

refracted ray directions rather than angles. So let **a** denote the unit vector in the direction of an incident ray upon a so-called polydomain sample; see Fig. 8.9.

Let $\mathbf{m}_1$ denote the (unit) direction of the ray after refraction into the sample. After some time this ray suffers a Bragg reflection, where the angle between $\mathbf{m}_1$ and the cholesteric axis $\mathbf{q}_0 = \mathbf{c}/|\mathbf{c}|$ (i.e., the associated unit vector) is denoted by r, and the reflected ray has unit direction $\mathbf{m}_2$. This ray propagates to the sample interface and then exits the sample, refracting into the direction $\mathbf{m}_3$. At the first interface, we have the first vector form of Snell's Law (here we assume outside the sample is air):

$$\mathbf{a} \times \boldsymbol{\nu} = \overline{n}(\mathbf{m}_1 \times \boldsymbol{\nu}) \tag{8.58}$$

where $\boldsymbol{\nu}$ is the unit normal perpendicular to the sample and $\overline{n}$ is as usual the average refractive index. We will only consider here the case when $\mathbf{q}_0$ lies in the plane of incidence. At the exit interface, another vector Snell's Law holds:

$$\overline{n}(\mathbf{m}_2 \times \boldsymbol{\nu}) = \mathbf{m}_3 \times \boldsymbol{\nu} \tag{8.59}$$

In a first approximation of wavelength, we may write

$$\lambda \approx \frac{\lambda_0}{\overline{n}}, \qquad \text{assuming} \qquad |n_e - n_o| \ll \overline{n} \tag{8.60}$$

where λ_0 denotes the wavelength in vacuum. Now, the angle r is related to the first refracted angle θ_t^1 and the exit refracted angle θ_t^2 by

$$2r = \theta_t^1 + \theta_t^2$$

Our goal is to determine a formula for the vacuum wavelength, not in terms of angles, but rather just in terms of the incident ray **a** and the exit refracted ray $\mathbf{m}_3$. By simple trigonometry and geometry, as well as the Snell Laws (8.58)-(8.59), we find that

$$\begin{aligned}
\cos(\theta_t^1+\theta_t^2) &= \cos(\theta_t^1)\cos(\theta_t^2)-\sin(\theta_t^1)\sin(\theta_t^2)\\
&= (\mathbf{m}_1\cdot\boldsymbol{\nu})(\mathbf{m}_2\cdot\boldsymbol{\nu})-||\mathbf{m}_1\times\boldsymbol{\nu}||\,||\mathbf{m}_2\times\boldsymbol{\nu}||\\
&= \left(\sqrt{1-||\mathbf{m}_1\times\boldsymbol{\nu}||^2}\right)\left(\sqrt{1-||\mathbf{m}_2\times\boldsymbol{\nu}||^2}\right)-\frac{1}{\overline{n}}||\mathbf{m}_1\times\boldsymbol{\nu}||\,||\mathbf{m}_3\times\boldsymbol{\nu}||\\
&= \left(\sqrt{1-\frac{1}{\overline{n}^2}||\mathbf{a}\times\boldsymbol{\nu}||^2}\right)\left(\sqrt{1-\frac{1}{\overline{n}^2}||\mathbf{m}_3\times\boldsymbol{\nu}||^2}\right)-\frac{1}{\overline{n}^2}||\mathbf{a}\times\boldsymbol{\nu}||\,||\mathbf{m}_3\times\boldsymbol{\nu}||
\end{aligned} \tag{8.61}$$

Now, since $\cos(\theta_t^1+\theta_t^2)=\cos(2r)=-1+\cos^2(r)$, we find that

$$\cos(r)=\pm\frac{1}{\sqrt{2}}\sqrt{1+\cos(\theta_t^1+\theta_t^2)}=\frac{1}{\sqrt{2}}\sqrt{1+\cos(\theta_t^1+\theta_t^2)}$$

where we have taken the plus sign by the Bragg Law $\cos(r)=m\lambda/p$ with m a positive integer. Thus, combining (8.61) and the Bragg Law with (8.60)

$$p\cos(r)=m\lambda\approx\frac{m\lambda_0}{\overline{n}}$$

we obtain

$$\begin{aligned}
\lambda_0 &= \frac{p\overline{n}}{m}\cos(r)\\
&= \frac{p\overline{n}}{\sqrt{2}m}\sqrt{1+\left(\sqrt{1-\frac{1}{\overline{n}^2}||\mathbf{a}\times\boldsymbol{\nu}||^2}\right)\left(\sqrt{1-\frac{1}{\overline{n}^2}||\mathbf{m}_3\times\boldsymbol{\nu}||^2}\right)-\frac{1}{\overline{n}^2}||\mathbf{a}\times\boldsymbol{\nu}||\,||\mathbf{m}_3\times\boldsymbol{\nu}||}
\end{aligned} \tag{8.62}$$

Note that (8.62), written in terms of all the angles $r, \theta_t^1, \theta_t^2$, reduces to the so-called Fergason equation; see, e.g., equation (6.39) in [98] or equation (2) in [227]. Moreover, if the polydomain sample is just a monodomain, this agrees with equation (1) in [325] for the angular dependence of the wavelength (with $m=1$). Indeed, in this case λ_0 reduces to, in the case $m=1$,

$$\begin{aligned}
\lambda_0 &= \frac{p\overline{n}}{\sqrt{2}}\sqrt{1+\sqrt{1-\frac{1}{\overline{n}^2}\sin^2(\theta_i)}}\\
&= p\overline{n}\cos\left(\frac{1}{2}\arcsin\left(\frac{\sin(\theta_i)}{\overline{n}}\right)\right)
\end{aligned} \tag{8.63}$$

where θ_i denotes the angle of incidence between $\mathbf{a}$ and $\boldsymbol{\nu}$, and we have used the identity

$$\cos\left(\frac{1}{2}\arcsin(x)\right) = \frac{\sqrt{1+\sqrt{1-x^2}}}{\sqrt{2}}$$

Let us call Θ the Bragg angle

$$\Theta = \arcsin\left(\frac{1}{\overline{n}}\sin(\theta_i)\right)$$

As in [325], equations (2) and (3), for the monodomain, one can obtain similar formulas in terms of the Bragg angle for the ordinary and extraordinary rays:

$$\lambda_e = p\overline{n}_e\cos(\Theta_e) \tag{8.64}$$

$$\lambda_o = p\overline{n}_o\cos(\Theta_o) \tag{8.65}$$

where $\Theta_{e,o}$ denote the Bragg angles for the extraordinary and ordinary rays and $\overline{n}_e, \overline{n}_o$ denote the average refractive indices along each Bragg angle, respectively. Let us determine these average refractive indices and Bragg angles.

Now, the average refractive index $\overline{n}_o$ is just the ordinary refractive index (see equation (8) in [325]), so the Bragg angle Θ_o is just given by Snell's Law

$$n_o\sin(\Theta_o) = \sin(\theta_i)$$

Remark 8.9 We could also introduce the Bragg direction $\mathbf{m}_B$, so that in vector form we have the Snell Law

$$\mathbf{a}\times\boldsymbol{\nu} = n_o\mathbf{m}_B\times\boldsymbol{\nu}$$

where **a** as usual denotes the incident direction vector. As seen in Chap. 4, we may alternatively write this as

$$\mathbf{a} - n_o\mathbf{m}_B = \kappa\boldsymbol{\nu}, \qquad \kappa\in\mathbb{R}$$

for some constant κ. In fact, by taking dot products with $\boldsymbol{\nu}$, we find that

$$\kappa = (\mathbf{a}\cdot\boldsymbol{\nu}) - n_o\sqrt{1-\frac{1}{n_o^2}\left(1-(\mathbf{a}\cdot\boldsymbol{\nu})^2\right)}$$

so that

$$\mathbf{m}_B = \frac{\mathbf{a} - \left[(\mathbf{a}\cdot\boldsymbol{\nu}) - n_o\sqrt{1-\frac{1}{n_o^2}\left(1-(\mathbf{a}\cdot\boldsymbol{\nu})^2\right)}\right]\boldsymbol{\nu}}{n_o}$$

Writing

$$\cos(\Theta_o) = \cos\left(\arcsin\left(\frac{1}{n_o}\sin(\theta_i)\right)\right) = \sqrt{1 - \frac{1}{n_o^2}\left(1 - (\mathbf{a}\cdot\boldsymbol{\nu})^2\right)}$$

we find the band edge wavelength for the ordinary ray from (8.65) written only in terms of the incident ray and unit normal

$$\lambda_o = pn_o\sqrt{1 - \frac{1}{n_o^2}\left(1 - (\mathbf{a}\cdot\boldsymbol{\nu})^2\right)} \tag{8.66}$$

Now, for the extraordinary ray, the refractive index can vary between n_e and the effective refractive index $n_{\mathrm{eff},p}$ as seen in Chap. 3:

$$n_{\mathrm{eff},p} = \frac{n_e n_o}{\sqrt{n_e^2\sin^2(\Theta_e) + n_o^2\cos^2(\Theta_e)}}$$

The average refractive index $\overline{n}_e$ is then just obtained as an integral average over the azimuthal angles ϕ of the liquid crystal molecules of $n_e^2\cos^2(\phi) + n_{\mathrm{eff},p}^2\sin^2(\phi)$, which becomes

$$\overline{n}_e = \sqrt{\frac{n_e^2 + n_{\mathrm{eff},p}^2}{2}}$$

see in particular equation (5) in [325]. Now, Snell's Law in vector form (4.32) and scalar form (4.33) have been written earlier, which in this context gives (in scalar form)

$$\sin(\theta_i) = \overline{n}_e\sin(\Theta_e)$$

Plugging in the value of $\overline{n_e}$ into the above Snell's Law, we find that

$$\overline{n}_e^2\sin^2(\Theta_e) = \frac{n_e^2}{2}\sin^2(\Theta_e)\left(1 + \frac{1}{1 + \left(\frac{n_e^2}{n_o^2} - 1\right)\sin^2(\theta_e)}\right) \tag{8.67}$$

and solving for $\sin^2(\Theta_e)$ from (8.67) yields

$$\sin^2(\Theta_e) = \frac{\sin^2(\theta_i)}{n_e^2} - \frac{n_o^2}{n_e^2(n_e^2 - n_o^2)} + \sqrt{\frac{n_o^4}{(n_e^2 - n_o^2)^2} + \frac{\sin^4(\theta_i)}{n_e^4}}$$

Solving for Θ_e then yields

$$\Theta_e = \arcsin\left(\sqrt{\frac{1-(\mathbf{a}\cdot\boldsymbol{\nu})^2}{n_e^2} - \frac{n_o^2}{n_e^2-n_o^2} + \sqrt{\frac{\left(1-(\mathbf{a}\cdot\boldsymbol{\nu})^2\right)^2}{n_e^4} + \frac{n_o^4}{(n_e^2-n_o^2)^2}}}\right) \tag{8.68}$$

which, if written out in terms of angles, is precisely equation (7) in [325]. As with the ordinary ray, this leads to the band edge wavelength for the extraordinary ray from (8.64), again written with only the incident ray **a** and the unit normal $\boldsymbol{\nu}$:

$$\lambda_e = p\sqrt{\frac{n_e^2 + \dfrac{n_e^2 n_o^2}{f(\mathbf{a},\boldsymbol{\nu})(n_e^2-n_o^2)+n_o^2}}{2}} \times \sqrt{1-\left(\frac{1-(\mathbf{a}\cdot\boldsymbol{\nu})^2}{n_e^2} - \frac{n_o^2}{n_e^2-n_o^2} + \sqrt{\frac{\left(1-(\mathbf{a}\cdot\boldsymbol{\nu})^2\right)^2}{n_e^4} + \frac{n_o^4}{(n_e^2-n_o^2)^2}}\right)} \tag{8.69}$$

with Θ_e given by (8.68) and where

$$f(\mathbf{a},\boldsymbol{\nu}) = \frac{1-(\mathbf{a}\cdot\boldsymbol{\nu})^2}{n_e^2} - \frac{n_o^2}{n_e^2-n_o^2} + \sqrt{\frac{\left(1-(\mathbf{a}\cdot\boldsymbol{\nu})^2\right)^2}{n_e^4} + \frac{n_o^4}{(n_e^2-n_o^2)^2}}$$

Remark 8.10 Writing

$$\overline{n}_e = \sqrt{\frac{n_e^2 + \dfrac{n_e^2 n_o^2}{f(\mathbf{a},\boldsymbol{\nu})(n_e^2-n_o^2)+n_o^2}}{2}}$$

$$= \sqrt{\frac{n_e^2 + \dfrac{n_e^2 n_o^2}{\left(\dfrac{1-(\mathbf{a}\cdot\boldsymbol{\nu})^2}{n_e^2} - \dfrac{n_o^2}{n_e^2-n_o^2} + \sqrt{\dfrac{\left(1-(\mathbf{a}\cdot\boldsymbol{\nu})^2\right)^2}{n_e^4} + \dfrac{n_o^4}{(n_e^2-n_o^2)^2}}\right)(n_e^2-n_o^2)+n_o^2}}{2}} \tag{8.70}$$

we can proceed as in Remark 8.9, using the vector Snell Law

$$\mathbf{a} \times \boldsymbol{\nu} = \overline{n}_e \mathbf{m}_{B,e} \times \boldsymbol{\nu}$$

to find the Bragg direction $\mathbf{m}_{B,e}$ for the extraordinary ray, just in terms of the incident ray $\mathbf{a}$ and the normal vector $\boldsymbol{\nu}$:

$$\mathbf{m}_{B,e} = \frac{\mathbf{a} - \left[(\mathbf{a} \cdot \boldsymbol{\nu}) - \overline{n}_e \sqrt{1 - \frac{1}{\overline{n}_e^2} \left(1 - (\mathbf{a} \cdot \boldsymbol{\nu})^2\right)} \right] \boldsymbol{\nu}}{\overline{n}_e}$$

with $\overline{n}_e$ given by (8.70).

Remark 8.11 In the isotropic limit that $n_e \rightarrow n_o := n$, then from (8.67) (and equation (5) in [325]), we find that $\overline{n}_e = |n_e|$. If $n_e > 0$, then Θ_e is then just easily read off from the isotropic, scalar Snell Law. However, as mentioned in Remark 4.4, for metamaterials one could have $n_e < 0$. Thus, a natural question is then: Can one obtain "negative" Bragg reflection [83], analogous to negative refraction [333]? What could be said about the band edge wavelengths in such a case?

Chapter 9
Conclusions and Outlook

9.1 Brief Summary

Both Fermat's Principle and Snell's Law are key in the various optics problems we have illustrated. We have seen that Snell's Law is intimately related with the Weierstrass–Erdmann corner condition, recall equation (4.8). We have restricted the various optics problems we have considered to the case of rather simple directors. More sophisticated materials, such as cholesteric liquid crystals or biaxial nematics, require a more careful study—for instance, the effective refractive index for biaxial nematics (recall equation (3.14)) is not homogeneous in $\mathbf{v}$, so one cannot immediately formulate the parametric variational integral because this will now depend on the parameterization chosen.

Moreover, we have seen that the Finsler geometry that often appears in the literature for liquid crystals (see, e.g., [443, 359, 360]) really just reduces to the Riemannian case for the extraordinary ray since the Finsler function is just the effective refractive index, whose Hessian is independent of direction $\mathbf{v}$. The regularity of the director $\mathbf{d}$ often imposes restrictions, for instance, on the type of extremals one can find. Broken extremals, in particular, are expected in geometric optics, as there is typically a surface discontinuity in which rays pass. When $\mathbf{d} \in H^1$ (or BV or $W^{1,1}$), even defining $\mathbf{d}(\mathbf{x})$ is a delicate matter, let alone showing lower semicontinuity of the appropriate functional (with the goal of showing existence of a minimizer in an appropriate class of functions). Absolute instruments are also important devices for optical imaging. However, there are a number of challenges that need to be overcome in order to obtain an absolute instrument using liquid crystals. In particular, if a suitable theory for velocity dependent potentials in the context of superintegrable systems can be developed, this will allow for the determination of various integrals of motion, paving the way to understand separability of the corresponding Hamiltonian and ultimately leading to an absolute instrument. In order to relax some of the constraints required by

E. Stachura, *Mathematical Methods in Liquid Crystal Optics and Lens Design*,
Springer Tracts in Modern Physics 294,
https://doi.org/10.1007/978-3-031-46614-4_9

an absolute instrument, in the next section we propose a weaker notion: a "nearly absolute instrument."

Chiral media are also particularly noteworthy, in part due to the special phenomenon of selective reflection. One parameter in particular we have not taken into account is temperature. Indeed, experiments show that the pitch p actually depends on temperature [254]. However, there is still some uncertainty about the theory on *how* the pitch depends on temperature [382]; see, in particular, [417, 292, 261, 291]. Thus, if one is to include temperature effects, for instance, in a Bragg–Snell law in a polydomain, then a clear mathematical model for how pitch depends on temperature is necessary to start from.[1]

In the following section we enumerate a few open directions in detail.

9.2 Open Directions

In the various problems discussed below for nematic liquid crystals, it is assumed that the nematic is uniaxial. Similar questions persist if the nematic is biaxial, but the complicated nature of the effective refractive index (3.14) and the a and b rays would make things more delicate, yet somehow also more interesting due to recent interest in biaxial nematics, both theoretically and in a more applied sense [248, 188, 245, 144, 124, 302, 440, 430, 301, 282, 439].

9.2.1 Snell's Law

In the case of homogeneous, isotropic media, much analysis has been done related to Snell's Law from a rigorous mathematical perspective [155]. More recently, some of these ideas have also been extended to anisotropic media [159], as well as negative refractive index materials [168, 395] and metasurfaces [164, 163]. The key feature in all these problems is that light rays have linear trajectories, which is not necessarily the case for inhomogeneous media (such as liquid crystals). In light of Remark 4.3, it would be interesting to see if the analysis in [159] would also work in a biaxial nematic liquid crystal. The ordinary ray should not cause any issue, but it is not clear what norm one can impose on the a and b rays to obtain Snell's Law as in Equation (3.2) in [159].

A typical problem begins with a vector form of Snell's Law. Then, one can ask what are the surfaces, interfaces between two media, which reflect or refract radiation in a uniform manner? A next question is typically: Given a weak formulation of a particular imaging problem (defined using conservation of energy),

[1] Some contrasting results on how pitch depends on temperature have arisen due to different types of approximations used [382].

what is the corresponding PDE? As has been shown in many of the above references, often the PDE is of Monge–Ampère type

$$\det\left(D^2\phi(x) + A(x, \nabla_x\phi(x))\right) = B(x, \nabla_x\phi(x))$$

A natural next step is to study such problems when one material is inhomogeneous (say, a liquid crystal). In particular, what would the correct formulation of such imaging problems look like? Moreover, is the resulting PDE still going to be of Monge–Ampère type? One answer has been found in [159], where an optimal mass transport connection has been made when one can associate a norm with an anisotropic material. For instance, let us apply this result to the case of a uniaxial nematic liquid crystal with uniform director. Then we found in Remark 4.3 that the norm to associate with the extraordinary ray is

$$N(\mathbf{v}) = n_{\mathrm{eff},p}(\mathbf{v})|\mathbf{v}|$$

where $n_{\mathrm{eff},p}$ is given by (3.9). Then, we assume that $\mathbf{m}$ satisfies the following condition:[2]

$$\{\mathbf{a} : \mathbf{a} \cdot \nabla N(\mathbf{m}) > 1 \text{ and } \mathbf{m} \cdot \mathbf{a} < 1\} \neq \emptyset \tag{9.1}$$

which is a condition from [159] necessary for obtaining a surface of uniform refraction, and the appropriate cost function turns out to be of the form[3]

$$c(\mathbf{a}, \mathbf{m}) = \log\left(\mathbf{a} \cdot \nabla N(\mathbf{m}) - 1\right),$$

compare with Eq. (1.1) in Chap. 1. Note that, since

$$\nabla N(\mathbf{m}) = -\frac{(n_e^2 - n_o^2)n_o^2 n_e^2(\mathbf{m}\cdot\mathbf{d})}{\left(n_o^2(1-(\mathbf{m}\cdot\mathbf{d})^2) + n_e^2(\mathbf{m}\cdot\mathbf{d})^2\right) n_{\mathrm{eff},p}(\mathbf{m})}\mathbf{d} + n_{\mathrm{eff},p}(\mathbf{m})\frac{\mathbf{m}}{|\mathbf{m}|}$$

the first condition in (9.1) depends on the director $\mathbf{d}$, and in particular, the angle between the incident angle $\mathbf{a}$ and $\mathbf{d}$.

There is a wealth of problems to be considered here (especially ones with nonuniform director), all which start from Snell's Law and the Fermat principle.

---

[2] Essentially this condition says that $n_{\mathrm{eff},p}^2 \geq 1$, so in some sense we are assuming the liquid crystal is denser than air.

[3] Recall that $\mathbf{a}$ is the incident ray direction.

9.2.2 Director Design

In Sect. 6.3 we designed a director field **d** in order to obtain a linear ray trajectory. Unsurprisingly, we found that uniform fields did the job. A linear trajectory is somehow the simplest possible trajectory we could ask for, so it is certainly an interesting question to determine the director fields for other trajectories (elliptical, parabolic, etc.). One possible goal of this trajectory design problem is to optical cloaking [64] using liquid crystals [359], one benefit of which includes the ability to manipulate light without the use of metamaterials. Indeed, negative refracting materials can be absorptive and narrowband [394, 243], so there is certainly motivation to manipulate light without the use of negatively refracting materials such as metamaterials. In particular, the flexibility afforded by liquid crystals allows for easy tuning of the optical axis by external factors such as using stress or electric/magnetic fields [340, 380]. In turn this flexibility allows for easier beam steering, which should be attractive from the point of view of optical cloaking.

However, mathematically speaking, the ray equations (6.31) are already quite complicated even for a trajectory which traces part of a circle. Indeed, in this case, the ray equations in 2D become, writing $\mathbf{x}(t) = (x(t), y(t))$,

$$\begin{aligned} d_1(x, y)\left(d_2(x, y) - 2t(t^2-1)\frac{\partial d_1}{\partial y} + \sqrt{1-t^2}\left(t^2\frac{\partial d_2}{\partial y} + (t^2-1)\frac{\partial d_1}{\partial x}\right)\right) \\ = -t^2\sqrt{1-t^2}d_2(x, y)\left(\frac{\partial d_2}{\partial x} - \frac{\partial d_1}{\partial y}\right), \end{aligned}$$

$$\begin{aligned} \frac{d_2(x, y)^2}{\sqrt{1-t^2}} - \frac{n_e^2}{(n_e^2 - n_o^2)\sqrt{1-t^2}} = (t^2-1)d_1(x, y)\left(\frac{\partial d_1}{\partial y} - \frac{\partial d_2}{\partial x}\right) + \\ d_2(x, y)\left(t^2\frac{\partial d_2}{\partial y} - (t^2-1)\frac{\partial d_1}{\partial x} - 2t\sqrt{1-t^2}\frac{\partial d_2}{\partial x}\right) \end{aligned}$$

9.2.3 Absolute Instruments

Is separability of the action (characteristic function) a requirement for superintegrability in systems analogous to absolute instruments? No such proof seems to exist, see [413]. More precisely: If the Hamilton–Jacobi equation for the Hamilton characteristic function S_0, given by

$$\frac{(\nabla S_0)^2}{2m} + V(x, y, z) = E \tag{9.2}$$

is separable in an orthogonal curvilinear coordinate system (q_1, q_2, q_3), then we can write

$$S_0 = S_1(q_1) + S_2(q_2) + S_3(q_3) \tag{9.3}$$

Is this necessary to obtain superintegrability and thus the design of an absolute instrument?

Moreover, we have seen that the study of absolute instruments is similar to finding bound systems which are superintegrable. An important notion for superintegrability is that of Hamiltonian separability.

It would be interesting, but likely quite difficult, to find explicit integrals of motion (linear and quadratic) via (7.34) and (7.35) for the system (7.32)–(7.33). If one can conclude anything about the general Hamiltonian (7.30), the coefficients A, B, C, D, E, W can be identified, and thus constraints on the director can be obtained in order to obtain linear and quadratic integrals of motion for the Hamiltonian (7.28). This in turn could address the question that was addressed in [288] except now for potentials which are quadratic in momenta. Namely, does the existence of second order integrals of motion allow for separation of variables in some coordinate system, assuming say the corresponding Hamilton–Jacobi equation separates there?

9.2.4 Lens Design

Here we will see another use for Snell's Law. On-axis spherical aberration results from a difference in optical ray paths. We can define on-axis spherical aberration by

$$a(Q) = PQI - POI \tag{9.4}$$

where PQI and POI are the optical paths from the source to the focal point, see, e.g., [353, Figure 4] and Fig. 9.1:

More concretely, suppose a light ray emanates from the origin O in 2D. Let (r, z) denote cylindrical coordinates, and suppose that a surface interface separating two distinct media, one of which being a liquid crystal, is described by $(r, z(r))$.

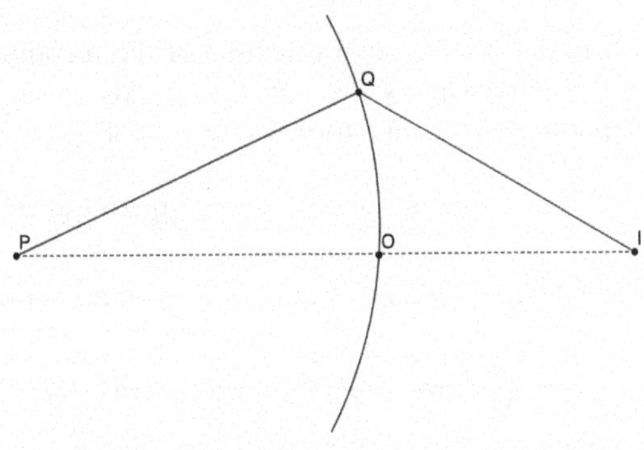

Fig. 9.1 Illustration of a lens to define on-axis spherical aberration. A source is placed at the point P, and the focal point is the point I

Consider a reference ray which emanates from O along the z-axis and strikes the interface at the point $(a, 0)$. The ray then continues throughout the lens until a final point $(b, 0)$. The total optical path length of this ray is given by

$$\mathrm{OPL}_{\mathrm{ref}} = a + \int_{(a,0)}^{(b,0)} n_{\mathrm{eff}}(s)ds \tag{9.5}$$

Consider another ray emanating from O with direction $\boldsymbol{x}$ and striking the surface at a point $P = (r_0, z(r_0))$; the surface is designed so that this ray is also refracted into the point $(b, 0)$. The total optical path length of this second ray is given by

$$\mathrm{OPL}_{\mathrm{gen}} = \sqrt{z(r_0)^2 + r_0^2} + \int_{(r_0,z(r_0))}^{(b,0)} n_{\mathrm{eff}}(s)ds \tag{9.6}$$

This yields the optical path difference (OPD)

$$\begin{aligned}\mathrm{OPD} = F(P) &:= \mathrm{OPL}_{\mathrm{gen}} - \mathrm{OPL}_{\mathrm{ref}}\\ &= \sqrt{z(r_0)^2 + r_0^2} + \int_{(r_0,z(r_0))}^{(b,0)} n_{\mathrm{eff}}(s)ds - \left(a + \int_{(a,0)}^{(b,0)} n_{\mathrm{eff}}(s)ds\right)\end{aligned} \tag{9.7}$$

Aberration occurs when $F(P) \neq 0$. The goal is to minimize $F(P)$.

One question to address in such a minimization problem is the following: What is the shape of the interface and the director such that the rays refracted from O are refracted into the point $(b, 0)$?

The starting point is the following vector form of Snell's Law at the interface, for the ray passing through the point P:

$$\begin{aligned}&\sqrt{n_o^2(\mathbf{m}\cdot\mathbf{d}(P))^2 + n_e^2(1-(\mathbf{m}\cdot\mathbf{d}(P))^2)}\,\mathbf{m} - \boldsymbol{a} = \lambda\boldsymbol{\nu}(P) -\\ &\qquad \frac{(n_o^2 - n_e^2)(\mathbf{m}\cdot\mathbf{d}(P))}{\sqrt{n_o^2(\mathbf{m}\cdot\mathbf{d}(P))^2 + n_e^2(1-(\mathbf{m}\cdot\mathbf{d}(P))^2)}}\mathbf{d}(P)\end{aligned} \tag{9.8}$$

where $\boldsymbol{a}$ is the incident direction of the incoming ray, $\mathbf{m}$ is the refracted direction of the extraordinary ray, and $\lambda \in \mathbb{R}$. Also, $\boldsymbol{\nu}$ is the unit normal at the first surface. Suppose that the director satisfies the following strong anchoring condition at P:

$$\mathbf{d}(P) \parallel \boldsymbol{\nu}(P) \tag{9.9}$$

Then, taking cross product in (9.8) with $\boldsymbol{\nu}$ yields

$$\sqrt{n_o^2(\mathbf{m}\cdot\mathbf{d}(P))^2 + n_e^2(1-(\mathbf{m}\cdot\mathbf{d}(P))^2)}\,\mathbf{m}\times\boldsymbol{\nu}(P) = \boldsymbol{a}\times\boldsymbol{\nu}(P)$$

which means there exists some $\widetilde{\lambda}$ such that

$$\sqrt{n_o^2(\mathbf{m}\cdot\mathbf{d}(P))^2+n_e^2(1-(\mathbf{m}\cdot\mathbf{d}(P))^2)}\mathbf{m}-\boldsymbol{a}=\widetilde{\lambda}\boldsymbol{\nu}(P) \tag{9.10}$$

Now, let $Q=(b,0)$, and let $X(t)$ be a curve on the interface surface, call it Γ. Suppose that the curve hits the point P at $t=t_0$, i.e., $X(t_0)=P$. Then from Snell's Law (9.10), we see that the tangent vector $X'(t)$ must satisfy (since $\boldsymbol{a}=X/|X|$)

$$X'(t)\cdot\left(\sqrt{(n_o^2-n_e^2)\left(\frac{Q-X(t)}{|Q-X(t)|}\cdot\mathbf{d}(X(t_0))\right)^2+n_e^2}\,\frac{Q-X(t)}{|Q-X(t)|}-\frac{X(t)}{|X(t)|}\right)=0 \tag{9.11}$$

for $X(t)$ near P, i.e., t near t_0. Thus the problem is to solve (9.11). Note that if the effective index was just 1, the above equation would become

$$|X(t)|'+|Q-X(t)|'=0$$

and after integrating one obtains $F(X)=|X|+|Q-X|=b$ for some constant b. The complicated nature of the effective index in terms of the director $\mathbf{d}$ leads to a nontrivial differential equation now for $X(t)$.

This surface design question is a natural extension of similar problems in the setting of isotropic media, see, for instance, [157, 155, 168].

To end discussions on lens design, recall that in Sect. 6.2 we have considered only a very simple case, which essentially assumes the liquid crystal layer is isotropic. It remains to consider the full inhomogeneous and anisotropic liquid crystal layer. This problem could be simplified by using the Huygens method as outlined in [228]. In this method, the propagation of light rays within the liquid crystal is assumed negligible in one direction; essentially, the thickness of the liquid crystal layer is assumed to tend to zero. Thus one can associate an effective refractive index that just varies in one spatial direction rather than two. The reason we suggest this method is that it can be viewed as an extension of Snell's Law to liquid crystals; as Snell's Law was instrumental in the lens design problem in Sect. 6.2, this seems a natural fit.

The obvious difficulty is that the trajectories of the rays are no longer linear in such a general medium. As such the function $D(\boldsymbol{x})$ should be defined in terms of the optical path length of the ray passing through the lens.

9.2.5 Nearly Absolute Instruments

Instead of requiring perfect stigmatic imaging as in absolute instruments, one could allow for a small amount of wave aberrations so that $W(x,y;x_0)\neq 0$. The goal of

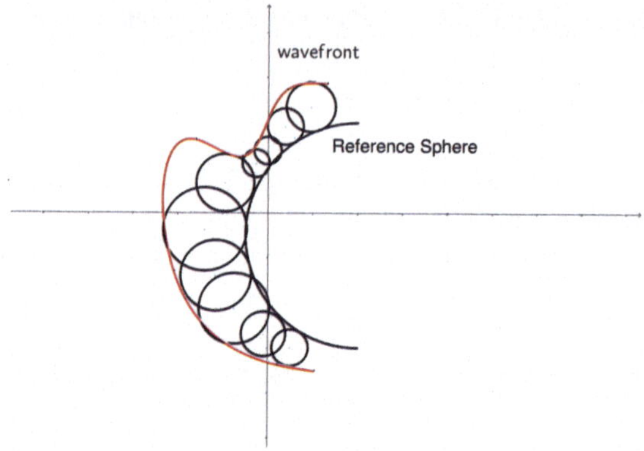

Fig. 9.2 Huygens-like construction of the wavefront

defining such instruments is to allow for more flexibility (only minimal aberrations rather than none), which will hopefully lead to a richer class of optical instruments for (almost) stigmatic imaging, and we illustrate one possible way forward below.

We start from the setup in [349] which we recall now. This all relies on a Huygens-like interpretation of the wavefront as an envelope of spheres with varying radius which is described by W; see Fig. 9.2. Recall that Huygens' principle states that each point on a wavefront acts as a new source of propagating wavelets [269].

Let us define W to be the path length between points on the aberrated wavefront and the reference sphere, connected along the direction of the propagation of the aberrated ray (see Fig. 6.1). Mathematically, we take the domain of W to be the reference sphere, which we denote S^2, the unit sphere in $\mathbb{R}^3$. The typical parameterization in terms of exit pupil coordinates as discussed previously in Sect. 6.1.3 then corresponds to one parameterization of S^2. As discussed in [349], a key property of W is that its gradient in magnitude must be smaller than 1. Indeed, the fact that W is single valued means that one can find an open ball around any point on S^2 that does not contain wavefront points. In 1D, if x is a point on the reference sphere and the norm of the gradient of W is larger than 1, $|dW/dx| > 1$, then this implies one of the balls completely contains the other. This is impossible since by definition, both balls do not contain wavefront points.

This leads to the following definition of a class of optical path difference functions:

Definition 9.1 An optical path difference function $W : S^2 \to \mathbb{R}$ is a C^2 function such that $|\nabla W| \leq 1$.

Note that the gradient above is taken on the sphere. Using spherical coordinates (ρ, θ, ϕ), we can write the gradient as

$$\nabla_{S^2} W = \frac{\partial W}{\partial \theta} \mathbf{e}_\theta + \frac{1}{\sin(\phi)} \frac{\partial W}{\partial \phi} \mathbf{e}_\phi$$

where $\mathbf{e}_\theta$ and $\mathbf{e}_\phi$ are basis vectors in $\mathbb{R}^3$.

Now, what if W is given as in (6.4)? Then since B^* is a point on the reference sphere, we can consider $W = W(B^*)$ as being defined on the sphere. How does this W then fit into the previous framework? We write

$$\int_{A^*}^{B^*} n_{\text{eff}} ds = \int_0^1 n_{\text{eff}}(\mathbf{x}(s))ds, \qquad |\dot{\mathbf{x}}(s)| = 1$$

where $\mathbf{x}(0) = A^*$ and $\mathbf{x}(1) = B^*$, and we have used the arclength parameterization. Recall that $\mathbf{x}(t)$ denotes the ray trajectory. We let $\mathbf{z} = (x, y, z)$ and consider

$$W(\mathbf{z}) = \int_{A \to \mathbf{z}} n_{\text{eff}}(s)ds$$

along the path of the ray. Then the gradient of W in the direction of a vector $\mathbf{u}$ is given by

$$\nabla W(\mathbf{z})[\mathbf{u}] = \frac{d}{dt} W(\mathbf{z} + t\mathbf{u})\bigg|_{t=0}$$

again considered along the path of the ray. Note that, typically in the optics community, the gradient and Hessian of W are computed with finite difference or other approximation methods and are typically done in homogeneous, isotropic media, see, e.g., [84] and the references therein.

Now, since there is no notion of path independence here, it matters which direction $\mathbf{u}$ we take. A natural choice is to take $\mathbf{u} = \dot{\mathbf{x}}$. Then we consider a parameterization of the ray from $(x, y, z) = \mathbf{x}(1)$ to $\mathbf{x}(1) + t\mathbf{u}$: Let us take $\mathbf{y}(s) = \mathbf{x}(1) + s\dot{\mathbf{x}}(t)$ for $0 < s < t$. Then $|\dot{\mathbf{y}}(s)| = 1$ and so

$$\frac{\partial W}{\partial \mathbf{u}} = \lim_{t \to 0} \frac{1}{t} \int_0^t n_{\text{eff}}(\mathbf{x}(1) + s\dot{\mathbf{x}}(t))ds = \frac{d}{dt} \int_0^t n_{\text{eff}}(\mathbf{x}(1) + s\dot{\mathbf{x}}(t))ds \bigg|_{t=0}$$

We find that

$$\frac{d}{dt} \int_0^t n_{\text{eff}}(\mathbf{x}(1) + s\dot{\mathbf{x}}(t))ds = n_{\text{eff}}(\mathbf{x}(1) + t\dot{\mathbf{x}}(t)) + \int_0^t \frac{d}{dt} n_{\text{eff}}(\mathbf{x}(1) + s\dot{\mathbf{x}}(t))ds$$

Let $\mathbf{r}(t) = \mathbf{x}(1) + s\dot{\mathbf{x}}(t)$. Then

$$\begin{aligned}\frac{\partial n_{\text{eff}}(\mathbf{r}(t))}{\partial t} &= 2\left[n_e^2 \dot{\mathbf{r}}\ddot{\mathbf{r}} + (n_o^2 - n_e^2)(\dot{\mathbf{r}} \cdot \mathbf{d}(\mathbf{r})\,(\ddot{\mathbf{r}} \cdot \mathbf{d}(\mathbf{r}) + \dot{\mathbf{r}} \cdot (J_d \dot{\mathbf{r}}))\right] \\ &= 2n_e^2 s^2 \ddot{\mathbf{x}}(t)\dddot{\mathbf{x}}(t) + 2(n_o^2 - n_e^2) \times \\ &\quad [s\ddot{\mathbf{x}}(t) \cdot \mathbf{d}(\mathbf{x}(1) + s\dot{\mathbf{x}}(t))\,(s\dddot{\mathbf{x}}(t) \cdot \mathbf{d}(\mathbf{x}(1) + s\dot{\mathbf{x}}(t) + s\ddot{\mathbf{x}}(t) \cdot (J_d s\ddot{\mathbf{x}}(t)))]\end{aligned}$$

Integrating with respect to s from 0 to t and setting $t = 0$ yield

$$\int_0^t \frac{d}{dt} n_{\text{eff}}(\mathbf{x}(1) + s\dot{\mathbf{x}}(t))ds \Big|_{t=0} = 0$$

and so we finally have

$$\frac{\partial W}{\partial \dot{\mathbf{x}}} = n_{\text{eff}}(\mathbf{x}(1))$$

i.e.,

$$\nabla_{\mathbf{u}} W(x, y, z) = n_{\text{eff}}(x, y, z) \tag{9.12}$$

Since the director $\mathbf{d}$ depends on $\mathbf{x}(1) = (x, y, z)$, we see from the previous calculations that we require $\mathbf{d}$ to be continuously differentiable (at least). Moreover, the growth condition on ∇W translates to a size constraint on the effective refractive index (when viewed as a function defined on the reference sphere).

Remark 9.1 What does smallness of the effective index mean? Well, in the isotropic limit $n_e = n_o = n$, we see the effective index is small if n is small. Since we can also write

$$n_{\text{eff}} = n_e \sqrt{|\mathbf{v}|^2 + \epsilon(\mathbf{v} \cdot \mathbf{d}(\mathbf{x}))^2}, \qquad \epsilon = \frac{n_o^2 - n_e^2}{n_e^2}$$

we again obtain a small effective index if n_e is small and small birefringence is considered.

To be clear, since we are working with wave aberrations W, we should specify that we are defining nearly absolute instruments in the sense that we allow for small *wave* aberrations. We could also define nearly absolute instruments by allowing for small *ray* aberrations (sometimes called *geometric* aberrations). Indeed, we can define a nearly absolute instrument by allowing ϵ_x, ϵ_y to be nonzero but small. Thus we can *define* a nearly absolute instrument as one whose wave aberration function W is small or whose ray aberrations ϵ_x, ϵ_y are small (either a nearly absolute instrument in the wave sense or the ray sense).

In general, if $W \in C^2(S^2)$, then it is natural to expect that $\epsilon_x, \epsilon_y \in C^1(S^2)$ in light of (6.3).

Clearly, classical absolute instruments as defined in [48, 415] are specific instances of nearly absolute instruments in ray sense with $\epsilon_x = \epsilon_y = 0$.

This opens the door to study a much larger class of instruments than absolute instruments. For instance, in [415, 414, 413] absolute instruments are constructed in the sense of eliminating *ray* aberrations. A natural question to ask is whether one could construct nearly absolute instruments similar to those mentioned above by merely *minimizing* ray aberrations. This should in principle lead to a much richer class of instruments, since it is typically a tall task to eliminate all ray aberrations.

Moreover, it is known that the only devices that can produce perfect stigmatic images in both the wave and ray sense are Pendry's slab [334] and Maxwell's fisheye surrounded by a mirror [243, 265]. It would be interesting to determine if the class of instruments which can image "almost perfectly" in both senses can be enlarged by allowing this additional flexibility of being "nearly" stigmatic in both the wave and ray senses.

Additionally, in [413] the close relationship between absolute instruments and the Hamilton–Jacobi equation was observed. By using the same mechanical-optical analogy, what would the corresponding PDE look like for nearly absolute instruments? This would perhaps lead to more flexibility in looking at the maximally superintegrable systems as discussed in Sect. 7.2, which we have seen to be challenging in the context of liquid crystals.

Finally, we should mention there may be other notions of nearly absolute instruments. Indeed, we have defined optical path difference functions in Definition 9.1 which required such functions to be twice continuously differentiable on the (reference) sphere with small norm of the gradient. We saw this led to a size constraint on the effective refractive index, but this may be an undesirable effect. Definition 9.1 arose from a Huygens-like interpretation of the wavefront. However, taking a different approach could lead to a different optical path difference function and therefore a different notion of nearly absolute instrument.

9.2.6 *Optical Transfer Function*

There is also an interesting connection to the theory of oscillatory integrals in harmonic analysis. The optical transfer function (OTF) is a quantitative measure of the image quality of an optical system [48, 381]. However, frequently this function can only be found approximately. Here we briefly discuss how the theory of oscillatory integrals can be used to obtain asymptotics for the OTF.

The effect of aberrations on image formation can be described using the pupil function [71]:

$$P(x, y) = A(x, y)e^{i\frac{2\pi}{\lambda}W(x,y)} \tag{9.13}$$

where $A(x, y)$ is the aperture or amplitude transmittance function of the system and W is the wave aberration function. The coordinates (x, y) are as in Fig. 6.8.

Assuming the exit pupil is perfectly circular, the aperture function $A(x, y)$ is given by

$$A(x, y) = \begin{cases} 1 & \text{if } (x, y) \in B(0, r) \\ 0 & \text{if } (x, y) \notin B(0, r) \end{cases} \tag{9.14}$$

where $B(0, r)$ is the exit pupil. Then we can define the Optical Transfer Function as the convolution

$$\mathrm{OTF}(u, v) = \frac{1}{S}(P * P)(u, v) = \frac{1}{S}\int_{\mathbb{R}^2} P(x, y)\overline{P(x - u, y - v)}dxdy \qquad (9.15)$$

where S is the normalization factor

$$S = \int_{\mathbb{R}^2} |P(x, y)|^2 dxdy$$

Consider a circularly symmetric $W(x_0, \rho)$ given by (6.10) in a not necessarily uniformly circular pupil. Thus we need to consider the phase function

$$\Phi(x, y) = W(x - u, y - v) - W(x, y)$$

so that

$$P(x, y)\overline{P(x - u, y - v)} = A(x, y)A(x - u, y - v)\exp\left(i\frac{2\pi}{\lambda}\Phi(x, y)\right)$$

By a suitable change of variables ($\widetilde{x} = x - u/2, \widetilde{y} = y - v/2$), we may write

$$\Phi(x, y) = W\left(x - \frac{u}{2}, y - \frac{v}{2}\right) - W\left(x + \frac{u}{2}, y + \frac{v}{2}\right) \qquad (9.16)$$

Given the radial nature of W, we see that $\Phi(0, 0) = 0$ and that $\nabla\Phi(0, 0) = 0$. We then consider the OTF as an oscillatory integral with parameter λ:

$$\mathrm{OTF}(\lambda; (u, v)) = \int_{-\infty}^{\infty}\int_{-\infty}^{\infty} \mathcal{A}(x, y; u, v)\exp\left(i\frac{2\pi}{\lambda}\Phi(x, y; u, v)\right)dydx \qquad (9.17)$$

where $\mathcal{A}(x, y; u, v) = A(x, y)A(x - u, y - v)$.

Recall that the theory of oscillatory integrals is concerned with integrals of the form

$$I(\tau, y) = \int_{\mathbb{R}^n} e^{i\tau f(x,y)}\phi(x, y)dx, \qquad y \in \mathbb{R}^k \qquad (9.18)$$

for large values of the parameter τ [17]. Here this means considering small wavelengths λ. Usually it is assumed that $\phi \in C^\infty(\mathbb{R}^n \times \mathbb{R}^k)$.

A typical estimate (see [18]) is the following: For every $\epsilon > 0$, the integral (9.18) satisfies

$$|I(\tau, y)| \leq O(1/\tau)^{\frac{n}{2} - \beta_j - \epsilon} \qquad \text{as } \tau \to \infty \qquad (9.19)$$

where the β_j depend on the degeneracies of the critical points. For the simplest type of singularities, the β's have been computed in [19]. Thus, in order to understand the decay of (9.17), we need to understand the singularities of the phase function Φ. In fact, if one can show that all critical points of Φ are simple or parabolic, then Theorem 6.17 in [17] (see also [92]) implies an estimate for (9.18) of the form

$$|I(\tau, y)| \le c\tau^{-n/2} \sum_{(x,y)\in\Sigma\cap\text{supp}(\phi)} |\det f''_{xx}(x, y)|^{-1/2}$$

where Σ is the set of critical points and f_{xx} denotes the matrix of second derivatives of f. In terms of the optical transfer function, this would become

$$|\text{OTF}(\lambda; (u, v))| \le C\lambda^{-1} \sum_{(x,y,u,v)\in\Sigma\cap\text{supp}(\mathcal{A})} |\det \Phi''_{xy}(x, y, u, v)|^{-1/2} \qquad (9.20)$$

The estimate (9.20) would generalize the result from [381] as the pupil is nonuniformly circular and field curvature and piston are included (although there is no piston contribution to the estimate—which makes sense since this term in the W expansion only involves x_0). Thus, a detailed study of the critical points for an appropriate "optical" phase function leads to asymptotics for the optical transfer function for an optical system.

9.2.7 *Controlling Light with Topological Solitons*

In light of the scalar Snell's Law obtained in [186], there is also the natural question of determining surfaces of uniform refraction for a soliton embedded into a liquid crystal background. Indeed, we have spent a good amount of time describing singular configurations in this text; these correspond to singular topological defects where the director is not defined. There are also nonsingular configurations where the[4] director is everywhere continuous but cannot be smoothly deformed to a uniform field without introducing singularities. These configurations correspond to topological solitons; one and two dimensional topological solitons are characterized by elements of the first and second fundamental homotopy groups, respectively, and behave like particles. Indeed, these nonsingular structures can also be characterized by maps $\phi : S^2 \to S^2$ [440]. For example, 2D baby skrymions (or just "skrymions") are the topological solitons characterized by elements of the second homotopy groups $\pi_2(S^2) = \mathbb{Z}$.

A starting point would be a basic ray tracing problem, such as the following. An incident ray having unit direction **a** strikes a liquid crystal background which has director $\mathbf{d}_1$. After entering the liquid crystal background, the ray eventually

[4] We may also say "a" director if in a cholesteric, for example.

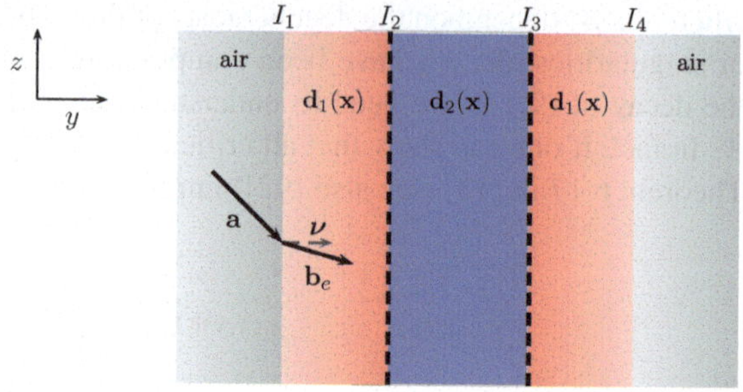

Fig. 9.3 Generalized ray tracing problem

encounters a 1D soliton having director $\mathbf{d}_2$; see Fig. 9.3. The question would be to determine the trajectory of the ray as it passes through the soliton and the liquid crystal background.

There are four interfaces which the ray encounters; thus at each interface we need to impose an appropriate version of Snell's Law in order to determine the direction of the refracted ray. As these materials are birefringent, two rays will be produced: an ordinary and an extraordinary ray (this is the ray denoted by $\mathbf{b}_e$ in Fig. 9.3). Note though that the result in [186] does not make the geometric optics assumption as we have done here, and moreover, they have made the assumption that the ray only refracts once it strikes the soliton layer. There is a plethora of analysis that can be done here, especially related to various types of solitons that can be embedded in the liquid crystal. In particular, by considering a 3D version of this problem, we would allow for the inclusion of recently discovered topological objects to be embedded within a liquid crystal, including so-called hopfions [2, 424, 209] and topological colloids [369, 346]. Moreover, one can even look to cholesteric liquid crystals for soliton-like structures such as skyrmions, hopfions,[5] and heliknotons [440].

9.2.8 Cholesterics

In Chap. 8 we saw various versions of Snell's Law appearing in the context of selective reflection and Bragg reflection. The periodic nature of cholesteric liquid crystals, and the associated mathematical techniques such as Floquet theory, would be brand new to the world of designing so-called surfaces of uniform refraction and the associated mathematical problems [157, 155, 280].

We also commented (recall Remark 8.2) on discotic liquid crystals, for which a chiral mesophase does exist. More generally, one might wonder about the distinction between disc-like molecules and rod-like molecules as seen throughout this book. We have assumed rod-like molecules throughout, but the geometric optics analysis

[5] In fact, cholesteric liquid crystals were where Hopf-like solitons were discovered experimentally [387].

we have seen (especially, in Chap. 4) does not see the underlying molecular structure: only the director. Since discotic molecules and rod-like molecules both have a director (recall that for discotic molecules, the director is the average orientation of the unit normals to the disc-like molecules), our analysis does not see the distinction between disc-like and rod-like molecules. However, this is worth exploring further, as there are distinct nematic mesophases for discotic molecules—does this affect the appropriate version of Snell's Law?

Also in Chap. 8, we computed the band edge wavelengths for the ordinary and extraordinary rays undergoing Bragg reflection. However, this was done under the assumption of small birefringence; when there is large birefringence, one must take into account multiple reflections [98], so more applications of Snell's Law will be required.

In general it seems there remains much to be said on the mathematical side for optical problems in cholesteric liquid crystals. In particular, it is certainly not clear that the tools from optimal mass transport theory [421, 7] can be used as in other reflection/refraction problems (recall the discussion from Chap. 1). The controllable nature of selective reflection and Bragg reflection are attractive [226, 279, 88] for the design of optical instruments to control light [326, 223, 183] and as such deserve a rigorous mathematical underpinning as well (likely based on geometrical optics [327]).

9.3 Conclusion

Motivated by a need to place various aspects of liquid crystal optics on a rigorous mathematical footing, we have provided details for a number of relevant topics, particularly related to variational problems in liquid crystal optics. Starting from Fermat's principle and then obtaining the ray equations using Riemannian geometry, we are always motivated by an underlying liquid crystal problem. We have further seen that broken extremals, while typically seen in mechanics (e.g., phase transition theory), also have applications in liquid crystal problems.

There is also the question of designing a director field **d** to obtain a particular ray trajectory. This can be seen as an inverse problem: What nematic environment will produce a given ray path?

The notion of absolute instruments is also a popular one due to their implications for imaging. By allowing additional flexibility and letting a small amount of aberrations exist, we have defined the so-called nearly absolute instruments. Again, the hope for such an idea is to expand the realm of constructible instruments to be used for (almost) stigmatic imaging. The key feature here is the wave aberration function.

Ultimately we hope this book serves as a starting point for a more rigorous look into liquid crystal optics. It is clear that there is much to do, and contributions can be made from many fields.

Appendix A
Appendix

A.1 Review of Manifolds and Tensor Calculus

A.1.1 Differential Geometry

Here we take the perspective of [110]. The idea of a manifold allows one to extend the ideas of calculus beyond Euclidean space $\mathbb{R}^n$. To define a differentiable manifold, consider a pair $\{(U_j, \phi_j)\}_j$, where U_j is an open subset of a set M and $\phi_j : U_j \rightarrow \mathbb{R}^n$. The set M is called a differentiable manifold if the following conditions are satisfied:

1. $\cup_j U_j = M$.
2. For each j, the image $\phi_j(U_j)$ is an open subset of $\mathbb{R}^n$ and ϕ_j is a homeomorphism[1] from U_j onto $\phi_j(U_j)$.
3. For any j, k such that $U_j \cap U_k \neq \emptyset$, the map

$$\phi_k \circ \phi_j^{-1} : \phi_j(U_j \cap U_k) \rightarrow \phi_k(U_j \cap U_k)$$

   is a differentiable map.

The family $\{(U_j, \phi_j)\}_j$ is called a differentiable coordinate neighborhood system of M. If $\{x^i\}$ denote coordinates in $\mathbb{R}^n$, setting $x_j^i = x^i \circ \phi_j$ we can write

$$\phi_j(p) = (x_j^1(p), \ldots, x_j^n(p))$$

[1] That is, ϕ_j is continuous and bijective, and the inverse ϕ_j^{-1} is continuous.

E. Stachura, *Mathematical Methods in Liquid Crystal Optics and Lens Design*,
Springer Tracts in Modern Physics 294,
https://doi.org/10.1007/978-3-031-46614-4

for a given p in the neighborhood (U_j, ϕ_j). The x_j^i are called coordinate functions in U_j. The tuple $(x_j^1, \ldots, x_j^n)$ is called a local coordinate system in (U_j, ϕ_j). Clearly, Euclidean space with the identity coordinate neighborhood system is a differentiable manifold.

Suppose $f : U \rightarrow \mathbb{R}$ is a function defined in some open subset U of a differentiable manifold M. We say f is differentiable if given a coordinate neighborhood (U_j, ϕ_j) with $U \cap U_j \neq \emptyset$, the function

$$f \circ \phi_j^{-1} : \phi_j(U \cap U_j) \rightarrow \mathbb{R}$$

is a differentiable function. Note that $U \cap U_j$ is an open subset of $\mathbb{R}^n$. Now, given a point p on a differentiable manifold M, let $D(p)$ denote the set of all differentiable functions defined in an open neighborhood of p. A differentiable function $v(t) : (-\epsilon, \epsilon) \rightarrow M$ is called a differentiable curve in M. Suppose that $v(0) = p$. The tangent vector to the curve v at $t = 0$ is a function $v'(0) : D(p) \rightarrow \mathbb{R}$ given by

$$v'(0)f = \left.\frac{d(f \circ v)}{dt}\right|_{t=0}, \quad f \in D(p)$$

A tangent vector at p is the tangent vector at $t = 0$ of some curve $v(t) : (-\epsilon, \epsilon) \rightarrow M$ with $v(0) = p$. The set of all tangent vectors to M at p is called the tangent vector space of M at p and denoted T_pM.

Let (U_j, ϕ_j) be a coordinate neighborhood containing the point p. Given $f \in D(p)$ defined near p, the function $f \circ \phi_j^{-1}$ is differentiable near the point $\phi_j(p) \in \mathbb{R}^n$, so the partial derivatives of $f \circ \phi_j^{-1}$ with respect to the coordinates x^i can be formed. Defining

$$\left(\frac{\partial}{\partial x_j^i}\right)_p (f) = \frac{\partial(f \circ \phi_j^{-1})}{\partial x^i}(\phi_j)(p)$$

produces a tangent vector to M at the point p and

$$\left\{\left(\frac{\partial}{\partial x_j^1}\right)_p, \ldots, \left(\frac{\partial}{\partial x_j^n}\right)_p\right\}$$

determines a basis in T_pM.

Now, to introduce a Riemannian manifold, we draw an analogy with surfaces in $\mathbb{R}^3$. In this setting, there is a natural way of measuring the lengths of tangent vectors: the inner product. To generalize this, suppose there is a given inner product g_x in the tangent space T_xM of a differentiable manifold. In particular, g_x is a bilinear form $g_x : T_xM \times T_xM \rightarrow \mathbb{R}$ which is symmetric and positive definite. On a surface in $\mathbb{R}^3$, this corresponds to the first fundamental form at the point x.

If we assume that given an arbitrary coordinate neighborhood U and its local coordinate system $(x^1, \dots, x^n)$, the function

$$g_{ij}(x) = g_x\left(\left(\frac{\partial}{\partial x^i}\right)_x, \left(\frac{\partial}{\partial x^j}\right)_x\right), \qquad 1 \le i, j \le n$$

is differentiable in U for all i, j. Then, the family of inner products $g = \{g_x\}_{x \in M}$ is called a Riemannian metric on M. A differentiable manifold together with a Riemannian metric is called a Riemannian manifold. The most basic example of a Riemannian manifold is the following. Let $M = \mathbb{R}^n$, and identify $\dfrac{\partial}{\partial x^i}$ with the basis vector $\mathbf{e}_i$. The metric is given by $g_{ij} = \delta_{ij}$, the Kronecker delta, so that

$$\langle \mathbf{e}_i, \mathbf{e}_j \rangle = \delta_{ij}$$

If $\alpha(t)$ is a differentiable curve on a Riemannian manifold, we can use the inner product to measure the norm of the velocity vector. Indeed, $|\alpha'(t)|$ is given by

$$|\alpha'(t)| = \sqrt{g_{\alpha(t)}(\alpha'(t), \alpha'(t))}$$

where the inner product $g_{\alpha(t)}$ is defined in each tangent space $T_{\alpha(t)}M$ for $t \in [a, b]$. The length of the curve $\alpha(t)$ is then the integral of this norm:

$$L(\alpha) = \int_a^b |\alpha'(t)| dt$$

A.1.2 Tensor Calculus

Let ds denote the infinitesimal distance between points x^i and $x^i + dx^i$ in some coordinate system. For Cartesian coordinates, we have

$$ds^2 = dx^2 + dy^2 + dz^2 = \delta_{ij} dx^i dx^j, \qquad x^i \in \{x, y, z\}$$

where as usual δ_{ij} denotes the Kronecker delta. The line element ds^2 can be written by changing coordinate systems from Cartesian to an arbitrary system $x^{i'}$ through the metric tensor:

$$ds^2 = g_{i'j'} dx^{i'} dx^{j'} \tag{A.1}$$

Note that this is the (squared) length of the vector dx^i. Indeed, given a vector $\mathbf{F}$ in $\mathbb{R}^3$, we can decompose $\mathbf{F}$ into its components F^i in the basis $\mathbf{e}_i = \{\mathbf{e}_x, \mathbf{e}_y, \mathbf{e}_z\}$, the unit vectors in each spatial direction, just by $\mathbf{F} = F^i \mathbf{e}_i$. Then we can write

$$|\mathbf{F}|^2 = \mathbf{F} \cdot \mathbf{F} = g_{ij} F^i F^j$$

Each summation above, over an upper and lower index, is called a contraction. Now, in general, we can take the scalar product of two vectors

$$\mathbf{F} \cdot \mathbf{G} = g_{ij} F^i G^j$$

In this expression, we use a lower index to define

$$G_i = g_{ij} G^j \tag{A.2}$$

which is contracted with F^i to get the scalar product $\mathbf{F} \cdot \mathbf{G}$. The G_i are the components of a covariant vector (one-form), and the "usual" vector components G^i form the components of a contravariant vector. Essentially (A.2) uses the metric tensor g_{ij} to lower the index of G^j to produce the covariant vector G_i.

Now we can talk about differentiation. Given a scalar function f depending on the coordinates x^i, it is clear how to take partial derivatives of f. Now, given a vector field $\mathbf{F}$, since it can be decomposed into basis vectors and components of $\mathbf{F}$, we can differentiate according to the Leibniz rule:

$$\frac{\partial}{\partial x^i}\mathbf{F} = \frac{\partial F^j}{\partial x^i}\mathbf{e}_j + F^j \frac{\partial \mathbf{e}_j}{\partial x^i}$$

Since the derivative of a vector is also a vector, we can expand

$$\frac{\partial \mathbf{e}_j}{\partial x^i} = \Gamma^k_{ji} \mathbf{e}_k \tag{A.3}$$

where the Γ^k_{ji} are the Christoffel symbols. Thus we find that

$$\frac{\partial}{\partial x^i}\mathbf{F} = \frac{\partial F^j}{\partial x^i}\mathbf{e}_j + \Gamma^k_{ji} F^j = \left(\frac{\partial F^j}{\partial x^i} + \Gamma^j_{ki} F^k\right)\mathbf{e}_j \tag{A.4}$$

which is the definition of covariant derivative of $\mathbf{F}$.

This idea also works on a Riemannian manifold, but first we need the correct notion of vector field, similar to $\mathbf{F}$ above. So let M be a smooth manifold and U some open subset of M (possibly all of M). A vector field $X = \{X(x)\}_{x \in U}$ is a correspondence that assigns to each $x \in U$ a tangent vector $X(x) \in T_x M$. What are the components of $X(x)$? Well, let $\{x^i\}_{i=1}^m$ be the local coordinate system; then, at each $x \in U$, $X(x)$ can uniquely be written as

$$X(x) = \sum_{i=1}^{m} \xi(x) \left(\frac{\partial}{\partial x^i}\right)\bigg|_x$$

The component functions $\xi^i(x)$ are called the components of X.

Now, let $\mathfrak{X}(M) = C^\infty(M)$ be the collection of all C^∞ vector fields in M.[2] A bilinear mapping is

$$\nabla : \mathfrak{X}(M) \times \mathfrak{X}(M) \to \mathfrak{X}(M)$$

such that

- $\nabla(fX, Y) = f\nabla(X, Y)$
- $\nabla(X, fY) = (Xf)Y + f\nabla(X, Y)$

for any $X, Y \in \mathfrak{X}(M)$ and function $f \in C^\infty(M)$ is called an affine connection in M. The term $\nabla(X, Y)$, sometimes written as $\nabla_X Y$, is the covariant derivative of Y by X. For example, if $M = \mathbb{R}^m$, then a smooth vector field Y is just a smooth function in $\mathbb{R}^m$: $Y = (f^1, \ldots, f^m)$. Then we can write

$$\nabla_X Y = (Xf^1, \ldots, Xf^m)$$

which defines a covariant derivative of Y by X. We can similarly define an analog of the Christoffel symbols Γ^i_{jk}. In this setting, the smooth functions Γ^k_{ij} are called the connection coefficients and are defined by

$$\nabla_{\partial_{x^i}} \frac{\partial}{\partial x_j} = \Gamma^k_{ij} \frac{\partial}{\partial x^k} \tag{A.5}$$

compare with Eq. (A.3). Note also the implied sum on k above. Given $X, Y \in \mathfrak{X}(M)$, we can write

$$X = \xi^i \frac{\partial}{\partial x^i}, \qquad Y = \eta^i \frac{\partial}{\partial x^i} \tag{A.6}$$

and the covariant derivative $\nabla_X Y$ is given by

$$\nabla_X Y = \left(\xi^i \left(\frac{\partial \eta^k}{\partial x^i} + \Gamma^k_{ij}\eta^j\right)\right) \frac{\partial}{\partial x^k}$$

compare with Eq. (A.4). The sum above is on k, i, j.

Now let us return to simple vectors and the Christoffel symbols. By interpreting the covariant derivative in terms of the metric tensor, we can get an expression for the Christoffel symbols in terms of the metric. Starting from the scalar product of vectors **F** and **G**, we take covariant derivative, which is just the partial derivative:

[2] A vector field is C^∞ smooth if its component functions $\xi^i(x)$ are C^∞ smooth.

$$\frac{\partial}{\partial x^j}(\mathbf{F}\cdot\mathbf{G}) = \frac{\partial}{\partial x^j}\left(g_{ij}F^iG^j\right) = \frac{\partial}{\partial x^j}(F_iG^j) = \frac{\partial F^i}{\partial x^j}G^i + F_i\frac{\partial G^i}{\partial x^j}$$

Rewriting this in terms of the covariant derivative of G^i, we find that

$$\frac{\partial}{\partial x^j}(\mathbf{F}\cdot\mathbf{G}) = F_i\left(\frac{\partial G^i}{\partial x^j} + \Gamma^i_{kj}G^k\right) + \left(\frac{\partial F^i}{\partial x^j} - \Gamma^k_{ij}F_k\right)G^j$$

after relabeling indices. We define the covariant derivative of the one form F_i as the quantity

$$\nabla_j F_i = \frac{\partial F^i}{\partial x^j} - \Gamma^k_{ij}F_k$$

Let us now apply this to find the covariant of the metric tensor. Ultimately we find that

$$\nabla_k g_{ij} = \frac{\partial}{\partial x^k}g_{ij} - \Gamma^\ell_{ik}g_{\ell j} - \Gamma^\ell_{jk}g_{i\ell}$$

It is a straightforward exercise to see that the covariant derivative of the metric is zero in Cartesian coordinates, which actually means that it vanishes in all coordinate systems,[3] so

$$\nabla_k g_{ij} = 0$$

Thus we find that (after permuting indices three times—see, e.g., [244, pg. 118])

$$\begin{aligned}
\frac{\partial}{\partial x^k}g_{ij} &= \Gamma^\ell_{ik}g_{\ell j} + \Gamma^\ell_{jk}g_{i\ell},\\
\frac{\partial}{\partial x^j}g_{ik} &= \Gamma^\ell_{ij}g_{\ell k} + \Gamma^\ell_{kj}g_{i\ell},\\
-\frac{\partial}{\partial x^i}g_{jk} &= -\Gamma^\ell_{ji}g_{\ell k} - \Gamma^\ell_{ki}g_{j\ell}
\end{aligned}$$

Adding the above equations together and using symmetry of the Christoffel symbols (namely, that $\Gamma^i_{jk} = \Gamma^i_{kj}$), we get

$$2g_{i\ell}\Gamma^\ell_{jk} = \frac{\partial}{\partial x^k}g_{ij} + \frac{\partial}{\partial x^j}g_{ik} - \frac{\partial}{\partial x^i}g_{jk}$$

Multiply this last equation by the inverse metric g^{mi} and contract (sum) on i to get

[3] If a tensor is zero in one coordinate system, it is zero in all of them [244].

$$\Gamma^i_{jk} = \frac{1}{2} g^{i\ell} \left(\frac{\partial}{\partial x^k} g_{\ell j} + \frac{\partial}{\partial x^j} g_{\ell k} - \frac{\partial}{\partial x^\ell} g_{jk} \right)$$

Finally, we see how to obtain a similar formula on a manifold. In our previous discussion, we only required a smooth manifold. However, now we consider M to be a Riemannian manifold, so that the metric g_{ij} is a Riemannian metric; that is, locally the metric is given by

$$g_{ij}(x) = g_x \left(\left(\frac{\partial}{\partial x^i} \right)\Big|_x , \left(\frac{\partial}{\partial x^j} \right)\Big|_x \right)$$

where g_x is a given inner product on the tangent space $T_x M$. A fundamental result in Riemannian geometry (see, e.g., [241]) says that among the affine connections on a Riemannian manifold M, there is a unique affine connection ∇ such that for any $X, Y, Z, \in \mathfrak{X}(M)$, we have

- $X_g(Y, Z) = g(\nabla_X Y, Z) + g(Y, \nabla_X Z)$
- $\nabla_X Y - \nabla_Y X = [X, Y]$

where the commutator $[X, Y]$ of X and Y is given by

$$[X, Y](x) = \left(\xi^i \frac{\partial \eta^j}{\partial x^i} - \eta \frac{\partial \xi^j}{\partial x^i} \right)(x) \left(\frac{\partial}{\partial x^j} \right)\Big|_x$$

where X, Y are given as in (A.6). This connection is called the Levi-Civita connection.

So let $\{x^i\}_{i=1}^m$ be a local coordinate system on M and Γ^k_{ij} the connection coefficients of this Levi-Civita connection. From (A.5) it is a simple exercise to show that

$$\Gamma^k_{ij} g_{k\ell} = \frac{1}{2} \left(\frac{\partial g_{i\ell}}{\partial x^i} + \frac{\partial g_{i\ell}}{\partial x^j} - \frac{\partial g_{ij}}{\partial x^\ell} \right)$$

Proceeding just like before (i.e., multiplying both sides by the inverse of g and contracting), we arrive at

$$\Gamma^k_{ij} = \frac{1}{2} g^{k\ell} \left(\frac{\partial g_{i\ell}}{\partial x^i} + \frac{\partial g_{i\ell}}{\partial x^j} - \frac{\partial g_{ij}}{\partial x^\ell} \right)$$

Thus, the connection coefficients of the Levi-Civita connection are precisely the Christoffel symbols.

A.2 Christoffel Symbols and Geodesics

We compute the Christoffel symbols $\Gamma^i_{j,k}$ for the metric (5.49) and find the following:

$$\Gamma^1_{1,1} = \frac{1}{n_e^2 n_o^2}(n_e^2-n_o^2)d_1\left((n_e^2-n_o^2)d_1\left(d_2\left(\frac{\partial d_1}{\partial x_2}-\frac{\partial d_2}{\partial x_1}\right)+d_3\left(\frac{\partial d_1}{\partial x_3}-\frac{\partial d_3}{\partial x_1}\right)\right)\right.$$
$$\left.-n_e^2\frac{\partial d_1}{\partial x_1}\right)$$

$$\Gamma^1_{2,1} = \frac{1}{2n_e^2 n_o^2}(n_e^2-n_o^2)\left(d_1\left((n_e^2-n_o^2)d_3^2\left(\frac{\partial d_1}{\partial x_2}-\frac{\partial d_2}{\partial x_1}\right)+2(n_e^2-n_o^2)d_2^2\left(\frac{\partial d_1}{\partial x_2}-\frac{\partial d_2}{\partial x_1}\right)\right.\right.$$
$$\left.+(n_e^2-n_o^2)d_2d_3\left(\frac{\partial d_1}{\partial x_3}-\frac{\partial d_3}{\partial x_1}\right)-2n_e^2\frac{\partial d_1}{\partial x_2}\right)$$
$$\left.+(n_e^2-n_o^2)d_3\left(d_2^2+d_3^2-1\right)\left(\frac{\partial d_3}{\partial x_2}-\frac{\partial d_2}{\partial x_3}\right)\right)\frac{\partial d_3}{\partial x_1}$$

$$\Gamma^1_{2,2} = -\frac{1}{n_e^2 n_o^2}(n_e^2-n_o^2)\left(d_1\left(n_e^2\frac{\partial d_2}{\partial x_2}-(n_e^2-n_o^2)d_2d_3\left(\frac{\partial d_2}{\partial x_3}-\frac{\partial d_3}{\partial x_2}\right)\right)\right.$$
$$\left.-d_2\left(\frac{\partial d_1}{\partial x_2}-\frac{\partial d_2}{\partial x_1}\right)\left((n_e^2-n_o^2)\left(d_2^2+d_3^2\right)-n_e^2\right)\right)$$

$$\Gamma^1_{3,1} = -\frac{1}{2n_e^2 n_o^2}(n_e^2-n_o^2)d_1\left((n_e^2-n_o^2)\left(d_2\left(d_3\left(\frac{\partial d_2}{\partial x_1}-\frac{\partial d_1}{\partial x_2}\right)+d_1\left(\frac{\partial d_2}{\partial x_3}-\frac{\partial d_3}{\partial x_2}\right)\right)\right.\right.$$
$$\left.\left.+\frac{\partial d_3}{\partial x_1}\left(d_2^2+2d_3^2\right)\right)+\frac{\partial d_3}{\partial x_3}\left(2n_e^2-(n_e^2-n_o^2)\left(d_2^2+2d_3^2\right)\right)\right)$$

$$\Gamma^1_{3,2} = -\frac{1}{2n_e^2 n_o^2}\left(n_o^2-n_e^2\right)\left(d_3\left(\frac{\partial d_1}{\partial x_2}-\frac{\partial d_2}{\partial x_1}\right)+d_2\left(\frac{\partial d_1}{\partial x_3}-\frac{\partial d_3}{\partial x_1}\right)\right.$$
$$\left.+d_1\left(\frac{\partial d_2}{\partial x_3}+\frac{\partial d_3}{\partial x_2}\right)\right)\times\left((n_e^2-n_o^2)\left(d_2^2+d_3^2\right)-n_e^2\right)+2\left(n_e^2-n_o^2\right)^2 d_1\frac{\partial d_2}{\partial x_3}d_2^2$$
$$+2\left(n_e^2-n_o^2\right)^2 d_1 d_3^2\frac{\partial d_3}{\partial x_2}$$

$$\Gamma^1_{3,3} = \frac{1}{n_e^2 n_o^2}(n_e^2 - n_o^2)\left(d_3\left((n_e^2 - n_o^2)d_2^2\left(\frac{\partial d_1}{\partial x_3} - \frac{\partial d_3}{\partial x_1}\right) - n_e^2\frac{\partial d_1}{\partial x_3}\right.\right.$$
$$\left. -(n_e^2-n_o^2)d_1 d_2\left(\frac{\partial d_2}{\partial x_3}-\frac{\partial d_3}{\partial x_2}\right)+n_e^2\frac{\partial d_3}{\partial x_1}\right)+(n_e^2-n_o^2)d_3^3\left(\frac{\partial d_1}{\partial x_3}-\frac{\partial d_3}{\partial x_1}\right)$$
$$\left. -n_e^2 d_1\frac{\partial d_3}{\partial x_3}\right)$$

$$\Gamma^2_{1,1} = \frac{1}{n_e^2 n_o^2}(n_e^2 - n_o^2)\left(d_1\left((n_e^2 - n_o^2)d_2^2\left(\frac{\partial d_1}{\partial x_2} - \frac{\partial d_2}{\partial x_1}\right) + n_o^2\left(\frac{\partial d_1}{\partial x_2} - \frac{\partial d_2}{\partial x_1}\right)\right.\right.$$
$$\left.\left. +(n_e^2 - n_o^2)d_2 d_3\left(\frac{\partial d_1}{\partial x_3} - \frac{\partial d_3}{\partial x_1}\right)\right) - n_e^2\frac{\partial d_1}{\partial x_1}d_2\right)$$

$$\Gamma^2_{2,1} = \frac{1}{2n_e^2 n_o^2}(n_e^2 - n_o^2)d_2\left((n_e^2 - n_o^2)\left(\frac{\partial d_1}{\partial x_2}\left(d_3^2 - 2\right) + d_1 d_3\left(\frac{\partial d_2}{\partial x_3} - \frac{\partial d_3}{\partial x_2}\right)\right)\right.$$
$$+2(n_e^2 - n_o^2)d_2^2\left(\frac{\partial d_1}{\partial x_2} - \frac{\partial d_2}{\partial x_1}\right) + (n_e^2 - n_o^2)d_2 d_3\left(\frac{\partial d_1}{\partial x_3} - \frac{\partial d_3}{\partial x_1}\right)$$
$$\left. +\frac{\partial d_2}{\partial x_1}\left((n_o - n_e)(n_e + n_o)d_3^2 - 2n_o^2\right)\right)$$

$$\Gamma^2_{2,2}=\frac{1}{n_e^2 n_o^2}(n_e^2-n_o^2)d_2\left((n_e^2-n_o^2)d_2\left(d_1\left(\frac{\partial d_2}{\partial x_1}-\frac{\partial d_1}{\partial x_2}\right)+d_3\left(\frac{\partial d_2}{\partial x_3}-\frac{\partial d_3}{\partial x_2}\right)\right)\right.$$
$$\left. -n_e^2\frac{\partial d_2}{\partial x_2}\right)$$

$$\Gamma^2_{3,1}=-\frac{1}{2n_e^2 n_o^2}(n_e^2-n_o^2)\left(\left(d_3\left(\frac{\partial d_2}{\partial x_1}-\frac{\partial d_1}{\partial x_2}\right)+d_2\left(\frac{\partial d_1}{\partial x_3}+\frac{\partial d_3}{\partial x_1}\right)+d_1\left(\frac{\partial d_2}{\partial x_3}-\frac{\partial d_3}{\partial x_2}\right)\right)\right.$$
$$\left. \times\left((n_e^2 - n_o^2)d_2^2 + n_o^2\right) + 2(n_e^2 - n_o^2)\frac{\partial d_1}{\partial x_3}d_1^2 d_2 + 2(n_e^2 - n_o^2)d_2 d_3^2\frac{\partial d_3}{\partial x_1}\right)$$

$$\Gamma^2_{3,2}=-\frac{1}{2n_e^2 n_o^2}(n_e^2-n_o^2)d_2\left((n_e^2 - n_o^2)d_1\left(d_3\left(\frac{\partial d_1}{\partial x_2} - \frac{\partial d_2}{\partial x_1}\right) + d_2\left(\frac{\partial d_1}{\partial x_3} - \frac{\partial d_3}{\partial x_1}\right)\right)\right.$$
$$\left. +\frac{\partial d_2}{\partial x_3}\left((n_e^2 - n_o^2)\left(d_2^2 - d_3^2\right) + n_e^2 + n_o^2\right) + (n_e^2 - n_o^2)\frac{\partial d_3}{\partial x_2}\left(-d_2^2 + d_3^2 + 1\right)\right)$$

$$\Gamma^2_{3,3} = -\frac{1}{n_e^2 n_o^2}(n_e^2 - n_o^2)\left((n_e^2 - n_o^2)d_1 d_2 d_3\left(\frac{\partial d_1}{\partial x_3} - \frac{\partial d_3}{\partial x_1}\right)\right.$$
$$\left. - d_3\left(\frac{\partial d_2}{\partial x_3} - \frac{\partial d_3}{\partial x_2}\right)\left((n_o - n_e)(n_e + n_o)d_2^2 - n_o^2\right) + n_e^2 d_2 \frac{\partial d_3}{\partial x_3}\right)$$

$$\Gamma^3_{1,1} = \frac{1}{n_e^2 n_o^2}(n_e^2 - n_o^2)\left(d_1\left((n_e^2 - n_o^2)d_2 d_3\left(\frac{\partial d_1}{\partial x_2} - \frac{\partial d_2}{\partial x_1}\right) + (n_e^2 - n_o^2)d_3^2\left(\frac{\partial d_1}{\partial x_3} - \frac{\partial d_3}{\partial x_1}\right)\right.\right.$$
$$\left.\left. + n_o^2\frac{\partial d_1}{\partial x_3} - n_o^2\frac{\partial d_3}{\partial x_1}\right) - n_e^2\frac{\partial d_1}{\partial x_1}d_3\right)$$

$$\Gamma^3_{2,1} = -\frac{1}{2n_e^2 n_o^2}(n_e^2 - n_o^2)\left(\left(d_3\left(\frac{\partial d_1}{\partial x_2} + \frac{\partial d_2}{\partial x_1}\right) + d_2\left(\frac{\partial d_3}{\partial x_1} - \frac{\partial d_1}{\partial x_3}\right) + d_1\left(\frac{\partial d_3}{\partial x_2} - \frac{\partial d_2}{\partial x_3}\right)\right)\right.$$
$$\left. \times\left((n_e^2 - n_o^2)d_3^2 + n_o^2\right) + 2(n_e^2 - n_o^2)\frac{\partial d_1}{\partial x_2}d_1^2 d_3 + 2(n_e^2 - n_o^2)d_2^2\frac{\partial d_2}{\partial x_1}d_3\right)$$

$$\Gamma^3_{2,2} = \frac{1}{n_e^2 n_o^2}(n_e^2 - n_o^2)\left(d_2\left(-(n_e^2 - n_o^2)d_1 d_3\left(\frac{\partial d_1}{\partial x_2} - \frac{\partial d_2}{\partial x_1}\right) + (n_e^2 - n_o^2)d_3^2\left(\frac{\partial d_2}{\partial x_3} - \frac{\partial d_3}{\partial x_2}\right)\right.\right.$$
$$\left.\left. + n_o^2\frac{\partial d_2}{\partial x_3} - n_o^2\frac{\partial d_3}{\partial x_2}\right) - n_e^2\frac{\partial d_2}{\partial x_2}d_3\right)$$

$$\Gamma^3_{3,1} = \frac{1}{2n_e^2 n_o^2}(n_e^2 - n_o^2)d_3\left((n_e^2 - n_o^2)\left(d_2\left(d_3\left(\frac{\partial d_1}{\partial x_2} - \frac{\partial d_2}{\partial x_1}\right) + d_1\left(\frac{\partial d_3}{\partial x_2} - \frac{\partial d_2}{\partial x_3}\right)\right)\right.\right.$$
$$\left.\left. + \frac{\partial d_1}{\partial x_3}\left(d_2^2 + 2d_3^2 - 2\right)\right) + \frac{\partial d_3}{\partial x_1}\left(-(n_e^2 - n_o^2)\left(d_2^2 + 2d_3^2\right) - 2n_o^2\right)\right)$$

$$\Gamma^3_{3,2} = -\frac{1}{2n_e^2 n_o^2}(n_e^2 - n_o^2)d_3\left((n_e^2 - n_o^2)d_1\left(d_3\left(\frac{\partial d_1}{\partial x_2} - \frac{\partial d_2}{\partial x_1}\right) + d_2\left(\frac{\partial d_1}{\partial x_3} - \frac{\partial d_3}{\partial x_1}\right)\right)\right.$$
$$\left. + (n_e^2 - n_o^2)\frac{\partial d_2}{\partial x_3}\left(d_2^2 - d_3^2 + 1\right) + \frac{\partial d_3}{\partial x_2}\left(-(n_e^2 - n_o^2)\left(d_2^2 - d_3^2\right) + n_e^2 + n_o^2\right)\right)$$

$$\Gamma^3_{3,3} = -\frac{1}{n_e^2 n_o^2}(n_e^2 - n_o^2)d_3\left((n_e^2 - n_o^2)d_1 d_3\left(\frac{\partial d_1}{\partial x_3} - \frac{\partial d_3}{\partial x_1}\right) + (n_e^2 - n_o^2)d_2 d_3\left(\frac{\partial d_2}{\partial x_3} - \frac{\partial d_3}{\partial x_2}\right)\right.$$
$$\left. + n_e^2\frac{\partial d_3}{\partial x_3}\right)$$

From here we can calculate the geodesic equation, see (5.52). This leads to the functions F_1, F_2, F_3 in the system (5.53), which are given by

$$
\begin{aligned}
F_1(\mathbf{d}, x_1, x_2, x_3) = & (n_e^2-n_o^2)\Bigg((x_2'd_2+x_3'd_3)\Bigg(x_2'\Bigg(\frac{\partial d_1}{\partial x_2}-\frac{\partial d_2}{\partial x_1}\Bigg)+x_3'\Bigg(\frac{\partial d_1}{\partial x_3}-\frac{\partial d_3}{\partial x_1}\Bigg)\Bigg)\\
& \times\Big((n_e^2-n_o^2)\Big(d_2^2+d_3^2\Big)-n_e^2\Big)-d_1\Bigg(-(n_e^2-n_o^2)d_2d_3\Bigg(x_3'\Bigg(x_1'\Bigg(\frac{\partial d_1}{\partial x_2}-\frac{\partial d_2}{\partial x_1}\Bigg)+x_3'\frac{\partial d_3}{\partial x_2}\Bigg)\\
& +x_1'x_2'\frac{\partial d_1}{\partial x_3}+\Big((x_2')^2-(x_3')^2\Big)\frac{\partial d_2}{\partial x_3}+(x_2')^2\Bigg(-\frac{\partial d_3}{\partial x_2}\Bigg)-x_1'x_2'\frac{\partial d_3}{\partial x_1}\Bigg)\\
& -(n_e^2-n_o^2)d_2^2\Bigg(x_2'\Bigg(2x_1'\Bigg(\frac{\partial d_1}{\partial x_2}-\frac{\partial d_2}{\partial x_1}\Bigg)+x_3'\Bigg(\frac{\partial d_3}{\partial x_2}-\frac{\partial d_2}{\partial x_3}\Bigg)\Bigg)+x_1'x_3'\frac{\partial d_1}{\partial x_3}\\
& -x_1'x_3'\frac{\partial d_3}{\partial x_1}\Bigg)-(n_e^2-n_o^2)d_3^2\Bigg(x_2'\Bigg(x_1'\Bigg(\frac{\partial d_1}{\partial x_2}-\frac{\partial d_2}{\partial x_1}\Bigg)+x_3'\Bigg(\frac{\partial d_2}{\partial x_3}-\frac{\partial d_3}{\partial x_2}\Bigg)\Bigg)\\
& +2x_1'x_3'\frac{\partial d_1}{\partial x_3}-2x_1'x_3\frac{\partial d_3}{\partial x_1}\Bigg)\\
& +2n_e^2x_1'x_3'\frac{\partial d_1}{\partial x_3}+2n_e^2x_1'x_2'\frac{\partial d_1}{\partial x_2}+n_e^2(x_1')^2\frac{\partial d_1}{\partial x_1}+n_e^2x_2'x_3'\frac{\partial d_2}{\partial x_3}+n_e^2(x_2')^2\frac{\partial d_2}{\partial x_2}\\
& +n_e^2(x_3')^2\frac{\partial d_3}{\partial x_3}+n_e^2x_2'x_3'\frac{\partial d_3}{\partial x_2}\Bigg)-\Bigg(x_1'(n_e^2-n_o^2)\Big(d_2^2+d_3^2-1\Big)\Bigg(d_2\Bigg(x_1'\Bigg(\frac{\partial d_1}{\partial x_2}-\frac{\partial d_2}{\partial x_1}\Bigg)\\
& +x_3'\Bigg(\frac{\partial d_3}{\partial x_2}-\frac{\partial d_2}{\partial x_3}\Bigg)\Bigg)+d_3\Bigg(x_1'\Bigg(\frac{\partial d_1}{\partial x_3}-\frac{\partial d_3}{\partial x_1}\Bigg)+x_2'\Bigg(\frac{\partial d_2}{\partial x_3}-\frac{\partial d_3}{\partial x_2}\Bigg)\Bigg)\Bigg)\Bigg)\Bigg)
\end{aligned}
\tag{A.7}
$$

and

$$
\begin{aligned}
F_2(\mathbf{d}, x_1, x_2, x_3) = & (n_e^2-n_o^2)\Bigg((n_e^2-n_o^2)\Bigg(x_1'x_3'\frac{\partial d_1}{\partial x_3}+x_2'\Bigg(x_3'\Bigg(\frac{\partial d_3}{\partial x_2}-\frac{\partial d_2}{\partial x_3}\Bigg)\\
& +2x_1'\Bigg(\frac{\partial d_1}{\partial x_2}-\frac{\partial d_2}{\partial x_1}\Bigg)\Bigg)\\
& -x_1'x_3'\frac{\partial d_3}{\partial x_1}\Bigg)d_2^3+(n_e^2-n_o^2)d_3\Bigg(-\frac{\partial d_3}{\partial x_2}(x_2')^2+x_1'\frac{\partial d_1}{\partial x_3}x_2'-x_1'\frac{\partial d_3}{\partial x_1}x_2'\\
& +\Big(x_2^2-x_3^2\Big)\frac{\partial d_2}{\partial x_3}+x_3'\Bigg(x_3'\frac{\partial d_3}{\partial x_2}+x_1'\Bigg(\frac{\partial d_1}{\partial x_2}-\frac{\partial d_2}{\partial x_1}\Bigg)\Bigg)\Bigg)d_2^2\\
& -(n_e^2-n_o^2)d_1\Bigg(-\frac{\partial d_1}{\partial x_2}(x_1')^2+\frac{\partial d_2}{\partial x_1}(x_1')^2+x_3'\frac{\partial d_2}{\partial x_3}x_1'-x_3'\frac{\partial d_3}{\partial x_2}x_1'+x_2'x_3'\frac{\partial d_1}{\partial x_3}\\
& +(x_2')^2\frac{\partial d_1}{\partial x_2}-(x_2')^2\frac{\partial d_2}{\partial x_1}-x_2'x_3'\frac{\partial d_3}{\partial x_1}\Bigg)d_2^2\\
& -\Bigg((x_3')^2\frac{\partial d_3}{\partial x_3}n_e^2+2x_1'x_2'\frac{\partial d_1}{\partial x_2}n_e^2+(x_2')^2\frac{\partial d_2}{\partial x_2}n_e^2+x_2'x_3'\frac{\partial d_3}{\partial x_2}n_e^2+(x_1')^2\frac{\partial d_1}{\partial x_1}n_e^2\\
& +\Big(2n_e^2-n_o^2\Big)x_1'x_3'\frac{\partial d_1}{\partial x_3}
\end{aligned}
$$

$$
\begin{aligned}
&+\Big(n_e^2+n_o^2\Big)\,x_2'x_3'\frac{\partial d_2}{\partial x_3}-2n_o^2x_1'x_2'\frac{\partial d_1}{\partial x_2}-n_o^2x_2'x_3'\frac{\partial d_3}{\partial x_2}+2n_o^2x_1'x_2'\frac{\partial d_2}{\partial x_1}\\
&+n_o^2x_1'x_3'\frac{\partial d_3}{\partial x_1}\bigg)\,d_2+(n_e^2-n_o^2)d_1d_3\bigg(\Big(({x_1'})^2-({x_3'})^2\Big)\frac{\partial d_1}{\partial x_3}\\
&+x_2'\bigg(x_1'\bigg(\frac{\partial d_2}{\partial x_3}-\frac{\partial d_3}{\partial x_2}\bigg)+x_3'\bigg(\frac{\partial d_2}{\partial x_1}-\frac{\partial d_1}{\partial x_2}\bigg)\bigg)+\Big(({x_3'})^2-({x_1'})^2\Big)\frac{\partial d_3}{\partial x_1}\bigg)\,d_2\\
&+(n_e^2-n_o^2)d_3^2\bigg(2x_1'x_3'\frac{\partial d_1}{\partial x_3}+x_2'\bigg(x_3'\bigg(\frac{\partial d_2}{\partial x_3}-\frac{\partial d_3}{\partial x_2}\bigg)+x_1'\bigg(\frac{\partial d_1}{\partial x_2}-\frac{\partial d_2}{\partial x_1}\bigg)\bigg)\\
&-2x_1'x_3'\frac{\partial d_3}{\partial x_1}\bigg)\,d_2+n_o^2d_1x_1'\bigg(x_3'\bigg(\frac{\partial d_3}{\partial x_2}-\frac{\partial d_2}{\partial x_3}\bigg)+x_1'\bigg(\frac{\partial d_1}{\partial x_2}-\frac{\partial d_2}{\partial x_1}\bigg)\bigg)\\
&+n_o^2d_3x_3'\bigg(x_3'\bigg(\frac{\partial d_3}{\partial x_2}-\frac{\partial d_2}{\partial x_3}\bigg)+x_1'\bigg(\frac{\partial d_1}{\partial x_2}-\frac{\partial d_2}{\partial x_1}\bigg)\bigg)\bigg)\bigg)
\end{aligned}
\tag{A.8}
$$

and

$$
\begin{aligned}
&F_3(\mathbf{d},x_1,x_2,x_3)=(n_e^2-n_o^2)\bigg((n_e^2-n_o^2)\bigg(2x_1'x_3'\frac{\partial d_1}{\partial x_3}+x_2'\bigg(x_3'\bigg(\frac{\partial d_2}{\partial x_3}-\frac{\partial d_3}{\partial x_2}\bigg)\\
&+x_1'\bigg(\frac{\partial d_2}{\partial x_2}-\frac{\partial d_2}{\partial x_1}\bigg)\bigg)-2x_1'x_3'\frac{\partial d_3}{\partial x_1}\bigg)\,d_3^3+(n_e^2-n_o^2)d_1\bigg(\Big(({x_1'})^2-({x_3'})^2\Big)\frac{\partial d_1}{\partial x_3}\\
&+x_2'\bigg(x_1'\bigg(\frac{\partial d_2}{\partial x_3}-\frac{\partial d_3}{\partial x_2}\bigg)+x_3'\bigg(\frac{\partial d_2}{\partial x_1}-\frac{\partial d_2}{\partial x_2}\bigg)\bigg)+\Big(({x_3'})^2-({x_1'})^2\Big)\frac{\partial d_3}{\partial x_1}\bigg)\,d_3^2\\
&+(n_e^2-n_o^2)d_2\bigg(-\frac{\partial d_3}{\partial x_2}({x_2'})^2+x_1'\frac{\partial d_1}{\partial x_3}x_2'-x_1'\frac{\partial d_3}{\partial x_1}x_2'+\Big(({x_2'})^2-({x_3'})^2\Big)\frac{\partial d_2}{\partial x_3}\\
&+x_3'\bigg(x_3'\frac{\partial d_3}{\partial x_2}+x_1'\bigg(\frac{\partial d_2}{\partial x_2}-\frac{\partial d_2}{\partial x_1}\bigg)\bigg)\bigg)\,d_3^2-\bigg(({x_3'})^2\frac{\partial d_3}{\partial x_3}n_e^2+2x_1'x_2'\frac{\partial d_2}{\partial x_2}n_e^2\\
&+({x_2'})^2\frac{\partial d_2}{\partial x_2}n_e^2+x_2'x_3'\frac{\partial d_3}{\partial x_2}n_e^2+({x_1'})^2\frac{\partial d_1}{\partial x_1}n_e^2+2(n_e^2-n_o^2)x_1'x_3'\frac{\partial d_1}{\partial x_3}\\
&+(n_e^2-n_o^2)x_2'x_3'\frac{\partial d_2}{\partial x_3}-n_o^2x_1'x_2'\frac{\partial d_2}{\partial x_2}+n_o^2x_2'x_3'\frac{\partial d_3}{\partial x_2}+n_o^2x_1'x_2'\frac{\partial d_2}{\partial x_1}+2n_o^2x_1'x_3'\frac{\partial d_3}{\partial x_1}\bigg)\,d_3\\
&+(n_e^2-n_o^2)d_2^2\bigg(x_1'x_3'\frac{\partial d_1}{\partial x_3}+x_2'\bigg(x_3'\bigg(\frac{\partial d_3}{\partial x_2}-\frac{\partial d_2}{\partial x_3}\bigg)+2x_1'\bigg(\frac{\partial d_2}{\partial x_2}-\frac{\partial d_2}{\partial x_1}\bigg)\bigg)\\
&-x_1'x_3'\frac{\partial d_3}{\partial x_1}\bigg)\,d_3-(n_e^2-n_o^2)d_1d_2\bigg(-\frac{\partial d_2}{\partial x_2}({x_1'})^2+\frac{\partial d_2}{\partial x_1}({x_1'})^2+x_3'\frac{\partial d_2}{\partial x_3}x_1\\
&-x_3'\frac{\partial d_3}{\partial x_2}x_1'+x_2'x_3'\frac{\partial d_1}{\partial x_3}+({x_2'})^2\frac{\partial d_2}{\partial x_2}-({x_2'})^2\frac{\partial d_2}{\partial x_1}-x_2'x_3'\frac{\partial d_3}{\partial x_1}\bigg)\,d_3
\end{aligned}
$$

$$
\begin{aligned}
&+n_o^2 d_1 x_1' \left(x_2' \left(\frac{\partial d_2}{\partial x_3} - \frac{\partial d_3}{\partial x_2} \right) + x_1' \left(\frac{\partial d_1}{\partial x_3} - \frac{\partial d_3}{\partial x_1} \right) \right) \\
&+n_o^2 d_2 x_2' \left(x_2' \left(\frac{\partial d_2}{\partial x_3} - \frac{\partial d_3}{\partial x_2} \right) + x_1' \left(\frac{\partial d_1}{\partial x_3} - \frac{\partial d_3}{\partial x_1} \right) \right) \Bigg) \Bigg)
\end{aligned} \tag{A.9}
$$

A Mathematica notebook with these calculations, based on a notebook developed in [178], can be found in the supplementary materials.[4]

A.3 Calculating Integrals of Motion

Recall in Sect. 7.2 we were seeking integrals of motion. We had the Hamilton equations of motion in (7.31). After solving the first two equations for p_x, p_y, we found

$$
p_x = \frac{-((1+2B(x,y))D(x,y)) + \dot{x}(1+2B(x,y)) - \dot{y}C(x,y) + C(x,y)E(x,y)}{1+2B(x,y)+2A(x,y)(1+2B(x,y))-C(x,y)^2},
$$

$$
p_y = \frac{-C(x,y)\dot{x} + C(x,y)D(x,y) + (1+2A(x,y))\dot{y} - E(x,y)(1+2A(x,y))}{1+2B(x,y)+2A(x,y)(1+2B(x,y)) - C(x,y)^2}
$$

and obtained

$$
\begin{aligned}
\ddot{x} &= \dot{p_x} + 2\left((A_x\dot{x} + A_y\dot{y})\, p_x + A\dot{p_x} \right) + (C_x\dot{x} + C_y\dot{y})\, p_y + C(x,y)\dot{p_y} + D_x\dot{x} + D_y\dot{y} \\
&= P_1\dot{x}^2 + P_2\dot{y}^2 + P_3\dot{x} + P_4\dot{y} + P_5\dot{x}\dot{y} + P_6
\end{aligned}
$$

for some functions $P_1(x, y), \ldots, P_6(x, y)$ which can be found explicitly through the Mathematica script on the following page.

For the y equation,

$$
\begin{aligned}
\ddot{y} &= \dot{p_y} + 2\left((B_x\dot{x} + B_y\dot{y})\, p_y + B\dot{p_y} \right) + (C_x\dot{x} + C_y\dot{y})\, p_x + C(x,y)\dot{p_x} + E_x\dot{x} + E_y\dot{y} \\
&= Q_1\dot{x}^2 + Q_2\dot{y}^2 + Q_3\dot{x} + Q_4\dot{y} + Q_5\dot{x}\dot{y} + Q_6
\end{aligned}
$$

for some functions $Q_1(x, y)$, ..., $Q_6(x, y)$ which again can be computed explicitly through the Mathematica script on the following page.

[4] It was not practical to include a PDF output of the Mathematica notebook here as in the next section due to the length of the equations involved.

```
(*Below is a script to explicitly compute the Hamilton
 equations of motion from equation (6.31). The coefficients A, B,
c,d,W are as described in Section 6.2.2, and the coefficients Aex,
Cex, etc. below denote A_x, C_x, etc. in the text. Moreover,
clearly terms like doty denote \dot{y} and terms like pxdot denote \dot{px}.*)


In[ ]:= Px = (-((1 + 2 B) d) + dotx + 2 B dotx + c (-doty + e)) / (1 + 2 B + A (2 + 4 B) - c^2);

In[ ]:= Py = (c (d - dotx) + (1 + 2 A) (doty - e)) / (1 + 2 B + A (2 + 4 B) - c^2);

In[ ]:= pxdot = - (Aex * Px^2 + Bex * Py^2 + Cex * Px * Py + dex * Px + eex * Py + Wex);

In[ ]:= pydot = - (Ay * Py^2 + By * Py^2 + Cy * Px * Py + dy * Px + ey * Py + Wy);

In[ ]:= xdoubledot = pxdot + 2 * ((Aex * dotx + Ay * doty) * Px + A * pxdot) +
          (Cex * dotx + Cy * doty) * Py + c * pydot + dex * dotx + dy * doty;

In[ ]:= ydoubledot = pydot + 2 * ((Bex * dotx + By * doty) * Px + B * pydot) +
          (Cex * dotx + Cy * doty) * Px + c * pxdot + eex * dotx + ey * doty;

In[ ]:= FullSimplify[Collect[FullSimplify[xdoubledot], {dotx, doty}]]

In[ ]:= FullSimplify[Collect[FullSimplify[ydoubledot], {dotx, doty}]]
```

Reference

1. Milton Abramowitz and Irene A. Stegun, editors. *Handbook of mathematical functions with formulas, graphs, and mathematical tables*, volume 55. National Bureau of Standards Applied Mathematics Series, 1972.
2. Paul J. Ackerman and Ivan I. Smalyukh. Diversity of knot solitons in liquid crystals manifested by linking of preimages in torons and hopfions. *Physical Review X*, 7(1):011006, 2017.
3. Aditya Agrawal, Paul Luchette, Peter Palffy-Muhoray, Sibani Lisa Biswal, Walter G. Chapman, and Rafael Verduzco. Surface wrinkling in liquid crystal elastomers. *Soft Matter*, 8(27):7138–7142, 2012.
4. S. Alfonso, C. Alberdi, J.M. Diñeiro, M. Berrogui, B. Hernandez, and C. Sáenz. Complex unitary vectors for the direction of propagation and for the polarization of electromagnetic waves in absorbing isotropic media. *JOSA A*, 21(9):1776–1784, 2004.
5. Les Allen, Marco W. Beijersbergen, R. J. C. Spreeuw, and J.P. Woerdman. Orbital angular momentum of light and the transformation of Laguerre-Gaussian laser modes. *Physical review A*, 45(11):8185, 1992.
6. Luigi Ambrosio, Nicola Fusco, and Diego Pallara. *Functions of bounded variation and free discontinuity problems*. Courier Corporation, 2000.
7. Luigi Ambrosio and Nicola Gigli. A user's guide to optimal transport. *Modelling and Optimisation of Flows on Networks: Cetraro, Italy 2009, Editors: Benedetto Piccoli, Michel Rascle*, 2062:1–155, 2013.
8. Luigi Ambrosio and Epifanio G. Virga. A boundary-value problem for nematic liquid crystals with a variable degree of orientation. *Archive for rational mechanics and analysis*, 114:335–347, 1991.
9. Aboutaleb Amiri, Romain Mueller, and Amin Doostmohammadi. Unifying polar and nematic active matter: emergence and co-existence of half-integer and full-integer topological defects. *Journal of Physics A: Mathematical and Theoretical*, 55(9):094002, 2022.
10. H. Ammari and J.-C. Nédélec. Small chirality behaviour of solutions to electromagnetic scattering problems in chiral media. *Mathematical methods in the applied sciences*, 21(4):327–359, 1998.
11. Habib Ammari, Brian Fitzpatrick, Hyeonbae Kang, Matias Ruiz, Sanghyeon Yu, and Hai Zhang. *Mathematical and computational methods in photonics and phononics*, volume 235. American Mathematical Society, 2018.
12. Habib Ammari, Hyeonbae Kang, and Hyundae Lee. *Layer potential techniques in spectral analysis*. Number 153. American Mathematical Soc., 2009.

E. Stachura, *Mathematical Methods in Liquid Crystal Optics and Lens Design*,
Springer Tracts in Modern Physics 294,
https://doi.org/10.1007/978-3-031-46614-4

13. Habib Ammari and Jean-Claude Nédélec. Time-harmonic electromagnetic fields in chiral media. *Modern mathematical methods in diffraction theory and its applications in engineering*, 1(997), 1997.
14. Dong An, Wei Wang, and Pingwen Zhang. On equilibrium configurations of nematic liquid crystals droplet with anisotropic elastic energy. *Research in the Mathematical Sciences*, 4(1):7, 2017.
15. Denis Andrienko. Introduction to liquid crystals. *Journal of Molecular Liquids*, 267:520–541, 2018.
16. Tilo Arens, Felix Hagemann, Frank Hettlich, and Andreas Kirsch. The definition and measurement of electromagnetic chirality. *Mathematical Methods in the Applied Sciences*, 41(2):559–572, 2018.
17. V. I. Arnold, A. N. Varchenko, and S. M. Gusein-Zade. *Singularities of Differentiable Maps: Volume II, Monodromy and Asymptotic Integrals*. Birkhauser Basel, 1988.
18. Vladimir I. Arnol'd. Remarks on the stationary phase method and coxeter numbers. *Russian Mathematical Surveys*, 28(5):19, 1973.
19. Vladimir Igorevich Arnol'd. Integrals of rapidly oscillating functions and singularities of projections of Lagrangian manifolds. *Funktsional'nyi Analiz i ego Prilozheniya*, 6(3):61–62, 1972.
20. Ahmed H. Arnous, Malik Zaka Ullah, Mir Asma, Seithuti P. Moshokoa, Mohammad Mirzazadeh, Anjan Biswas, and Milivoj Belic. Nematicons in liquid crystals by modified simple equation method. *Nonlinear Dynamics*, 88:2863–2872, 2017.
21. S. N. Aronishidze, V. E. Dmitrienko, D. G. Khoshtariya, and G.S. Chilaya. Circular dichroism in absorbing mixtures with a cholesteric structure. *Soviet Journal of Experimental and Theoretical Physics Letters*, 32:17, 1980.
22. Viktar S. Asadchy, Mohammad Sajjad Mirmoosa, Ana Diaz-Rubio, Shanhui Fan, and Sergei A. Tretyakov. Tutorial on electromagnetic nonreciprocity and its origins. *Proceedings of the IEEE*, 108(10):1684–1727, 2020.
23. Gaetano Assanto. *Nematicons: spatial optical solitons in nematic liquid crystals*. John Wiley & Sons, 2012.
24. T.J. Atherton and J. R. Sambles. Numerical simulation of a twisted nematic cell constructed from micropatterned substrates. *Molecular Crystals and Liquid Crystals*, 475(1):3–11, 2007.
25. Vijay Kumar Baliyan, Kwang-Un Jeong, and Shin-Woong Kang. Dichroic-dye-doped short pitch cholesteric liquid crystals for the application of electrically switchable smart windows. *Dyes and Pigments*, 166:403–409, 2019.
26. John M. Ball. Mathematics and liquid crystals. *Molecular Crystals and Liquid Crystals*, 647(1):1–27, 2017.
27. John M. Ball and Apala Majumdar. Nematic liquid crystals: from Maier-Saupe to a continuum theory. *Molecular crystals and liquid crystals*, 525(1):1–11, 2010.
28. John M. Ball and Arghir Zarnescu. Orientability and energy minimization in liquid crystal models. *Archive for rational mechanics and analysis*, 202(2):493–535, 2011.
29. David Bao, S.-S. Chern, and Zhongmin Shen. *An introduction to Riemann-Finsler geometry*, volume 200. Springer Science & Business Media, 2000.
30. Gang Bao and Peijun Li. *Maxwell's Equations in Periodic Structures*. Springer, 2022.
31. J. Baudry, S. Pirkl, and P. Oswald. Topological properties of singular fingers in frustrated cholesteric liquid crystals. *Physical Review E*, 57(3):3038, 1998.
32. Patricia Bauman, Daniel Phillips, and Jinhae Park. Existence of solutions to boundary value problems for smectic liquid crystals. *Discrete Contin. Dyn. Syst. Ser. S*, 8(2):243–257, 2015.
33. S. Belli, S. Dussi, Marjolein Dijkstra, and R. H. H. G. van Roij. Density functional theory for chiral nematic liquid crystals. *Physical Review E*, 90(2):020503, 2014.
34. V. A. Belyakov and V. E. Dmitrienko. Optics of absorbing cholsteric liquid crystals. *Fizika Tverdogo Tela*, 18(10):2880–2886, 1976.
35. V. A. Belyakov, Vladimir E. Dmitrienko, and V.P. Orlov. Optics of cholesteric liquid crystals. *Soviet Physics Uspekhi*, 22(2):64, 1979.

36. Sergio Benenti, C. Chanu, and G. Rastelli. Variable separation for natural Hamiltonians with scalar and vector potentials on Riemannian manifolds. *Journal of Mathematical Physics*, 42(5):2065–2091, 2001.
37. Marcel Berger. *A panoramic view of Riemannian geometry*. Springer, 2003.
38. Dwight W. Berreman. Optics in stratified and anisotropic media: 4× 4-matrix formulation. *Josa*, 62(4):502–510, 1972.
39. Dwight W. Berreman. Numerical modelling of twisted nematic devices. *Philosophical Transactions of the Royal Society of London. Series A, Mathematical and Physical Sciences*, 309(1507):203–216, 1983.
40. Dwight W. Berreman and Terry J. Scheffer. Reflection and transmission by single-domain cholesteric liquid crystal films: theory and verification. *Molecular Crystals and Liquid Crystals*, 11(4):395–405, 1970.
41. M. V. Berry. Optical vortices evolving from helicoidal integer and fractional phase steps. *Journal of Optics A: Pure and Applied Optics*, 6(2):259, 2004.
42. Josée Bérubé and Pavel Winternitz. Integrable and superintegrable quantum systems in a magnetic field. *Journal of Mathematical Physics*, 45(5):1959–1973, 2004.
43. Arthur L. Besse. *Manifolds all of whose geodesics are closed*, volume 93. Springer Science & Business Media, 1978.
44. Iwo Bialynicki-Birula. New solutions of the Dirac, Maxwell, and Weyl equations from the fractional Fourier transform. *Physical Review D*, 103(8):085001, 2021.
45. J.S. Biggins, M. Warner, and K. Bhattacharya. Elasticity of polydomain liquid crystal elastomers. *Journal of the Mechanics and Physics of Solids*, 60(4):573–590, 2012.
46. Corinne Binet, Michel Mitov, and Alain Boudet. Bragg reflections in cholesteric liquid crystals: from selectivity to broadening and reciprocally. *Molecular Crystals and Liquid Crystals Science and Technology. Section A. Molecular Crystals and Liquid Crystals*, 339(1):111–123, 2000.
47. Charlotte E. Boott, Andy Tran, Wadood Y. Hamad, and Mark J. MacLachlan. Cellulose nanocrystal elastomers with reversible visible color. *Angewandte Chemie*, 132(1):232–237, 2020.
48. Max Born and Emil Wolf. *Principles of optics: electromagnetic theory of propagation, interference and diffraction of light*. Pergamon Press, 1986.
49. Anne Boutet de Monvel and Dimitri Shepelsky. Direct and inverse scattering problem for a stratified nonreciprocal chiral medium. *Inverse Problems*, 13(2):239–251, 1997.
50. Helmut R. Brand, Harald Pleiner, and Philippe Martinoty. Selected macroscopic properties of liquid crystalline elastomers. *Soft Matter*, 2(3):182–189, 2006.
51. Michelle T. Brannum, Aubrey M. Steele, Maxwell C. Venetos, LaShanda T. J. Korley, Gary E. Wnek, and Timothy J. White. Light control with liquid crystalline elastomers. *Advanced Optical Materials*, 7(6):1801683, 2019.
52. E. Brasselet. Liquid crystal defects: Nonlinear optics and optical vortices. In *2010 International Conference on Advanced Optoelectronics and Lasers*, pages 115–117. IEEE, 2010.
53. Etienne Brasselet. Singular optical manipulation of birefringent elastic media using nonsingular beams. *Optics letters*, 34(20):3229–3231, 2009.
54. Etienne Brasselet. Singular optics of liquid crystal defects. *Liquid Crystals: New Perspectives, P Pieranki and MH Godinho, Wiley*, pages 1–70, 2021.
55. A Bregar, TJ White, and M Ravnik. Refraction of light on flat boundary of liquid crystals or anisotropic metamaterials. *Liquid Crystals Reviews*, 5(1):53–68, 2017.
56. Haïm Brezis, Jean-Michel Coron, and Elliott H Lieb. Harmonic maps with defects. *Communications in mathematical physics*, 107(4):649–705, 1986.
57. Kolja Brix, Yasemin Hafizogullari, and Andreas Platen. Designing illumination lenses and mirrors by the numerical solution of Monge–Ampère equations. *JOSA A*, 32(11):2227–2236, 2015.
58. Kolja Brix, Yasemin Hafizogullari, and Andreas Platen. Solving the Monge–Ampère equations for the inverse reflector problem. *Mathematical Models and Methods in Applied Sciences*, 25(05):803–837, 2015.

59. Annalisa Buffa, Martin Costabel, and Dongwoo Sheen. On traces for $H(curl, \Omega)$ in Lipschitz domains. *Journal of Mathematical Analysis and Applications*, 276(2):845–867, 2002.
60. Giuseppe Buttazzo, Mariano Giaquinta, and Stefan Hildebrandt. *One-dimensional variational problems: an introduction*, volume 15. Oxford University Press, 1998.
61. L. A. Caffarelli and V. I. Oliker. Weak solutions of one inverse problem in geometric optics. *Journal of Mathematical Sciences*, 154:39–49, 2008.
62. Luis A. Caffarelli, Cristian E. Gutiérrez, and Qingbo Huang. On the regularity of reflector antennas. *Annals of mathematics*, pages 299–323, 2008.
63. Luis A Caffarelli and Qingbo Huang. Reflector problem in $\mathbb{R}^n$ endowed with non-Euclidean norm. *Archive for rational mechanics and analysis*, 193(2):445–473, 2009.
64. Wenshan Cai, Uday K Chettiar, Alexander V Kildishev, and Vladimir M Shalaev. Optical cloaking with metamaterials. *Nature photonics*, 1(4):224–227, 2007.
65. Maria-Carme Calderer, Dmitry Golovaty, Fang Hua Lin, and Chun Liu. Time evolution of nematic liquid crystals with variable degree of orientation. *SIAM journal on mathematical analysis*, 33(5):1033–1047, 2002.
66. Christophe Caloz, Andrea Alu, Sergei Tretyakov, Dimitrios Sounas, Karim Achouri, and Zoé-Lise Deck-Léger. Electromagnetic nonreciprocity. *Physical Review Applied*, 10(4):047001, 2018.
67. Miguel Camacho-Lopez, Heino Finkelmann, Peter Palffy-Muhoray, and Michael Shelley. Fast liquid-crystal elastomer swims into the dark. *Nature materials*, 3(5):307–310, 2004.
68. Wenyi Cao, P. Palffy-Muhoray, B. Taheri, A. Marino, and G. Abbate. Lasing thresholds of cholesteric liquid crystals lasers. *Molecular Crystals and Liquid Crystals*, 429(1):101–110, 2005.
69. Constantin Carathéodory. *Variationsrechnung und partielle Differentialgleichungen erster Ordnung*, volume 1. B. G. Teubner, Berlin, 1935.
70. C. Caroli and E. Dubois-Violette. Energy of a disinclination line in an anisotropic cholesteric liquid crystal. *Solid State Communications*, 7(11):799–802, 1969.
71. Ana Belen Castano-Fernandez, Andrei Martinez-Finkelshtein, and D Robert Iskander. A semi-analytic approach to calculating the Strehl ratio for a circularly symmetric system. part 1: static wavefront. *Optica Applicata*, 48(2), 2018.
72. Lamberto Cesari. *Optimization—theory and applications: problems with ordinary differential equations*, volume 17. Springer Science & Business Media, 2012.
73. Michel Cessenat. *Mathematical methods in electromagnetism: linear theory and applications*. World Scientific, 1996.
74. S. Chandrasekhar. *Liquid Crystals*. Cambridge University Press, 1977.
75. S. Chandrasekhar. Discotic liquid crystals. a brief review. *Liquid Crystals*, 14(1):3–14, 1993.
76. S. Chandrasekhar and G. S. Ranganath. Discotic liquid crystals. *Reports on Progress in Physics*, 53(1):57, 1990.
77. S. Chandrasekhar, G. S. Ranganath, and K. A. Suresh. Dynamical theory of reflexion from cholesteric liquid crystals. *Pramana-Journal of Physics*, 1:341–352, 1975.
78. Peter C. Y. Chang, J. G. Walker, and K. I. Hopcraft. Ray tracing in absorbing media. *Journal of Quantitative Spectroscopy and Radiative Transfer*, 96(3-4):327–341, 2005.
79. L. Lawrence Chapoy, Bjorn Marcher, and Knud H. Rasmussen. The morphology of liquid-crystalline polymers and the possible consequences for their rheological behaviour. *Liquid Crystals*, 3(12):1611–1636, 1988.
80. Jing-huei Chen and T. C. Lubensky. Landau-Ginzburg mean-field theory for the nematic to smectic-c and nematic to smectic-a phase transitions. *Physical Review A*, 14(3):1202, 1976.
81. Peng Chen, Bing-Yan Wei, Wei Hu, and Yan-Qing Lu. Liquid-crystal-mediated geometric phase: from transmissive to broadband reflective planar optics. *Advanced Materials*, 32(27):1903665, 2020.
82. Quan-Ming Chen, Chun-Ting Xu, Xiao Liang, and Wei Hu. Helical structure endows liquid crystal planar optics with a customizable working band. *Advanced Quantum Technologies*, 6(2):2200153, 2023.

83. Xi Chen, Li-Gang Wang, and Chun-Fang Li. Transmission gap, Bragg-like reflection, and Goos-Hänchen shifts near the Dirac point inside a negative-zero-positive index metamaterial slab. *Physical Review A*, 80(4):043839, 2009.
84. Yu-Bin Chen and Psang Dain Lin. Second-order derivatives of optical path length of ray with respect to variable vector of source ray. *Applied optics*, 51(22):5552–5562, 2012.
85. Zhigang Chen, Mordechai Segev, and Demetrios N Christodoulides. Optical spatial solitons: historical overview and recent advances. *Reports on Progress in Physics*, 75(8):086401, 2012.
86. Zhiming Chen, Qiang Du, and Jun Zou. Finite element methods with matching and non-matching meshes for Maxwell equations with discontinuous coefficients. *SIAM Journal on Numerical Analysis*, 37(5):1542–1570, 2000.
87. Carmen Chicone. *Ordinary Differential Equations with Applications*. Texts in Applied Mathematics. Springer Science & Business Media, 2006.
88. Su Seok Choi, Stephen M/ Morris, Wilhelm T. S. Huck, and Harry J. Coles. Liquid crystals: electrically tuneable liquid crystal photonic bandgaps. *Advanced Materials*, 21(38-39), 2009.
89. P. E. Cladis and M. Kleman. Non-singular disclinations of strength s=+ 1 in nematics. *Journal de Physique*, 33(5-6):591–598, 1972.
90. P.E. Cladis, W. Van Saarloos, P. L. Finn, and A. R. Kortan. Dynamics of line defects in nematic liquid crystals. *Physical review letters*, 58(3):222, 1987.
91. Harry Coles and Stephen Morris. Liquid-crystal lasers. *Nature Photonics*, 4(10):676–685, 2010.
92. Y. Colin de Verdière. Nombre de points entiers dans une famille homothétique de domaines de $\mathbb{R}$. 10(4):559–575, 1977.
93. Peter J. Collings and John W. Goodby. *Introduction to liquid crystals: chemistry and physics*. CRC Press, 2019.
94. P. Coullet, L. Gil, and F. Rocca. Optical vortices. *Optics Communications*, 73(5):403–408, 1989.
95. Gianni Dal Maso. *An introduction to Γ-convergence*, volume 8. Springer Science & Business Media, 2012.
96. Antonio d'Alessandro and Rita Asquini. Light propagation in confined nematic liquid crystals and device applications. *Applied Sciences*, 11(18):8713, 2021.
97. Pierre Gilles de Gennes. An analogy between superconductors and smectics A. *Solid State Communications*, 10(9):753–756, 1972.
98. Pierre-Gilles De Gennes and Jacques Prost. *The physics of liquid crystals*. Number 83. Oxford university press, 1993.
99. Roberto De Leo, Cristian E. Gutiérrez, and Henok Mawi. On the numerical solution of the far field refractor problem. *Nonlinear Analysis*, 157:123–145, 2017.
100. A De Padua, Fernando Parisio-Filho, and Fernando Moraes. Geodesics around line defects in elastic solids. *Physics Letters A*, 238(2-3):153–158, 1998.
101. Guido De Philippis and Alessio Figalli. Sobolev regularity for Monge-Ampere type equations. *SIAM Journal on Mathematical Analysis*, 45(3):1812–1824, 2013.
102. Thierri De Vos and Marc Baus. A density functional study of a twisted nematic cell and its relation to the Frank–Oseen theory. *The Journal of chemical physics*, 128(19):194903, 2008.
103. Hill De Vries. Rotatory power and other optical properties of certain liquid crystals. *Acta Crystallographica*, 4(3):219–226, 1951.
104. Arno Deicke. Über die finsler-räume mit a i= 0. *Archiv der Mathematik*, 4:45–51, 1953.
105. Giovanni Di Fratta, Jonathan M. Robbins, Valeriy Slastikov, and Arghir Zarnescu. Half-integer point defects in the q-tensor theory of nematic liquid crystals. *Journal of Nonlinear Science*, 26:121–140, 2016.
106. Ingo Dierking. *Textures of liquid crystals*. John Wiley & Sons, 2003.
107. Ingo Dierking. Chiral liquid crystals: structures, phases, effects. *Symmetry*, 6(2):444–472, 2014.
108. J. M. Diñeiro, M. Berrogui, S. Alfonso, C. Alberdi, B. Hernández, and C. Sáenz. Complex unitary vectors for the direction of propagation and for the polarization of electromagnetic waves in uniaxial and absorbing dielectric media. *JOSA A*, 24(6):1767–1775, 2007.

109. Manfredo P. Do Carmo. *Differential geometry of curves and surfaces: revised and updated second edition*. Courier Dover Publications, 2016.
110. Manfredo Perdigao Do Carmo and J. Flaherty Francis. *Riemannian geometry*, volume 6. Springer, 1992.
111. J. W. Doane, N. A. Vaz, B.-G. Wu, and Slobodan Žumer. Field controlled light scattering from nematic microdroplets. *Applied Physics Letters*, 48(4):269–271, 1986.
112. B Dorizzi, B Grammaticos, A Ramani, and P Winternitz. Integrable Hamiltonian systems with velocity-dependent potentials. *Journal of mathematical physics*, 26(12):3070–3079, 1985.
113. Leonid L. Doskolovich, Dmitry A. Bykov, Evgeniy S. Andreev, Evgeni A. Bezus, and Vladimir Oliker. Designing double freeform surfaces for collimated beam shaping with optimal mass transportation and linear assignment problems. *Optics Express*, 26(19):24602–24613, 2018.
114. R. Dreher and G. Meier. Optical properties of cholesteric liquid crystals. *Physical Review A*, 8(3):1616, 1973.
115. R. Dreher, G. Meier, and A. Saupe. Selective reflection by cholesteric liquid crystals. *Molecular Crystals and Liquid Crystals*, 13(1):17–26, 1971.
116. David Dunmur and Tim Sluckin. *Soap, science, and flat-screen TVs: a history of liquid crystals*. Oxford University Press, 2014.
117. M. A. Dupertuis, M. Proctor, and B. Acklin. Generalization of complex Snell–Descartes and Fresnel laws. *JOSA A*, 11(3):1159–1166, 1994.
118. R. Echarri and M. T. Garea. Behaviour of the Poynting vector in uniaxial absorbent media. *Pure and Applied Optics: Journal of the European Optical Society Part A*, 3(6):931, 1994.
119. Shyunsuke Endo, Tadashi Kuribara, and Tadashi Akahane. A study of the anomalous transmission (Borrmann effect) for obliquely incident light in an absorbing single-domain cholesteric liquid crystal. *Japanese journal of applied physics*, 22(8A):L499, 1983.
120. Klaus-Jochen Engel, Rainer Nagel, and Simon Brendle. *One-parameter semigroups for linear evolution equations*, volume 194. Springer, 2000.
121. John H Erdmann, Slobodan Žumer, and J. William Doane. Configuration transition in a nematic liquid crystal confined to a small spherical cavity. *Physical review letters*, 64(16):1907, 1990.
122. J. L. Ericksen. Inequalities in liquid crystal theory. *The physics of Fluids*, 9(6):1205–1207, 1966.
123. Jerald L. Ericksen. Liquid crystals with variable degree of orientation. *Archive for Rational Mechanics and Analysis*, 113:97–120, 1991.
124. Jiahui Fang and Junyu Lin. Existence and blow up criterion for strong solutions to the compressible biaxial nematic liquid crystal flow. *Mathematical Methods in the Applied Sciences*, 46(6):6972–7012, 2023.
125. James L. Fergason. Cholesteric structure-1 optical properties. *Molecular Crystals and Liquid Crystals*, 1(2):293–307, 1966.
126. Ivan Fernandez-Corbaton, Martin Fruhnert, and Carsten Rockstuhl. Objects of maximum electromagnetic chirality. *Physical Review X*, 6(3):031013, 2016.
127. Alberto Fernández-Nieves, Vincenzo Vitelli, Andrew S. Utada, Darren R. Link, Manuel Márquez, David R. Nelson, and David A. Weitz. Novel defect structures in nematic liquid crystal shells. *Physical review letters*, 99(15):157801, 2007.
128. Alessio Figalli. *The Monge–Ampère equation and its applications*, volume 22 of *Zurich Lectures in Advanced Mathematics*. European Mathematical Society, 2017.
129. Heino Finkelmann, Sung Tae Kim, Antonio Munoz, Peter Palffy-Muhoray, and Bahman Taheri. Tunable mirrorless lasing in cholesteric liquid crystalline elastomers. *Advanced Materials*, 13(14):1069–1072, 2001.
130. Andrew Forbes. Structured light from lasers. *Laser & Photonics Reviews*, 13(11):1900140, 2019.
131. Frederick C. Frank. I. liquid crystals. on the theory of liquid crystals. *Discussions of the Faraday Society*, 25:19–28, 1958.

132. V. Fréedericksz and A. Repiewa. Theoretisches und experimentelles zur frage nach der natur der anisotropen flüssigkeiten. *Zeitschrift für Physik*, 42(7):532–546, 1927.
133. Augustin Jean Fresnel. *Mémoire sur la double réfraction*, volume 7. Chez Firmin Didot, Père et Fils, Libraires, 1827.
134. Henrik Freymond and Rainer Picard. The elusive Drude-Born-Fedorov model for chiral electromagnetic media. *arXiv preprint arXiv:1204.5350*, 2012.
135. G. Friedel. Mesomorphic states of matter. *Annales de Physique*, 18(273), 1922.
136. Sébastien Fumeron and Bertrand Berche. Introduction to topological defects: from liquid crystals to particle physics. *The European Physical Journal Special Topics*, pages 1–21, 2023.
137. Izrail Moiseevitch Gelfand and Richard A. Silverman. *Calculus of variations*. Courier Corporation, 2000.
138. Mariano Giaquinta and Stefan Hildebrandt. *Calculus of Variations I*, volume 310. Springer Science & Business Media, 2013.
139. Mariano Giaquinta and Stefan Hildebrandt. *Calculus of variations II*, volume 311. Springer Science & Business Media, 2013.
140. David Gilbarg and Neil S. Trudinger. *Elliptic partial differential equations of second order*, volume 224. Springer, 1977.
141. Andrew S Glassner. *An introduction to ray tracing*. Morgan Kaufmann, 1989.
142. Dmitry Golovaty, Michael Novack, and Peter Sternberg. A one-dimensional variational problem for cholesteric liquid crystals with disparate elastic constants. *Journal of Differential Equations*, 286:785–820, 2021.
143. Carlos Gomez-Reino, Maria Victoria Perez, and Carmen Bao. *Gradient-index optics: fundamentals and applications*. Springer Science & Business Media, 2002.
144. Weihua Gong and Junyu Lin. Existence of solutions to incompressible biaxial nematic liquid crystals flows. *Applicable Analysis*, 101(16):5872–5905, 2022.
145. Rafael G. González-Acuña and Héctor A. Chaparro-Romo. General formula for bi-aspheric singlet lens design free of spherical aberration. *Applied optics*, 57(31):9341–9345, 2018.
146. Rafael G González-Acuña, Héctor A Chaparro-Romo, and Julio C Gutiérrez-Vega. Analytic solution of the eikonal for a stigmatic singlet lens. *Physica Scripta*, 95(8):085201, 2020.
147. J. Richard Gott III. Gravitational lensing effects of vacuum strings-exact solutions. *The Astrophysical Journal*, 288:422–427, 1985.
148. Yury Grabovsky and Lev Truskinovsky. Roughening instability of broken extremals. *Archive for rational mechanics and analysis*, 200(1):183–202, 2011.
149. François Grandjean. Les propriétés optiques de certaines structures de liquides anisotropes. *Bulletin de Minéralogie*, 42(1):42–86, 1919.
150. Leon W. Green. Auf wiedersehensflachen. *Annals of Mathematics*, pages 289–299, 1963.
151. Pengfei Guan, Xu-Jia Wang, et al. On a Monge-Ampere equation arising in geometric optics. *J. Diff. Geom.*, 48(2):205–223, 1998.
152. Zhecun Guan, Ling Wang, and Jinhye Bae. Advances in 4D printing of liquid crystalline elastomers: materials, techniques, and applications. *Materials Horizons*, 9(7):1825–1849, 2022.
153. Tianyi Guo. *Using Light to Study Liquid Crystals and Using Liquid Crystals to Control Light*. PhD thesis, Kent State University, 2020.
154. Cristian E. Gutiérrez. Aspherical lens design. *JOSA A*, 30(9):1719–1726, 2013.
155. Cristian E Gutiérrez. Refraction problems in geometric optics. In *Fully Nonlinear PDEs in Real and Complex Geometry and Optics*, pages 95–150. Springer, 2014.
156. Cristian E. Gutiérrez. *The Monge-Ampere equation*. Birkhauser, 2nd edition, 2016.
157. Cristian E Gutiérrez and Qingbo Huang. The refractor problem in reshaping light beams. *Archive for rational mechanics and analysis*, 193(2):423–443, 2009.
158. Cristian E. Gutiérrez and Qingbo Huang. The near field refractor. *Annales de l'Institut Henri Poincaré C, Analyse non linéaire*, 31(4):655–684, 2014.
159. Cristian E Gutiérrez, Qingbo Huang, and Henok Mawi. Refractors in anisotropic media associated with norms. *Nonlinear Analysis*, 188:125–141, 2019.

160. Cristian E. Gutiérrez, Qingbo Huang, Quentin Mérigot, and Boris Thibert. Metasurfaces and optimal transport. *The SMAI Journal of computational mathematics*, 8:201–224, 2022.
161. Cristian E. Gutiérrez and Henok Mawi. The refractor problem with loss of energy. *Nonlinear Analysis: Theory, Methods, and Applications*, 82:12–46, 2013.
162. Cristian E. Gutiérrez and Henok Mawi. On the numerical solution of the near field refractor problem. *Applied Mathematics & Optimization*, 84(Suppl 2):1877–1902, 2021.
163. Cristian E. Gutiérrez and Luca Pallucchini. Reflection and refraction problems for metasurfaces related to Monge–Ampère equations. *JOSA A*, 35(9):1523–1531, 2018.
164. Cristian E Gutiérrez, Luca Pallucchini, and Eric Stachura. General refraction problems with phase discontinuities on nonflat metasurfaces. *JOSA A*, 34(7):1160–1172, 2017.
165. Cristian E. Gutiérrez and Ahmad Sabra. The reflector problem and the inverse square law. *Nonlinear Analysis: Theory, Methods & Applications*, 96:109–133, 2014.
166. Cristian E. Gutiérrez and Ahmad Sabra. Freeform lens design for scattering data with general radiant fields. *Archive for Rational Mechanics and Analysis*, 228:341–399, 2018.
167. Cristian E. Gutiérrez and Ahmad Sabra. Chromatic aberration in metalenses. *Advances in Applied Mathematics*, 124:102134, 2021.
168. Cristian E Gutiérrez and Eric Stachura. Uniform refraction in negative refractive index materials. *JOSA A*, 32(11):2110–2122, 2015.
169. Cristian E Gutiérrez and Eric Stachura. Metamaterial lens design. *JOSA A*, 33(10):2020–2026, 2016.
170. Cristian E. Gutierrez and Federico Tournier. Surfaces refracting and reflecting collimated beams. *JOSA A*, 28(9):1860–1863, 2011.
171. Cristian E. Gutiérrez and Federico Tournier. Regularity for the near field parallel refractor and reflector problems. *Calculus of Variations and Partial Differential Equations*, 54(1):917–949, 2015.
172. A. S. Gutman. Modified Luneberg lens. *Journal of Applied Physics*, 25(7):855–859, 1954.
173. Felix Hagemann. *Reconstructing the shape and measuring chirality of obstacles in electromagnetic scattering*. PhD thesis, Dissertation, Karlsruhe, Karlsruher Institut für Technologie (KIT), 2019, 2019.
174. Piort Hajlasz. Sobolev spaces on metric-measure spaces. *Contemporary Mathematics*, 338:173–218, 2003.
175. Brian C. Hall. *Lie groups, Lie algebras, and representations*. Springer, 2013.
176. Yucen Han and Apala Majumdar. Pattern formation for nematic liquid crystals—modelling, analysis, and applications. *Research in Mathematics of Materials Science*, pages 399–431, 2022.
177. Yucen Han and Apala Majumdar. Multistability for a reduced nematic liquid crystal model in the exterior of 2d polygons. *Journal of Nonlinear Science*, 33(2):24, 2023.
178. James B. Hartle. *Gravity: an introduction to Einstein's general relativity*. American Association of Physics Teachers, 2003.
179. Allen Hatcher. *Algebraic topology*. Cambridge University Press, 2002.
180. Qiguang He, Zhijian Wang, Yang Wang, Adriane Minori, Michael T Tolley, and Shengqiang Cai. Electrically controlled liquid crystal elastomer–based soft tubular actuator with multimodal actuation. *Science advances*, 5(10):eaax5746, 2019.
181. Sailing He. A time-harmonic green's function technique and wave propagation in a stratified nonreciprocal chiral slab with multiple discontinuities. *Journal of mathematical physics*, 33(12):4103–4110, 1992.
182. Sailing He and Yidong Hu. Electromagnetic scattering from a stratified bi-isotropic (nonreciprocal chiral) slab: numerical computations. *IEEE transactions on antennas and propagation*, 41(8):1057–1062, 1993.
183. Ziqian He, Fangwang Gou, Ran Chen, Kun Yin, Tao Zhan, and Shin-Tson Wu. Liquid crystal beam steering devices: principles, recent advances, and future developments. *Crystals*, 9(6):292, 2019.
184. Duvan Henao, Apala Majumdar, and Adriano Pisante. Uniaxial versus biaxial character of nematic equilibria in three dimensions. *Calculus of Variations and Partial Differential Equations*, 56(2):55, 2017.

185. G. Heppke, H. Kitzerow, F. Oestreicher, S. Quentel, and A. Ranft. Electrooptic effect in a non-polar nematic discotic liquid crystal. *Molecular Crystals and Liquid Crystals Letters*, 6(3):71–79, 1988.
186. Andrew J Hess, Guilhem Poy, Jung-Shen B Tai, Slobodan Žumer, and Ivan I Smalyukh. Control of light by topological solitons in soft chiral birefringent media. *Physical Review X*, 10(3):031042, 2020.
187. Yucheng Hu, Yang Qu, and Pingwen Zhang. On the disclination lines of nematic liquid crystals. *Communications in Computational Physics*, 19(2):354–379, 2016.
188. Jinrui Huang and Junyu Lin. Orientability and asymptotic convergence of q-tensor flow of biaxial nematic liquid crystals. *Calculus of Variations and Partial Differential Equations*, 61(5):173, 2022.
189. Matjaž Humar, Malte C Gather, and Seok-Hyun Yun. Cellular dye lasers: lasing thresholds and sensing in a planar resonator. *Optics express*, 23(21):27865–27879, 2015.
190. Radu Ignat, Luc Nguyen, Valeriy Slastikov, and Arghir Zarnescu. Stability of the melting hedgehog in the Landau–de Gennes theory of nematic liquid crystals. *Archive for Rational Mechanics and Analysis*, 215:633–673, 2015.
191. Radu Ignat, Luc Nguyen, Valeriy Slastikov, and Arghir Zarnescu. Instability of point defects in a two-dimensional nematic liquid crystal model. In *Annales de l'Institut Henri Poincaré C, Analyse non linéaire*, volume 33, pages 1131–1152. Elsevier, 2016.
192. Radu Ignat, Luc Nguyen, Valeriy Slastikov, and Arghir Zarnescu. Stability of point defects of degree ± 1 2±1 2 in a two-dimensional nematic liquid crystal model. *Calculus of Variations and Partial Differential Equations*, 55:1–33, 2016.
193. Yo Inoue, Hiroyuki Yoshida, Kenta Inoue, Yusuke Shiozaki, Hitoshi Kubo, Akihiko Fujii, and Masanori Ozaki. Tunable lasing from a cholesteric liquid crystal film embedded with a liquid crystal nanopore network. *Advanced Materials*, 23(46):5498–5501, 2011.
194. Andreas D. Ioannidis, Gerhard Kristensson, and Ioannis G. Stratis. On the well-posedness of the Maxwell system for linear bianisotropic media. *SIAM Journal on Mathematical Analysis*, 44(4):2459–2473, 2012.
195. Reginald William James. *The Optical Principles of the Diffraction of X-rays*, volume 2. G. Bell & Sons, 1954.
196. Seung-Yeol Jeon, Beijun Shen, Nicholas A. Traugutt, Zeyu Zhu, Lichen Fang, Christopher M. Yakacki, Thao D. Nguyen, and Sung Hoon Kang. Synergistic energy absorption mechanisms of architected liquid crystal elastomers. *Advanced Materials*, 34(14):2200272, 2022.
197. John D. Joannopoulos, Steven G. Johnson, Joshua N. Winn, and Robert D. Meade. Molding the flow of light. *Princet. Univ. Press. Princeton, NJ*, 2008.
198. Frank Jochmann. Regularity of weak solutions of Maxwell's equations with mixed boundary-conditions. *Mathematical methods in the applied sciences*, 22(14):1255–1274, 1999.
199. A Joets and R Ribotta. A geometrical model for the propagation of rays in an anisotropic inhomogeneous medium. *Optics communications*, 107(3-4):200–204, 1994.
200. Steven G. Johnson and John D. Joannopoulos. *Photonic crystals: the road from theory to practice*. Springer Science & Business Media, 2001.
201. Leonid V. Kantorovich. On the translocation of masses. *Journal of mathematical sciences*, 133(4):1381–1382, 2006.
202. Leonid Vasilevich Kantorovich and S. G. Rubinshtein. On a space of totally additive functions. *Vestnik of the St. Petersburg University: Mathematics*, 13(7):52–59, 1958.
203. Leonid Vitalevich Kantorovich. On a problem of Monge. *Journal of Mathematical Sciences*, 4(133):1383–1383, 2006.
204. Aram Karakhanyan and Xu-Jia Wang. On the reflector shape design. *Journal of Differential Geometry*, 84(3):561–610, 2010.
205. Aram L. Karakhanyan. An inverse problem for the refractive surfaces with parallel lighting. *SIAM Journal on Mathematical Analysis*, 48(1):740–784, 2016.
206. Aram L. Karakhanyan and Ahmad Sabra. Refractor surfaces determined by near-field data. *Journal of Differential Equations*, 269(2):1278–1318, 2020.

207. Tosio Kato. *Perturbation theory for linear operators*, volume 132. Springer Science & Business Media, 2013.
208. Louis H. Kauffman. *Knots and physics*, volume 1. World scientific, 2001.
209. Noah et al. Kent. Creation and observation of hopfions in magnetic multilayer systems. *Nature communications*, 12(1):1–7, 2021.
210. Iam Choon Khoo. Nonlinear optics of liquid crystalline materials. *Physics Reports*, 471(5-6):221–267, 2009.
211. Iam-Choon Khoo. *Liquid crystals*. John Wiley & Sons, 2022.
212. Iam-Choon Khoo and Shin-Tson Wu. *Optics and nonlinear optics of liquid crystals*, volume 1. world scientific, 1993.
213. MinSu Kim and Francesca Serra. Tunable dynamic topological defect pattern formation in nematic liquid crystals. *Advanced Optical Materials*, 8(1):1900991, 2020.
214. Jun Kitagawa. An iterative scheme for solving the optimal transportation problem. *Calculus of Variations and Partial Differential Equations*, 51:243–263, 2014.
215. Georgy Kitavtsev, Jonathan M. Robbins, Valeriy Slastikov, and Arghir Zarnescu. Liquid crystal defects in the Landau–de Gennes theory in two dimensions—beyond the one-constant approximation. *Mathematical Models and Methods in Applied Sciences*, 26(14):2769–2808, 2016.
216. Charles Kittel and Paul McEuen. *Introduction to solid state physics*. John Wiley & Sons, 2018.
217. Maurice Kléman. Points, lignes, parois dans les fluides anisotropes et les solides cristallins. *Edition de Physique, France*, 1977.
218. Maurice Kléman. Effect of frustration in liquid crystals and polymers. *Physica Scripta*, 1987(T19B):565, 1987.
219. Maurice Kléman. Defects in liquid crystals. *Reports on Progress in Physics*, 52(5):555, 1989.
220. Maurice Kleman and Oleg D Laverntovich. *Soft matter physics: an introduction*. Springer Science & Business Media, 2007.
221. Maurice Kleman and Oleg D. Lavrentovich. Topological point defects in nematic liquid crystals. *Philosophical Magazine*, 86(25-26):4117–4137, 2006.
222. Maurice Kleman, Oleg D. Lavrentovich, and Yuriy A. Nastishin. Dislocations and disclinations in mesomorphic phases. *Dislocations in solids*, 12:147–271, 2004.
223. Junji Kobashi, Yoshinori Mohri, Hiroyuki Yoshida, and Masanori Ozaki. Circularly-polarized, large-angle reflective deflectors based on periodically patterned cholesteric liquid crystals. *Optical Data Processing and Storage*, 3(1):61–66, 2017.
224. Junji Kobashi, Hiroyuki Yoshida, and Masanori Ozaki. Planar optics with patterned chiral liquid crystals. *Nature Photonics*, 10(6):389–392, 2016.
225. Sergey A. Kochengin and Vladimir I. Oliker. Determination of reflector surfaces from near-field scattering data. *Inverse problems*, 13(2):363, 1997.
226. Victor I. Kopp, B. Fan, H. K. M. Vithana, and Azriel Z. Genack. Low-threshold lasing at the edge of a photonic stop band in cholesteric liquid crystals. *Optics letters*, 23(21):1707–1709, 1998.
227. Eileen M. Korenic, Stephen D. Jacobs, Sadeg M. Fare, and Le Li. Cholesteric liquid crystal flakes—a new form of domain. *Molecular Crystals and Liquid Crystals Science and Technology. Section A. Molecular Crystals and Liquid Crystals*, 317(1):197–219, 1998.
228. TC Kraan, T Van Bommel, and RAM Hikmet. Modeling liquid-crystal gradient-index lenses. *JOSA A*, 24(11):3467–3477, 2007.
229. Yury A Kravtsov and Yuri Ilich Orlov. *Geometrical optics of inhomogeneous media*, volume 38. Springer, 1990.
230. Ekkehart Kröner. Continuum theory of defects. *Physics of defects*, 35:282–315, 1981.
231. D. Krüerke, H.-S. Kitzerow, G. Heppke, and V. Vill. First observation of selective reflection and blue phases in chiral discotic liquid crystals. *Berichte der Bunsengesellschaft für physikalische Chemie*, 97(10):1371–1375, 1993.
232. Peter Kuchment. The mathematics of photonic crystals. In *Mathematical modeling in optical science*, pages 207–272. SIAM, 2001.

233. Akhlesh Lakhtakia. *Beltrami fields in chiral media*, volume 2. World Scientific, 1994.
234. Akhlesh Lakhtakia, Vijay K. Varadan, and Vasundara V. Varadan. *Time-harmonic electromagnetic fields in chiral media*. Springer, 1989.
235. Éamon Lalor. An analytical approach to the theory of internal conical refraction. *Journal of Mathematical Physics*, 13(4):449–454, 1972.
236. Xavier Lamy. Uniaxial symmetry in nematic liquid crystals. In *Annales de l'Institut Henri Poincaré C, Analyse non linéaire*, volume 32, pages 1125–1144. Elsevier, 2015.
237. Lev Davidovich Landau, J. S. Bell, M. J. Kearsley, L.P. Pitaevskii, E. M. Lifshitz, and J.B. Sykes. *Electrodynamics of continuous media*, volume 8. Elsevier, 2013.
238. Oleg D. Lavrentovich. Ferroelectric nematic liquid crystal, a century in waiting. *Proceedings of the National Academy of Sciences*, 117(26):14629–14631, 2020.
239. Oleg D. Lavrentovich, Paolo Pasini, Claudio Zannoni, and Slobodan Zumer. *Defects in liquid crystals: Computer simulations, theory and experiments*, volume 43. Springer Science & Business Media, 2001.
240. Jonathan Leach, Eric Yao, and Miles J. Padgett. Observation of the vortex structure of a non-integer vortex beam. *New Journal of Physics*, 6(1):71, 2004.
241. John Lee. *Introduction to topological manifolds*, volume 202. Springer Science & Business Media, 2010.
242. O. Lehmann. über fliessende kristalle (on flowing crystals),. *Zeitschrift für Physikalische Chemie*, 4:462, 1889.
243. Ulf Leonhardt. Perfect imaging without negative refraction. *New Journal of Physics*, 11(9):093040, 2009.
244. Ulf Leonhardt and Thomas Philbin. *Geometry and light: the science of invisibility*. Courier Corporation, 2010.
245. Sirui Li, Chenchen Wang, and Jie Xu. Well-posedness of frame hydrodynamics for biaxial nematic liquid crystals. *arXiv preprint arXiv:2202.08998*, 2022.
246. Fang-Hua Lin. Nonlinear theory of defects in nematic liquid crystals; phase transition and flow phenomena. *Communications on Pure and Applied Mathematics*, 42(6):789–814, 1989.
247. Fang Hua Lin. On nematic liquid crystals with variable degree of orientation. *Communications on pure and applied mathematics*, 44(4):453–468, 1991.
248. Junyu Lin, Yimei Li, and Changyou Wang. On static and hydrodynamic biaxial nematic liquid crystals. *arXiv preprint arXiv:2006.04207*, 2020.
249. Tsung-Hsien Lin, Yi-Jan Chen, Chun-Hui Wu, Andy Y-G Fuh, J-H Liu, and P-C Yang. Cholesteric liquid crystal laser with wide tuning capability. *Applied Physics Letters*, 86(16):161120, 2005.
250. Ismo Lindell, Ari Sihvola, Sergei Tretyakov, and Ari J. Viitanen. *Electromagnetic waves in chiral and bi-isotropic media*. Artech House, 1994.
251. Ismo V. Lindell, Ari H. Sihvola, and Ari J. Viitanen. Plane-wave reflection from a bi-isotropic (nonreciprocal chiral) interface. *Microwave and Optical Technology Letters*, 5(2):79–81, 1992.
252. Ismo V. Lindell and A. J. Vitanen. Duality transformations for general bi-isotropic(nonreciprocal chiral) media. *IEEE transactions on antennas and propagation*, 40(1):91–95, 1992.
253. Jacques Louis Lions and Enrico Magenes. *Non-homogeneous boundary value problems and applications: Vol. 1*, volume 181. Springer Science & Business Media, 2012.
254. J. D. Litster, J. Als-Nielsen, R.J. Birgeneau, S.S. Dana, D. Davidov, F. Garcia-Golding, M. Kaplan, C.R. Safinya, and R. Schaetzing. High resolution x-ray and light scattering studies of bilayer smectic a compounds. *Le Journal de Physique Colloques*, 40(C3):C3–339, 1979.
255. Chun Liu and M. Carme Calderer. Liquid crystal flow: dynamic and static configurations. *SIAM Journal on Applied Mathematics*, 60(6):1925–1949, 2000.
256. Jung-Ping Liu, Hon-Fai Yau, Zhen Ye, Chao-Hsien Kuo, and BoWen Ke. Amphoteric refraction exhibited by anisotropic crystals. *Optik*, 118(3):147–152, 2007.
257. Wanguo Liu, Hai Hu, Fenghua Liu, and Hua Zhao. Manipulating light trace in a gradient-refractive-index medium: a Lagrangian optics method. *Optics express*, 27(4):4714–4726, 2019.

258. Alex Lonergan, David McNulty, and Colm O'Dwyer. Tetrahedral framework of inverse opal photonic crystals defines the optical response and photonic band gap. *Journal of Applied Physics*, 124(9):095106, 2018.
259. Teresa Lopez-Leon and Alberto Fernandez-Nieves. Drops and shells of liquid crystal. *Colloid and Polymer Science*, 289:345–359, 2011.
260. T. C. Lubensky, David Pettey, Nathan Currier, and Holger Stark. Topological defects and interactions in nematic emulsions. *Physical Review E*, 57(1):610, 1998.
261. G.R. Luckhurst, S. Romano, and H. B. Zewdie. Computer simulation studies of anisotropic systems. Part 25.—Chiral nematics, the Dzyaloshinsky–Moriya model. *Journal of the Chemical Society, Faraday Transactions*, 92(10):1781–1791, 1996.
262. Rudolf Karl Luneburg. *Mathematical theory of optics*. University of California press, 1964.
263. Hailu Luo, Wei Hu, Xunong Yi, Haiying Liu, and Jing Zhu. Amphoteric refraction at the interface between isotropic and anisotropic media. *Optics communications*, 254(4-6):353–360, 2005.
264. Xi-Nan Ma, Neil S Trudinger, and Xu-Jia Wang. Regularity of potential functions of the optimal transportation problem. *Archive for rational mechanics and analysis*, 177(2):151–183, 2005.
265. Yun Gui Ma, Sahar Sahebdivan, CK Ong, Tomáš Tyc, and Ulf Leonhardt. Evidence for subwavelength imaging with positive refraction. *New Journal of Physics*, 13(3):033016, 2011.
266. Thomas Machon and Gareth P. Alexander. Umbilic lines in orientational order. *Physical Review X*, 6(1):011033, 2016.
267. Ernest H. MacMillan. On the hydrodynamics of biaxial nematic liquid crystals part 1: General theory. *Archive for rational mechanics and analysis*, 117(3):193–239, 1992.
268. Ernest H. MacMillan. On the hydrodynamics of biaxial nematic liquid crystals part 2: Steady-state analysis. *Archive for Rational Mechanics and Analysis*, 117(3):241–294, 1992.
269. Juan A. Madrid. A geometrical approach to time evolving wave fronts. *Geophysical Journal International*, 172(3):1117–1122, 2008.
270. Wilhelm Maier and Alfred Saupe. Eine einfache molekular-statistische theorie der nematischen kristallinflüssigen phase. Teil ii. *Zeitschrift für Naturforschung A*, 15(4):287–292, 1960.
271. Apala Majumdar. Equilibrium order parameters of nematic liquid crystals in the Landau-de Gennes theory. *European Journal of Applied Mathematics*, 21(2):181–203, 2010.
272. Apala Majumdar. The radial-hedgehog solution in Landau–de Gennes' theory for nematic liquid crystals. *European Journal of Applied Mathematics*, 23(1):61–97, 2012.
273. Apala Majumdar and Xingbin Pan. The Landau-de Gennes theory of nematic liquid crystals: uniaxiality versus biaxiality. *Communications on Pure & Applied Analysis*, 11(3), 2012.
274. Apala Majumdar and Arghir Zarnescu. Landau–de Gennes theory of nematic liquid crystals: the Oseen–Frank limit and beyond. *Archive for rational mechanics and analysis*, 196(1):227–280, 2010.
275. A.S. Marathay. Matrix-operator description of the propagation of polarized light through cholesteric liquid crystals. *JOSA*, 61(10):1363–1372, 1971.
276. Antonella Marchesiello, Libor Šnobl, et al. Classical superintegrable systems in a magnetic field that separate in Cartesian coordinates. *SIGMA. Symmetry, Integrability and Geometry: Methods and Applications*, 16:015, 2020.
277. Ian Marquette. Supersymmetry as a method of obtaining new superintegrable systems with higher order integrals of motion. *Journal of mathematical physics*, 50(12):122102, 2009.
278. Jürg Marti. On integro-differential equations in Banach spaces. *Pacific Journal of Mathematics*, 20(1):99–108, 1967.
279. Yuko Matsuhisa, Yuhua Huang, Ying Zhou, Shin-Tson Wu, Ryotaro Ozaki, Yuuki Takao, Akihiko Fujii, and Masanori Ozaki. Low-threshold and high efficiency lasing upon band-edge excitation in a cholesteric liquid crystal. *Applied Physics Letters*, 90(9), 2007.
280. Henok Mawi. An overview of mathematical modeling of geometric optics problems involving refraction. *The Golden Anniversary Celebration of the National Association of Mathematicians*, 759:21, 2020.

281. Gina Mayonado, Shabbir M. Mian, Valentina Robbiano, and Franco Cacialli. Investigation of the Bragg-Snell law in photonic crystals. *BFY Proceedings*, pages 60–63, 2015.
282. Sophie McLauchlan, Yucen Han, Matthias Langer, and Apala Majumdar. The radial hedgehog solution in the Landau-de Gennes theory: Effects of the bulk potentials. *arXiv preprint arXiv:2306.15563*, 2023.
283. William Charles Hector McLean. *Strongly elliptic systems and boundary integral equations*. Cambridge university press, 2000.
284. E. J. McShane. Some Existence Theorems in the Calculus of Variations, I. Existence Theorems for Nonregular Problems. *Transactions of the American Mathematical Society*, 44(1):428–438, 1938.
285. E. J. McShane. Some Existence Theorems in the Calculus of Variations II. Existence Theorems For Isoperimetric Problems in the Plane. *Transactions of the American Mathematical Society*, 44(3):439–453, 1938.
286. E. J. McShane. Some existence theorems in the calculus of variations iii. existence theorems for nonregular problems. *Transactions of the American Mathematical Society*, 45(1):151–171, 1939.
287. E. J. McShane. Some Existence Theorems in the Calculus of Variations IV. Isoperimetric Problems in Non-Parametric Form. *Transactions of the American Mathematical Society*, 45(2):173–196, 1939.
288. Eric McSween and Pavel Winternitz. Integrable and superintegrable Hamiltonian systems in magnetic fields. *Journal of Mathematical Physics*, 41(5):2957–2967, 2000.
289. Sethna Meiboom, James P. Sethna, P. W. Anderson, and W.F. Brinkman. Theory of the blue phase of cholesteric liquid crystals. *Physical Review Letters*, 46(18):1216, 1981.
290. Yobani Mejia. Exact relations between wave aberration and the Sagitta difference, and between ray aberration and the slope difference. *Optik*, 123(8):726–730, 2012.
291. R. Memmer and F. Janssen. Computer simulation of chiral liquid crystal phases vi. a temperature induced helix inversion in a single component cholesteric liquid crystal. *Liquid crystals*, 24(6):805–809, 1998.
292. Reiner Memmer and F. Janssen. Computer simulation of chiral liquid crystal phases part 5 temperature dependence of the pitch of a cholesteric phase studied under self-determined boundary conditions. *Journal of the Chemical Society, Faraday Transactions*, 94(2):267–276, 1998.
293. Erich Merchand. *Gradient index optics*. Elsevier, 2012.
294. N. David Mermin. The topological theory of defects in ordered media. *Reviews of Modern Physics*, 51(3):591, 1979.
295. Robert B. Meyer. Piezoelectric effects in liquid crystals. *Physical Review Letters*, 22(18):918, 1969.
296. Robert B. Meyer. On the existence of even indexed disclinations in nematic liquid crystals. *The Philosophical Magazine: A Journal of Theoretical Experimental and Applied Physics*, 27(2):405–424, 1973.
297. Zongcheng Miao, Xianliang Chen, Yutong Zhang, Dong Wang, and Lei Wang. Bistable cholesteric liquid crystal films with excellent electro-optical performance and spacing stability for reflective displays. *ACS Applied Polymer Materials*, 5(1):476–484, 2022.
298. D. Michaelis, P. Schreiber, and A. Bräuer. Cartesian oval representation of freeform optics in illumination systems. *Optics letters*, 36(6):918–920, 2011.
299. Daniel S. Miller, Xiaoguang Wang, and Nicholas L. Abbott. Design of functional materials based on liquid crystalline droplets. *Chemistry of Materials*, 26(1):496–506, 2014.
300. Willard Miller Jr, Sarah Post, and Pavel Winternitz. Classical and quantum superintegrability with applications. *Journal of Physics A: Mathematical and Theoretical*, 46(42):423001, 2013.
301. Effran Mirzad Rafael, Luca Tonti, Daniel Corbett, Alejandro Cuetos, and Alessandro Patti. Dynamics of uniaxial-to-biaxial nematics switching in suspensions of hard cuboids. *Physics of Fluids*, 33(6), 2021.
302. Ryusei Momosaki, Kazunari Ashikawa, Moritsugu Sakamoto, Kohei Noda, Tomoyuki Sasaki, Nobuhiro Kawatsuki, and Hiroshi Ono. Incident angle dependence-reduced polarization

grating performance by using optically biaxial polymer liquid crystal. *Optics Letters*, 44(24):5929–5932, 2019.
303. Gaspard Monge. Mémoire sur la théorie des déblais et des remblais. *Mem. Math. Phys. Acad. Royale Sci.*, pages 666–704, 1781.
304. Peter Monk et al. *Finite element methods for Maxwell's equations*. Oxford University Press, 2003.
305. Duncan T. Moore. Gradient-index optics: a review. *Applied Optics*, 19(7):1035–1038, 1980.
306. Fernando Moraes. Geodesics around a dislocation. *Physics Letters A*, 214(3-4):189–192, 1996.
307. Samuel P. Morgan. General solution of the Luneberg lens problem. *Journal of Applied Physics*, 29(9):1358–1368, 1958.
308. Nigel J. Mottram and Christopher J. P. Newton. Introduction to q-tensor theory. *arXiv preprint arXiv:1409.3542*, 2014.
309. Urban Mur, Miha Ravnik, and David Seč. Controllable shifting, steering, and expanding of light beam based on multi-layer liquid-crystal cells. *Scientific Reports*, 12(1):1–13, 2022.
310. J. Nehring, A. R. Kmetz, and T. J. Scheffer. Analysis of weak-boundary-coupling effects in liquid-crystal displays. *Journal of Applied Physics*, 47(3):850–857, 1976.
311. Luc Nguyen and Arghir Zarnescu. Refined approximation for minimizers of a Landau-de Gennes energy functional. *Calculus of Variations and Partial Differential Equations*, 47:383–432, 2013.
312. Rajaram Nityananda, U. D. Kini, S. Chandrasekhar, and K. A. Suresh. Anomalous transmission (Borrmann effect) in absorbing cholesteric liquid crystals. 1975.
313. Idan Niv and Efi Efrati. Geometric frustration and compatibility conditions for two-dimensional director fields. *Soft matter*, 14(3):424–431, 2018.
314. Christian Ohm, Martin Brehmer, and Rudolf Zentel. Liquid crystalline elastomers as actuators and sensors. *Advanced materials*, 22(31):3366–3387, 2010.
315. Vladimir Oliker. On design of free-form refractive beam shapers, sensitivity to figure error, and convexity of lenses. *JOSA A*, 25(12):3067–3076, 2008.
316. Vladimir Oliker. Designing freeform lenses for intensity and phase control of coherent light with help from geometry and mass transport. *Archive for Rational Mechanics and Analysis*, 201(3):1013–1045, 2011.
317. Vladimir Oliker. The reflector problem for closed surfaces. In *Partial Differential Equations and Applications*, pages 265–270. Routledge, 2017.
318. Vladimir Oliker. Lightcurve inversion problem for objects with negative Gaussian curvature. *Advances in Applied Mathematics*, 148:102516, 2023.
319. Vladimir Oliker, Jacob Rubinstein, and Gershon Wolansky. Supporting quadric method in optical design of freeform lenses for illumination control of a collimated light. *Advances in Applied Mathematics*, 62:160–183, 2015.
320. Vladimir I. Oliker. Near radially symmetric solutions of an inverse problem in geometric optics. *Inverse Problems*, 3(4):743, 1987.
321. Lars Onsager. The effects of shape on the interaction of colloidal particles. *Annals of the New York Academy of Sciences*, 51(4):627–659, 1949.
322. C. W. Oseen. The theory of liquid crystals. *Transactions of the Faraday Society*, 29(140):883–899, 1933.
323. Patrick Oswald and Pawel Pieranski. *Nematic and cholesteric liquid crystals: concepts and physical properties illustrated by experiments*. CRC press, 2005.
324. Benjamin Outram. Long-pitch cholesterics. In *Liquid Crystals*. IOP Publishing, 2018.
325. Ryotaro Ozaki. Simple model for estimating band edge wavelengths of selective reflection from cholesteric liquid crystals for oblique incidence. *Physical Review E*, 100(1):012708, 2019.
326. Ryotaro Ozaki, Shunsuke Hashimura, Shinji Yudate, Kazunori Kadowaki, Hiroyuki Yoshida, and Masanori Ozaki. Optical properties of selective diffraction from Bragg-Berry cholesteric liquid crystal deflectors. *OSA Continuum*, 2(12):3554–3563, 2019.

327. Ryotaro Ozaki, Kohdai Saiki, Shunsuke Hashimura, Kazunori Kadowaki, Hiroyuki Yoshida, and Masanori Ozaki. Geometrical optics analysis of diffraction in patterned cholesteric liquid crystals. *ACS Applied Optical Materials*, 2023.
328. P. Palffy-Muhoray, E. C. Gartland, and J. R. Kelly. A new configurational transition in inhomogeneous nematics. *Liquid Crystals*, 16(4):713–718, 1994.
329. Peter Palffy-Muhoray. The diverse world of liquid crystals. *Physics today*, 60(9):54–60, 2007.
330. Peter Palffy-Muhoray and B. Taheri. Ultraviolet lasing in cholesteric liquid crystals. *Optics Letters*, 26(11):804–806, 2001.
331. Miha Papič, Urban Mur, Miha Ravnik, Igor Muševič, and Matjaž Humar. Vector beams generated by microlasers based on topological liquid-crystal structures. *arXiv preprint arXiv:2008.00929*, 2020.
332. Karthik Peddireddy, Venkata Subba R. Jampani, Shashi Thutupalli, Stephan Herminghaus, Christian Bahr, and Igor Muševič. Lasing and waveguiding in smectic a liquid crystal optical fibers. *Optics Express*, 21(25):30233–30242, 2013.
333. John B. Pendry. Negative refraction. *Contemporary Physics*, 45(3):191–202, 2004.
334. John Brian Pendry. Negative refraction makes a perfect lens. *Physical review letters*, 85(18):3966, 2000.
335. Volker Perlick. Fermat principle in finsler spacetimes. *General Relativity and Gravitation*, 38:365–380, 2006.
336. Sreekanth Perumbilavil, Armando Piccardi, Oleksandr Buchnev, Giuseppe Strangi, Martti Kauranen, and Gaetano Assanto. Spatial solitons to mold random lasers in nematic liquid crystals. *Optical Materials Express*, 8(12):3864–3878, 2018.
337. Gia Petriashvili, Kokhta Japaridze, Lali Devadze, Cisana Zurabishvili, Nino Sepashvili, Nino Ponjavidze, Maria P De Santo, Mario A Matranga, Ridha Hamdi, Federica Ciuchi, et al. Paper like cholesteric interferential mirror. *Optics Express*, 21(18):20821–20830, 2013.
338. Gregory Francis Pettis. *Hertzian dipoles and microstrip circuits on arbitrarily oriented biaxially anisotropic media*. PhD thesis, Syracuse University, 2008.
339. Armando Piccardi, Alessandro Alberucci, and Gaetano Assanto. Nematicons and their electro-optic control: Light localization and signal readdressing via reorientation in liquid crystals. *International Journal of Molecular Sciences*, 14(10):19932–19950, 2013.
340. Oleg P Pishnyak and Oleg D Lavrentovich. Electrically controlled negative refraction in a nematic liquid crystal. *Applied physics letters*, 89(25):251103, 2006.
341. P.H. Poulin and D. A. Weitz. Inverted and multiple nematic emulsions. *Physical Review E*, 57(1):626, 1998.
342. Philippe Poulin, Holger Stark, T. C. Lubensky, and D. A. Weitz. Novel colloidal interactions in anisotropic fluids. *Science*, 275(5307):1770–1773, 1997.
343. Guilhem Poy and Slobodan Žumer. Ray-based optical visualisation of complex birefringent structures including energy transport. *Soft matter*, 15(18):3659–3670, 2019.
344. E. Priestly. *Introduction to liquid crystals*. Springer Science & Business Media, 2012.
345. Roshdi Rashed. A pioneer in anaclastics: Ibn Sahl on burning mirrors and lenses. *Isis*, 81(3):464–491, 1990.
346. M. Ravnik, M. Stimulak, M. Cancula, and S. Zumer. Nematic defects and colloids as photonic elements. *Molecular Crystals and Liquid Crystals*, 619(1):61–69, 2015.
347. J. L. Rayces. Exact relation between wave aberration and ray aberration. *Optica Acta: International Journal of Optics*, 11(2):85–88, 1964.
348. F. Reinitzer. Beiträge zur kenntnis des cholesterins. *Monatshefte für Chemie*, 9:421–441, 1888.
349. John Restrepo, Pawel J Stoerck, and Ivo Ihrke. Ray and wave aberrations revisited: a Huygens-like construction yields exact relations. *JOSA A*, 33(2):160–171, 2016.
350. Gary Francis Roach, Ioannis G. Stratis, and Athanasios N. Yannacopoulos. *Mathematical analysis of deterministic and stochastic problems in complex media electromagnetics*. Princeton University Press, 2012.
351. Wilhelm Conrad Röntgen and J. Schneider. Ueber die compressibilität des wassers. *Annalen der Physik*, 269(4):644–660, 1888.

352. Carmelo Rosales-Guzmán, Bienvenu Ndagano, and Andrew Forbes. A review of complex vector light fields and their applications. *Journal of Optics*, 20(12):123001, 2018.
353. Miguel Ruphuy and Omar M. Ramahi. Spherical aberration in electrically thin flat lenses. *JOSA A*, 33(8):1531–1536, 2016.
354. W. V. T. Rusch and P. D. Potter. *Analysis of reflector antennas*. Academic Press, 2013.
355. Philip St. J. Russell, Timothy A. Birks, and F. Dominic Lloyd-Lucas. Photonic Bloch waves and photonic band gaps. *Confined Electrons and Photons: New Physics and Applications*, pages 585–633, 1995.
356. Alexander Ryabchun and Alexey Bobrovsky. Cholesteric liquid crystal materials for tunable diffractive optics. *Advanced Optical Materials*, 6(15):1800335, 2018.
357. D. G Sannikov and D. I. Sementsov. Interference of counterpropagating waves in a nonreciprocal chiral medium. *Technical Physics Letters*, 33:996–999, 2007.
358. José Sasián. *Introduction to aberrations in optical imaging systems*. Cambridge University Press, 2013.
359. Caio Sátiro and Fernando Moraes. Lensing effects in a nematic liquid crystal with topological defects. *The European Physical Journal E*, 20(2):173–178, 2006.
360. Caio Sátiro and Fernando Moraes. On the deflection of light by topological defects in nematic liquid crystals. *The European Physical Journal E*, 25(4):425–429, 2008.
361. Michelle Savescu, Stephen Johnson, Polina Sanchez, Qin Zhou, MF Mahmood, Essaid Zerrad, Anjan Biswas, and Milivoj Belic. Nematicons in liquid crystals. *Journal of Computational and Theoretical Nanoscience*, 12(11):4667–4673, 2015.
362. Toralf Scharf. *Polarized light in liquid crystals and polymers*. John Wiley & Sons, 2007.
363. T. J. Scheffer. Structures and energies of Grandjean-Cano liquid-crystal disclinations. *Physical Review A*, 5(3):1327, 1972.
364. Henrik Schlichtkrull. *Curves and Surfaces: Lecture Notes for Geometry 1*. Department of Mathematical Science, University of Copenhagen, 2010.
365. Richard Schoen and Karen Uhlenbeck. A regularity theory for harmonic maps. *Journal of Differential Geometry*, 17(2):307–335, 1982.
366. Bernd Schroers. Solvable models of magnetic skyrmions. In *Quantum Theory and Symmetries*, pages 535–544. Springer, 2021.
367. Jonathan V. Selinger. Interpretation of saddle-splay and the Oseen-Frank free energy in liquid crystals. *Liquid Crystals Reviews*, 6(2):129–142, 2018.
368. Robin L. B. Selinger, Badel L. Mbanga, and Jonathan V. Selinger. Modeling liquid crystal elastomers: actuators, pumps, and robots. In *Emerging Liquid Crystal Technologies III*, volume 6911, pages 82–86. SPIE, 2008.
369. Senyuk, B., Liu, Q., He, S. et al. Topological colloids. *Nature*, 493, 200–205, 2013.
370. Ravi Shanker. *Self-assembled photonic crystals infiltrated with nanoplatelets and nanotubes*. PhD thesis, University of Surrey (United Kingdom), 2015.
371. V. N. Shapovalov, V. G. Bagrov, and A. G. Meshkov. Separation of variables in the stationary Schrödinger equation. *Soviet Physics Journal*, 15(8):1115–1119, 1972.
372. Jie Shen, Tao Tang, and Li-Lian Wang. *Spectral methods: algorithms, analysis and applications*, volume 41. Springer Science & Business Media, 2011.
373. Yijie Shen, Xuejiao Wang, Zhenwei Xie, Changjun Min, Xing Fu, Qiang Liu, Mali Gong, and Xiaocong Yuan. Optical vortices 30 years on: OAM manipulation from topological charge to multiple singularities. *Light: Science & Applications*, 8(1):90, 2019.
374. Zhongmin Shen. *Lectures on Finsler geometry*. World Scientific, 2001.
375. Ping Sheng. Introduction to the elastic continuum theory of liquid crystals. *Introduction to Liquid Crystals*, pages 103–127, 1975.
376. Devanand K. Shenoy, D. Laurence Thomsen III, Amritha Srinivasan, Patrick Keller, and Banahalli R. Ratna. Carbon coated liquid crystal elastomer film for artificial muscle applications. *Sensors and Actuators A: Physical*, 96(2-3):184–188, 2002.
377. Baoming Shi, Yucen Han, and Lei Zhang. Nematic liquid crystals in a rectangular confinement: solution landscape, and bifurcation. *SIAM Journal on Applied Mathematics*, 82(5):1808–1828, 2022.

378. Valery Shibaev, Alexey Bobrovsky, and Natalia Boiko. Photoactive liquid crystalline polymer systems with light-controllable structure and optical properties. *Progress in Polymer Science*, 28(5):729–836, 2003.
379. Kevin E. Shopsowitz, Hao Qi, Wadood Y. Hamad, and Mark J. MacLachlan. Free-standing mesoporous silica films with tunable chiral nematic structures. *Nature*, 468(7322):422–425, 2010.
380. Vladlen G. Shvedov, Yana V. Izdebskaya, Yan Sheng, and Wieslaw Krolikowski. Magnetically controlled negative refraction of solitons in liquid crystals. *Applied Physics Letters*, 110(9):091107, 2017.
381. Aleksey N. Simonov and Michiel C. Rombach. Asymptotic behavior of the spatial frequency response of an optical system with defocus and spherical aberration. *JOSA A*, 27(12):2563–2573, 2010.
382. Robert A. Skutnik, Jan-Christoph Eichler, Marco G. Mazza, and Martin Schoen. The temperature dependence of the helical pitch in a cholesteric liquid crystal. *Molecular Physics*, 119(15-16):e1881638, 2021.
383. Tim Sluckin, David Dunmur, and Horst Stegemeyer. *Crystals that flow: Classic papers from the history of Liquid Crystals*. Taylor & Francis London, 2004.
384. Maarten Sluijter. *Ray-optics analysis of inhomogeneous optically anisotropic media*. PhD thesis, Delft University of Technology, 2010.
385. Maarten Sluijter, Dick K de Boer, and H Paul Urbach. Ray-optics analysis of inhomogeneous biaxially anisotropic media. *JOSA A*, 26(2):317–329, 2009.
386. Maarten Sluijter, Dick KG de Boer, and Joseph JM Braat. General polarized ray-tracing method for inhomogeneous uniaxially anisotropic media. *JOSA A*, 25(6):1260–1273, 2008.
387. Ivan I. Smalyukh. Knots and other new topological effects in liquid crystals and colloids. *Reports on Progress in Physics*, 83(10):106601, 2020.
388. Ivan I. Smalyukh, Daniel S. Kaputa, Aliaksandr V. Kachynski, Andrey N. Kuzmin, and Paras N. Prasad. Optical trapping of director structures and defects in liquid crystals using laser tweezers. *Optics Express*, 15(7):4359–4371, 2007.
389. David Ryan Smith, Jack J. Mock, A. F. Starr, and David Schurig. Gradient index metamaterials. *Physical Review E*, 71(3):036609, 2005.
390. Warren J. Smith. *Modern optical engineering: the design of optical systems*. McGraw-Hill Education, 2008.
391. A. Sommerfeld. *Optics, Lectures on theoretical physics, vol. IV.* Academic Press, 1959.
392. A. M. Somoza and P. Tarazona. Density functional theory of the elastic constants of a nematic liquid crystal. *Molecular Physics*, 72(4):911–926, 1991.
393. Myoung Hoon Song, Na Young Ha, Kazuhiro Amemiya, Byoungchoo Park, Yoichi Takanishi, Ken Ishikawa, Jeong Weon Wu, Suzushi Nishimura, Takehiro Toyooka, and Hideo Takezoe. Defect-mode lasing with lowered threshold in a three-layered hetero-cholesteric liquid-crystal structure. *Advanced materials*, 18(2):193–197, 2006.
394. Costas M. Soukoulis, Stefan Linden, and Martin Wegener. Negative refractive index at optical wavelengths. *Science*, 315(5808):47–49, 2007.
395. Eric Stachura. Existence of weak solutions to refraction problems in negative refractive index materials. *Nonlinear Analysis*, 157:76–103, 2017.
396. Eric Stachura. The time dependent Maxwell system with measurable coefficients in Lipschitz domains. *Journal of Mathematical Analysis and Applications*, 452(2):941–956, 2017.
397. Eric Stachura and Niklas Wellander. Quantitative trace estimates for the Maxwell system in Lipschitz domains. *Mathematical Methods in the Applied Sciences*, 44(13):10635–10652, 2021.
398. Jakob J. Stamnes and Glen S. Sithambaranathan. Reflection and refraction of an arbitrary electromagnetic wave at a plane interface separating an isotropic and a biaxial medium. *JOSA A*, 18(12):3119–3129, 2001.
399. Iain W. Stewart. *The static and dynamic continuum theory of liquid crystals: a mathematical introduction*. CRC Press, 2019.

400. Ioannis G. Stratis. Chirality notions and electromagnetic scattering: a mini review. *Complex Variables and Elliptic Equations*, 67(3):740–772, 2022.
401. Robert S. Strichartz. *A guide to distribution theory and Fourier transforms*. World Scientific Publishing Company, 2003.
402. K. A. Suresh. An experimental study of the anomalous transmission (Borrmann effect) in absorbing cholesteric liquid crystals. *Molecular Crystals and Liquid Crystals*, 35(3-4):267–273, 1976.
403. B. Taheri, A. F. Munoz, P. Palffy-Muhoray, and R. Twieg. Low threshold lasing in cholesteric liquid crystals. *Molecular Crystals and Liquid Crystals Science and Technology. Section A. Molecular Crystals and Liquid Crystals*, 358(1):73–82, 2001.
404. L. Tartar. On the characterization of traces of a Sobolev space used for Maxwell's equations. *Proceedings of a Meeting in Boreaux*, (6-7), 1997.
405. Jamie M. Taylor. γ-convergence of a mean-field model of a chiral doped nematic liquid crystal to the Oseen–Frank description of cholesterics. *Nonlinearity*, 33(6):3062, 2020.
406. Bernard D. H. Tellegen. The gyrator, a new electric network element. *Philips Res. Rep*, 3(2):81–101, 1948.
407. Gerald Teschl. *Ordinary differential equations and dynamical systems*, volume 140. American Mathematical Soc., 2012.
408. S. Thiberge, C. Chevallard, J. M. Gilli, and A. Buka. Critical radius of loop defects in homeotropic nematic liquid crystals. *Liquid crystals*, 26(8):1225–1234, 1999.
409. Donald L. Thomsen, Patrick Keller, Jawad Naciri, Roger Pink, Hong Jeon, Devanand Shenoy, and Banahalli R. Ratna. Liquid crystal elastomers with mechanical properties of a muscle. *Macromolecules*, 34(17):5868–5875, 2001.
410. Yu Tong, Yiwei Wang, and Pingwen Zhang. Defects around a spherical particle in cholesteric liquid crystals. *Numerical Mathematics: Theory, Methods and Applications*, 10(2):205–221, 2017.
411. Neil S. Trudinger. Recent developments in elliptic partial differential equations of Monge-Ampere type. In *International Congress of Mathematicians*, volume 3, pages 291–301. Citeseer, 2006.
412. Carsten Tschierske and Demetri J. Photinos. Biaxial nematic phases. *Journal of Materials Chemistry*, 20(21):4263–4294, 2010.
413. Tomáš Tyc and Aaron J Danner. Absolute optical instruments, classical superintegrability, and separability of the Hamilton-Jacobi equation. *Physical Review A*, 96(5):053838, 2017.
414. Tomáš Tyc, HL Dao, and Aaron J Danner. Absolute optical instruments without spherical symmetry. *Physical Review A*, 92(5):053827, 2015.
415. Tomáš Tyc, Lenka Herzánová, Martin Šarbort, and Klaus Bering. Absolute instruments and perfect imaging in geometrical optics. *New Journal of Physics*, 13(11):115004, 2011.
416. John Urbas. On the second boundary value problem for equations of Monge-Ampère type. *J. Reine Angew. Math.*, pages 115–124, 1997.
417. B. W. Van der Meer, G. Vertogen, A. J. Dekker, and J. G.J. Ypma. A molecular-statistical theory of the temperature-dependent pitch in cholesteric liquid crystals. *The Journal of Chemical Physics*, 65(10):3935–3943, 1976.
418. Andrii Varanytsia, Tianyi Guo, and Peter Palffy-Muhoray. Small footprint cholesteric liquid crystal laser. *Applied Optics*, 58(4):739–743, 2019.
419. Andrii Varanytsia, Hama Nagai, Kenji Urayama, and Peter Palffy-Muhoray. Tunable lasing in cholesteric liquid crystal elastomers with accurate measurements of strain. *Scientific reports*, 5(1):17739, 2015.
420. Alexander Vilenkin. Cosmic strings as gravitational lenses. *The Astrophysical Journal*, 282:L51–L53, 1984.
421. Cédric Villani. *Optimal transport: old and new*, volume 338. Springer, 2009.
422. Epifanio G. Virga. *Variational theories for liquid crystals*, volume 8. CRC Press, 1995.
423. Claes Waksjö and Stefan Rauch-Wojciechowski. How to find separation coordinates for the Hamilton–Jacobi equation: a criterion of separability for natural Hamiltonian systems. *Mathematical physics, analysis and geometry*, 6(4):301–348, 2003.

424. X. S. Wang, Alireza Qaiumzadeh, and Arne Brataas. Current-driven dynamics of magnetic hopfions. *Physical review letters*, 123(14):147203, 2019.
425. Xu-Jia Wang. On the design of a reflector antenna. *Inverse problems*, 12(3):351, 1996.
426. Xu-Jia Wang. On the design of a reflector antenna ii. *Calculus of Variations and Partial Differential Equations*, 20(3):329–341, 2004.
427. Xuewen Wang, Zhongquan Nie, Yao Liang, Jian Wang, Tao Li, and Baohua Jia. Recent advances on optical vortex generation. *Nanophotonics*, 7(9):1533–1556, 2018.
428. Yanyang Wang, Liang Liang, Hong Xin, and Lei Wu. Complex ray tracing in uniaxial absorbing media. *JOSA A*, 25(3):653–657, 2008.
429. Yanyang Wang, Peng Shi, Hong Xin, and Lei Wu. Complex ray tracing in biaxial anisotropic absorbing media. *Journal of Optics A: Pure and Applied Optics*, 10(7):075009, 2008.
430. Zhenming Wang, Thomas Raistrick, Aidan Street, Matthew Reynolds, Yanjun Liu, and Helen F Gleeson. Direct observation of biaxial nematic order in auxetic liquid crystal elastomers. *Materials*, 16(1):393, 2022.
431. Owies M Wani, Hao Zeng, and Arri Priimagi. A light-driven artificial flytrap. *Nature communications*, 8(1):15546, 2017.
432. M. Warner and E. M. Terentjev. Nematic elastomers—a new state of matter? *Progress in Polymer Science*, 21(5):853–891, 1996.
433. Mark Warner and Eugene Michael Terentjev. *Liquid crystal elastomers*, volume 120. Oxford university press, 2007.
434. Walter Thompson Welford. *Aberrations of optical systems*. Adam Hilger, Bristol, England, 1986.
435. Yishi Weng, Daming Xu, Yuning Zhang, Xiaohua Li, and Shin-Tson Wu. Polarization volume grating with high efficiency and large diffraction angle. *Optics Express*, 24(16):17746–17759, 2016.
436. Timothy J. White and Dirk J. Broer. Programmable and adaptive mechanics with liquid crystal polymer networks and elastomers. *Nature materials*, 14(11):1087–1098, 2015.
437. T. Wöhrle et al. Discotic liquid crystals. *Chemical Reviews*, 116(3):1139–1241, 2016.
438. Stefan Wojciechowski. Separability of an integrable case of the Hénon-Heiles system. *Physics Letters A*, 100(6):277–278, 1984.
439. Jin-Sheng Wu, Marina Torres Lazaro, Souvik Ghosh, Haridas Mundoor, Henricus H Wensink, and Ivan I Smalyukh. Unavoidable emergent biaxiality in chiral molecular-colloidal hybrid liquid crystals. *arXiv preprint arXiv:2304.07447*, 2023.
440. Jin-Sheng Wu and Ivan I. Smalyukh. Hopfions, heliknotons, skyrmions, torons and both abelian and nonabelian vortices in chiral liquid crystals. *Liquid Crystals Reviews*, pages 1–35, 2022.
441. Shin-Tson Wu and Deng-Ke Yang. *Reflective liquid crystal displays*. Wiley, 2001.
442. James C. Wyant and Katherine Creath. Basic wavefront aberration theory for optical metrology. *Applied optics and optical engineering*, 11(part 2):28–39, 1992.
443. Jie Xiang and Oleg D. Lavrentovich. Liquid crystal structures for transformation optics. *Molecular Crystals and Liquid Crystals*, 559(1):106–114, 2012.
444. Jie Xiang, Andrii Varanytsia, Fred Minkowski, Daniel A. Paterson, John M. D. Storey, Corrie T. Imrie, Oleg D. Lavrentovich, and Peter Palffy-Muhoray. Electrically tunable laser based on oblique heliconical cholesteric liquid crystal. *Proceedings of the National Academy of Sciences*, 113(46):12925–12928, 2016.
445. Michel Daoud Yacoub and Gustavo Fraidenraich. 96.33 a solution to the quartic equation. *The Mathematical Gazette*, 96(536):271–275, 2012.
446. D.-K. Yang, John L. West, L.-C. Chien, and J. William Doane. Control of reflectivity and bistability in displays using cholesteric liquid crystals. *Journal of applied physics*, 76(2):1331–1333, 1994.
447. Deng-Ke Yang, J. William Doane, Z. Yaniv, and J. Glasser. Cholesteric reflective display: drive scheme and contrast. *Applied physics letters*, 64(15):1905–1907, 1994.
448. Donghao Yang, Marouen Chemingui, Yu Wang, Xinzheng Zhang, Irena Drevensek-Olenik, Faheem Hassan, Qiang Wu, Yigang Li, Lotfi Saadaoui, and Jingjun Xu. Dual-wavelength

lasing with orthogonal circular polarizations generated in a single layer of a polymer–cholesteric liquid crystal superstructure. *Polymers*, 15(5):1226, 2023.
449. Fuzi Yang and JR Sambles. Critical angles for reflectivity at an isotropic-anisotropic boundary. *Journal of Modern Optics*, 40(6):1131–1142, 1993.
450. Ping Yang and K. N Liou. Light scattering by hexagonal ice crystals: comparison of finite-difference time domain and geometric optics models. *JOSA A*, 12(1):162–176, 1995.
451. Xiaomei Yao, Lei Zhang, and Jeff Z. Y. Chen. Defect patterns of two-dimensional nematic liquid crystals in confinement. *Physical Review E*, 105(4):044704, 2022.
452. Amnon Yariv and Pochi Yeh. *Optical waves in crystal propagation and control of laser radiation*. John Wiley and Sons, Inc., New York, NY, 1983.
453. Pochi Yeh and Claire Gu. *Optics of liquid crystal displays*, volume 67. John Wiley & Sons, 2009.
454. Hiroshi Yokoyama. Density-functional theory of surfacelike elasticity of nematic liquid crystals. *Physical Review E*, 55(3):2938, 1997.
455. Kōsaku Yosida. *Functional analysis*. Springer Science & Business Media, 2012.
456. Arghir Zarnescu. Topics in the q-tensor theory of liquid crystals. *Topics in mathematical modeling and analysis*, 7:187–252, 2012.
457. Arghir Zarnescu. Mathematical problems of nematic liquid crystals: between dynamical and stationary problems. *Philosophical Transactions of the Royal Society A*, 379(2201):20200432, 2021.
458. Qiwen Zhan. Cylindrical vector beams: from mathematical concepts to applications. *Advances in Optics and Photonics*, 1(1):1–57, 2009.
459. Xiuqin Zhan, Fa-Feng Xu, Zhonghao Zhou, Yongli Yan, Jiannian Yao, and Yong Sheng Zhao. 3d laser displays based on circularly polarized lasing from cholesteric liquid crystal arrays. *Advanced Materials*, 33(37):2104418, 2021.
460. Pei Zhang, Laurens T. de Haan, Michael G. Debije, and Albert P. H. J. Schenning. Liquid crystal-based structural color actuators. *Light: Science & Applications*, 11(1):248, 2022.
461. Yuzhen Zhao, Min Zhu, Yang Zhao, Zemin He, Huimin Zhang, Cheng Ma, Yongming Zhang, and Zongcheng Miao. Dye-coated polymer stabilized cholesteric liquid crystal films for bistable reflective displays. *Optical Materials*, 134:112991, 2022.
462. Rafael S. Zola, Hari Krishna Bisoyi, Hao Wang, Augustine M. Urbas, Timothy J. Bunning, and Quan Li. Dynamic control of light direction enabled by stimuli-responsive liquid crystal gratings. *Advanced Materials*, 31(7):1806172, 2019.

Index

E. Stachura, *Mathematical Methods in Liquid Crystal Optics and Lens Design*, Springer Tracts in Modern Physics 294, https://doi.org/10.1007/978-3-031-46614-4

SPRINGER NATURE

GPSR Compliance

The European Union's (EU) General Product Safety Regulation (GPSR) is a set of rules that requires consumer products to be safe and our obligations to ensure this.

If you have any concerns about our products, you can contact us on ProductSafety@springernature.com

In case Publisher is established outside the EU, the EU authorized representative is:

Springer Nature Customer Service Center GmbH
Europaplatz 3
69115 Heidelberg, Germany

The manufacturer's authorised representative in the EU is Springer Nature Customer Service Centre GmbH, Europaplatz 3, 69115 Heidelberg, Germany. If you have any concerns regarding our products, please contact ProductSafety@springernature.com

Printed and bound by CPI Group (UK) Ltd, Croydon, CR0 4YY
23/09/2026
02221546-0008